T0141990

Springer Optimization and Its Appli

VOLUME 108

Aims and Scope
Optimization has been expanding in all directions at an astonishing rate during the last few decades. New algorithmic and theoretical techniques have been developed, the diffusion into other disciplines has proceeded at a rapid pace, and our knowledge of all aspects of the field has grown even more profound. At the same time, one of the most striking trends in optimization is the constantly increasing emphasis on the interdisciplinary nature of the field. Optimization has been a basic tool in all areas of applied mathematics, engineering, medicine, economics, and other sciences.

The series *Springer Optimization and Its Applications* publishes undergraduate and graduate textbooks, monographs and state-of-the-art expository work that focus on algorithms for solving optimization problems and also study applications involving such problems. Some of the topics covered include nonlinear optimization (convex and nonconvex), network flow problems, stochastic optimization, optimal control, discrete optimization, multi-objective programming, description of software packages, approximation techniques and heuristic approaches.

More information about this series at http://www.springer.com/series/7393

Alexander J. Zaslavski

Numerical Optimization
with Computational Errors

 Springer

Alexander J. Zaslavski
Department of Mathematics
The Technion – Israel Institute
  of Technology
Haifa, Israel

ISSN 1931-6828   ISSN 1931-6836   (electronic)
Springer Optimization and Its Applications
ISBN 978-3-319-80917-5   ISBN 978-3-319-30921-7   (eBook)
DOI 10.1007/978-3-319-30921-7

Mathematics Subject Classification (2010): 47H09, 49M30, 65K10

Printed on acid-free paper

This Springer imprint is published by Springer Nature
The registered company is Springer International Publishing AG Switzerland

Preface

The book is devoted to the study of approximate solutions of optimization problems in the presence of computational errors. We present a number of results on the convergence behavior of algorithms in a Hilbert space, which are known as important tools for solving optimization problems and variational inequalities. According to the results known in the literature, these algorithms should converge to a solution. In this book, we study these algorithms taking into account computational errors which are always present in practice. In this case the convergence to a solution does not take place. We show that our algorithms generate a good approximate solution, if computational errors are bounded from above by a small positive constant. In practice it is sufficient to find a good approximate solution instead of constructing a minimizing sequence. On the other hand, in practice computations can induce numerical errors and if one uses optimization methods to solve minimization problems these methods usually provide only approximate solutions of the problems. Our main goal is, for a known computational error, to find out what an approximate solution can be obtained and how many iterates one needs for this.

This monograph contains 16 chapters. Chapter 1 is an introduction. In Chap. 2, we study the subgradient projection algorithm for minimization of convex and nonsmooth functions. The mirror descent algorithm is considered in Chap. 3. The gradient projection algorithm for minimization of convex and smooth functions is analyzed in Chap. 4. In Chap. 5, we consider its extension which is used for solving linear inverse problems arising in signal/image processing. The convergence of Weiszfeld's method in the presence of computational errors is discussed in Chap. 6. In Chap. 7, we solve constrained convex minimization problems using the extragradient method. Chapter 8 is devoted to a generalized projected subgradient method for minimization of a convex function over a set which is not necessarily convex. In Chap. 9, we study the convergence of a proximal point method in a Hilbert space under the presence of computational errors. Chapter 10 is devoted to the local convergence of a proximal point method in a metric space under the presence of computational errors. In Chap. 11, we study the convergence of a proximal point method to a solution of the inclusion induced by a maximal monotone operator, under the presence of computational errors. In Chap. 12, the convergence of the subgradient method for solving variational inequalities is proved under the presence of computational errors. The convergence of the subgradient method to a common solution of a finite family of variational inequalities and of a finite family of fixed point problems, under the presence of computational errors, is shown in Chap. 13. In Chap. 14, we study continuous subgradient method. Penalty methods are studied in Chap. 15. Chapter 16 is devoted to Newton's method. The results of Chaps. 2–6, 14, and 16 are new. The results of other chapters were obtained and published during the last 5 years.

The author believes that this book will be useful for researchers interested in the optimization theory and its applications.

Rishon LeZion, Israel   Alexander J. Zaslavski
October 19, 2015

Contents

Chapter 1
Introduction

In this book we study behavior of algorithms for constrained convex minimization problems in a Hilbert space. Our goal is to obtain a good approximate solution of the problem in the presence of computational errors. We show that the algorithm generates a good approximate solution, if the sequence of computational errors is bounded from above by a constant. In this section we discuss several algorithms which are studied in the book.

1.1 Subgradient Projection Method

In Chap. 2 we study the subgradient projection algorithm for minimization of convex and nonsmooth functions and for computing the saddle points of convex–concave functions, under the presence of computational errors. It should be mentioned that the subgradient projection algorithm is one of the most important tools in the optimization theory and its applications. See, for example, [1–3, 12, 30, 44, 51, 79, 89, 92, 95, 96, 105, 108, 109, 112] and the references mentioned therein.

We use this method for constrained minimization problems in Hilbert spaces equipped with an inner product denoted by $\langle \cdot, \cdot \rangle$ which induces a complete norm $\| \cdot \|$. For every $z \in R^1$ denote by $\lfloor z \rfloor$ the largest integer which does not exceed z: $\lfloor z \rfloor = \max\{i \in R^1 : i \text{ is an integer and } i \leq z\}$.

Let X be a Hilbert space. For each $x \in X$ and each $r > 0$ set

$$B_X(x, r) = \{y \in X : \|x - y\| \leq r\}.$$

For each $x \in X$ and each nonempty set $E \subset X$ set

$$d(x, E) = \inf\{\|x - y\| : y \in E\}.$$

© Springer International Publishing Switzerland 2016

A.J. Zaslavski, *Numerical Optimization with Computational Errors*, Springer Optimization and Its Applications 108, DOI 10.1007/978-3-319-30921-7_1

Let C be a nonempty closed convex subset of X, U be an open convex subset of X such that $C \subset U$ and let $f : U \to R^1$ be a convex function. For each $x \in U$ set [84]

$$\partial f(x) = \{l \in X : f(y) - f(x) \geq \langle l, y - x \rangle \text{ for all } y \in U\}$$

which is called the subdifferential of the function f at the point x [84].

Suppose that there exist $L > 0$, $M_0 > 0$ such that

$$C \subset B_X(0, M_0),$$

$$|f(x) - f(y)| \leq L\|x - y\| \text{ for all } x, y \in U.$$

It is not difficult to see that for each $x \in U$,

$$\emptyset \neq \partial f(x) \subset B_X(0, L).$$

For every nonempty closed convex set $D \subset X$ and every $x \in X$ there is a unique point $P_D(x) \in D$ satisfying

$$\|x - P_D(x)\| = \inf\{\|x - y\| : y \in D\}.$$

We consider the minimization problem

$$f(z) \to \min, \ z \in C.$$

Suppose that $\delta \in (0, 1]$ is a computational error produced by our computer system and that $\{a_k\}_{k=0}^{\infty} \subset (0, \infty)$.

Let us describe our algorithm.

Subgradient Projection Algorithm

Initialization: select an arbitrary $x_0 \in U$.

Iterative step: given a current iteration vector $x_t \in U$ calculate

$$\xi_t \in \partial f(x_t) + B_X(0, \delta)$$

and the next iteration vector $x_{t+1} \in U$ such that

$$\|x_{t+1} - P_C(x_t - a_t\xi_t)\| \leq \delta.$$

In Chap. 2 we prove the following result (see Theorem 2.4).

Theorem 1.1. *Let $\delta \in (0, 1]$, $\{a_k\}_{k=0}^{\infty} \subset (0, \infty)$ and let*

$$x_* \in C$$

satisfies

$$f(x_*) \leq f(x) \text{ for all } x \in C.$$

Assume that $\{x_t\}_{t=0}^{\infty} \subset U$, $\{\xi_t\}_{t=0}^{\infty} \subset X$,

$$\|x_0\| \leq M_0 + 1$$

and that for each integer $t \geq 0$,

$$\xi_t \in \partial f(x_t) + B_X(0, \delta)$$

and

$$\|x_{t+1} - P_C(x_t - a_t\xi_t)\| \leq \delta.$$

Then for each natural number T,

$$\sum_{t=0}^{T} a_t(f(x_t) - f(x_*))$$

$$\leq 2^{-1}\|x_* - x_0\|^2 + \delta(T+1)(4M_0 + 1)$$

$$+ \delta(2M_0 + 1)\sum_{t=0}^{T} a_t + 2^{-1}(L+1)^2\sum_{t=0}^{T} a_t^2.$$

Moreover, for each natural number T,

$$\max\left\{ f\left(\left(\sum_{t=0}^{T} a_t\right)^{-1}\sum_{t=0}^{T} a_t x_t\right) - f(x_*),\ \min\{f(x_t) : t = 0, \ldots, T\} - f(x_*)\right\}$$

$$\leq 2^{-1}\left(\sum_{t=0}^{T} a_t\right)^{-1}\|x_* - x_0\|^2 + \left(\sum_{t=0}^{T} a_t\right)^{-1}\delta(T+1)(4M_0 + 1)$$

$$+ \delta(2M_0 + 1) + 2^{-1}\left(\sum_{t=0}^{T} a_t\right)^{-1}(L+1)^2\sum_{t=0}^{T} a_t^2.$$

We are interested in an optimal choice of a_t, $t = 0, 1, \ldots$. Let T be a natural number and $A_T = \sum_{t=0}^{T} a_t$ be given. It is shown in Chap. 2 that the best choice is $a_i = (T+1)^{-1}A_T$, $i = 0, \ldots, T$.

Let T be a natural number and $a_t = a$, $t = 0, \ldots, T$. It is shown in Chap. 2 that best choice of $a > 0$ is

$$a = (2\delta(4M_0 + 1))^{1/2}(L+1)^{-1}.$$

Now we can think about the best choice of T. It is not difficult to see that it should be at the same order as $\lfloor \delta^{-1} \rfloor$.

1.2   The Mirror Descent Method

Let X be a Hilbert space equipped with an inner product $\langle \cdot, \cdot \rangle$ which induces a complete norm $\| \cdot \|$. We use the notation introduced in the previous section.

Let C be a nonempty closed convex subset of X, U be an open convex subset of X such that $C \subset U$ and let $f : U \to R^1$ be a convex function. Suppose that there exist $L > 0$, $M_0 > 0$ such that

$$C \subset B_X(0, M_0),$$

$$|f(x) - f(y)| \leq L\|x - y\| \text{ for all } x, y \in U.$$

It is not difficult to see that for each $x \in U$,

$$\emptyset \neq \partial f(x) \subset B_X(0, L).$$

For each nonempty set $D \subset X$ and each function $h : D \to R^1$ put

$$\inf(h, D) = \inf\{h(y) : y \in D\}$$

and

$$\text{argmin}\{h(y) : y \in D\} = \{y \in D : h(y) = \inf(h; D)\}.$$

In Chap. 3 we study the convergence of the mirror descent algorithm under the presence of computational errors. This method was introduced by Nemirovsky and Yudin for solving convex optimization problems [90]. Here we use a derivation of this algorithm proposed by Beck and Teboulle [19].

We consider the minimization problem

$$f(z) \to \min, \ z \in C.$$

Suppose that $\delta \in (0, 1]$ is a computational error produced by our computer system and that $\{a_k\}_{k=0}^{\infty} \subset (0, \infty)$. We describe the inexact version of the mirror descent algorithm.

Mirror Descent Algorithm
Initialization: select an arbitrary $x_0 \in U$.
Iterative step: given a current iteration vector $x_t \in U$ calculate

$$\xi_t \in \partial f(x_t) + B_X(0, \delta),$$

define

$$g_t(x) = \langle \xi_t, x \rangle + (2a_t)^{-1}\|x - x_t\|^2, \ x \in X$$

and calculate the next iteration vector $x_{t+1} \in U$ such that

$$B_X(x_{t+1}, \delta) \cap \operatorname{argmin}\{g_t(y) : y \in C\} \neq \emptyset.$$

Note that g_t is a convex bounded from below function on X which possesses a minimizer on C.

In Chap. 3 we prove the following result (see Theorem 3.1).

Theorem 1.2. *Let* $\delta \in (0, 1]$, $\{a_k\}_{k=0}^{\infty} \subset (0, \infty)$ *and let*

$$x_* \in C$$

satisfies

$$f(x_*) \leq f(x) \text{ for all } x \in C.$$

Assume that $\{x_t\}_{t=0}^{\infty} \subset U$, $\{\xi_t\}_{t=0}^{\infty} \subset X$,

$$\|x_0\| \leq M_0 + 1$$

and that for each integer $t \geq 0$,

$$\xi_t \in \partial f(x_t) + B_X(0, \delta)$$

and

$$B_X(x_{t+1}, \delta) \cap \operatorname{argmin}\{\langle \xi_t, v \rangle + (2a_t)^{-1}\|v - x_t\|^2 : v \in C\} \neq \emptyset.$$

Then for each natural number T,

$$\sum_{t=0}^{T} a_t(f(x_t) - f(x_*))$$

$$\leq 2^{-1}(2M_0 + 1)^2 + \delta(2M_0 + L + 2)\sum_{t=0}^{T} a_t$$

$$+\delta(T + 1)(8M_0 + 8) + 2^{-1}(L + 1)^2 \sum_{t=0}^{T} a_t^2.$$

Moreover, for each natural number T,

$$f\left(\left(\sum_{t=0}^{T} a_t\right)^{-1} \sum_{t=0}^{T} a_t x_t\right) - f(x_*), \ \min\{f(x_t) : t = 0, \dots, T\} - f(x_*)$$

$$\leq 2^{-1}(2M_0 + 1)^2 \left(\sum_{t=0}^{T} a_t\right)^{-1} + \delta(2M_0 + L + 2)$$

$$+\delta(T + 1)(8M_0 + 8) \left(\sum_{t=0}^{T} a_t\right)^{-1} + 2^{-1}(L + 1)^2 \sum_{t=0}^{T} a_t^2 \left(\sum_{t=0}^{T} a_t\right)^{-1}.$$

We are interested in an optimal choice of a_t, $t = 0, 1, \ldots$. Let T be a natural number and $A_T = \sum_{t=0}^{T} a_t$ be given. It is shown in Chap. 3 that the best choice is $a_t = (T + 1)^{-1} A_T$, $i = 0, \ldots, T$.

Let T be a natural number and $a_t = a$, $t = 0, \ldots, T$. It is shown in Chap. 3 that the best choice of $a > 0$

$$a = (16\delta(M_0 + 1))^{1/2}(L + 1)^{-1}.$$

If we think about the best choice of T, it is clear that it should be at the same order as $\lfloor \delta^{-1} \rfloor$.

1.3 Proximal Point Method

In Chap. 9 we analyze the behavior of the proximal point method in a Hilbert space which is an important tool in the optimization theory. See, for example, [9, 15, 16, 29, 31, 34, 36, 53, 55, 69, 70, 77, 81, 87, 103, 104, 106, 107, 111, 113] and the references mentioned therein.

Let X be a Hilbert space equipped with an inner product $\langle \cdot, \cdot \rangle$ which induces the norm $\| \cdot \|$.

For each function $g : X \to R^1 \cup \{\infty\}$ set

$$\inf(g) = \inf\{g(y) : y \in X\}.$$

Suppose that $f : X \to R^1 \cup \{\infty\}$ is a convex lower semicontinuous function and a is a positive constant such that

$$\mathrm{dom}(f) := \{x \in X : f(x) < \infty\} \neq \emptyset,$$

$$f(x) \geq -a \text{ for all } x \in X$$

and that

$$\lim_{\|x\| \to \infty} f(x) = \infty.$$

It is not difficult to see that the set

$$\mathrm{Argmin}(f) := \{z \in X : f(z) = \inf(f)\} \neq \emptyset.$$

Let a point

$$x^* \in \mathrm{Argmin}(f)$$

and let M be any positive number such that

$$M > \inf(f) + 4.$$

It is clear that there exists a number $M_0 > 1$ such that

$$f(z) > M + 4 \text{ for all } z \in X \text{ satisfying } \|z\| \ge M_0 - 1.$$

Evidently,

$$\|x^*\| < M_0 - 1.$$

Assume that

$$0 < \Lambda_1 < \Lambda_2 \le M_0^{-2}/2.$$

The following theorem is the main result of Chap. 9.

Theorem 1.3. *Let*

$$\lambda_k \in [\Lambda_1, \Lambda_2], \ k = 0, 1, \ldots,$$

$\Delta \in (0, 1]$, *a natural number L satisfy*

$$L > 2(4M_0^2 + 1)\Lambda_2 \Delta^{-1}$$

and let a positive number ϵ satisfy

$$\epsilon^{1/2}(L + 1)(2\Lambda_1^{-1} + 8M_0\Lambda_1^{-1/2}) \le 1 \text{ and } \epsilon(L + 1) \le \Delta/4.$$

Assume that a sequence $\{x_k\}_{k=0}^{\infty} \subset X$ satisfies

$$f(x_0) \le M$$

and

$$f(x_{k+1}) + 2^{-1}\lambda_k\|x_{k+1} - x_k\|^2 \le \inf(f + 2^{-1}\lambda_k\| \cdot -x_k\|^2) + \epsilon$$

for all integers $k \ge 0$. Then for all integers $k > L$,

$$f(x_k) \le \inf(f) + \Delta.$$

By this theorem, for a given $\Delta > 0$, we obtain $\xi \in X$ satisfying

$$f(\xi) \leq \inf(f) + \Delta$$

doing $\lfloor c_1 \Delta^{-1} \rfloor$ iterations with the computational error $\epsilon = c_2 \Delta^2$, where the constant $c_1 > 0$ depends only on M_0, Λ_2 and the constant $c_2 > 0$ depends only on $M_0, L, \Lambda_1, \Lambda_2$.

1.4 Variational Inequalities

In Chap. 12 we are interested in solving of variational inequalities. The studies of gradient-type methods and variational inequalities are important topics in optimization theory. See, for example, [3, 5, 12, 30, 31, 37–39, 44, 52, 54, 59, 68, 71–74, 93, 129] and the references mentioned therein.

Let $(X, \langle \cdot, \cdot \rangle)$ be a Hilbert space with an inner product $\langle \cdot, \cdot \rangle$ which induces a complete norm $\| \cdot \|$. For each $x \in X$ and each $r > 0$ set

$$B(x, r) = \{y \in X : \ \|x - y\| \leq r\}.$$

Let C be a nonempty closed convex subset of X.

Consider a mapping $f : X \rightarrow X$. We say that the mapping f is monotone on C if

$$\langle f(x) - f(y), x - y \rangle \geq 0 \text{ for all } x, y \in C.$$

We say that f is pseudo-monotone on C if for each $x, y \in C$ the inequality

$$\langle f(y), x - y \rangle \geq 0 \text{ implies that } \langle f(x), x - y \rangle \geq 0.$$

Clearly, if f is monotone on C, then f is pseudo-monotone on C. Denote by S the set of all $x \in C$ such that

$$\langle f(x), y - x \rangle \geq 0 \text{ for all } y \in C.$$

We suppose that

$$S \neq \emptyset.$$

For each $\epsilon > 0$ denote by S_ϵ the set of all $x \in C$ such that

$$\langle f(x), y - x \rangle \geq -\epsilon \|y - x\| - \epsilon \text{ for all } y \in C.$$

In Chap. 12, we present examples which provide simple and clear estimations for the sets S_ϵ in some important cases. These examples show that elements of S_ϵ can be considered as ϵ-approximate solutions of the variational inequality.

In Chap. 12, in order to solve the variational inequality (to find $x \in S$), we use the algorithm known in the literature as the extragradient method [75]. In each iteration of this algorithm, in order to get the next iterate x_{k+1}, two orthogonal projections onto C are calculated, according to the following iterative step. Given the current iterate x_k calculate $y_k = P_C(x_k - \tau_k f(x_k))$ and then

$$x_{k+1} = P_C(x_k - \tau_k f(y_k)),$$

where τ_k is some positive number. It is known that this algorithm generates sequences which converge to an element of S. In Chap. 12, we study the behavior of the sequences generated by the algorithm taking into account computational errors which are always present in practice. Namely, in practice the algorithm generates sequences $\{x_k\}_{k=0}^{\infty}$ and $\{y_k\}_{k=0}^{\infty}$ such that for each integer $k \geq 0$,

$$\|y_k - P_C(x_k - \tau_k f(x_k))\| \leq \delta$$

and

$$\|x_{k+1} - P_C(x_k - \tau_k f(y_k))\| \leq \delta,$$

with a constant $\delta > 0$ which depends only on our computer system. Surely, in this situation one cannot expect that the sequence $\{x_k\}_{k=0}^{\infty}$ converges to the set S. The goal is to understand what subset of C attracts all sequences $\{x_k\}_{k=0}^{\infty}$ generated by the algorithm. The main result of Chap. 12 (Theorem 12.2) shows that this subset of C is the set S_ϵ with some $\epsilon > 0$ depending on δ. The examples considered in Chap. 12 show that one cannot expect to find an attracting set smaller than S_ϵ, whose elements can be considered as approximate solutions of the variational inequality.

Chapter 2
Subgradient Projection Algorithm

In this chapter we study the subgradient projection algorithm for minimization of convex and nonsmooth functions and for computing the saddle points of convex–concave functions, under the presence of computational errors. We show that our algorithms generate a good approximate solution, if computational errors are bounded from above by a small positive constant. Moreover, for a known computational error, we find out what an approximate solution can be obtained and how many iterates one needs for this.

2.1 Preliminaries

The subgradient projection algorithm is one of the most important tools in the optimization theory and its applications. See, for example, [1–3, 12, 30, 44, 51, 79, 89, 92, 95, 96, 105, 108, 109, 112] and the references mentioned therein.

In this chapter we use this method for constrained minimization problems in Hilbert spaces equipped with an inner product denoted by $\langle \cdot, \cdot \rangle$ which induces a complete norm $\| \cdot \|$. For every $z \in R^1$ denote by $\lfloor z \rfloor$ the largest integer which does not exceed z: $\lfloor z \rfloor = \max\{i \in R^1 : i \text{ is an integer and } i \leq z\}$.

Let X be a Hilbert space. For each $x \in X$ and each $r > 0$ set

$$B_X(x, r) = \{y \in X : \|x - y\| \leq r\}.$$

For each $x \in X$ and each nonempty set $E \subset X$ set

$$d(x, E) = \inf\{\|x - y\| : y \in E\}.$$

Let C be a nonempty closed convex subset of X, U be an open convex subset of X such that $C \subset U$ and let $f : U \to R^1$ be a convex function. Recall that for each $x \in U$,

© Springer International Publishing Switzerland 2016
A.J. Zaslavski, *Numerical Optimization with Computational Errors*, Springer Optimization and Its Applications 108, DOI 10.1007/978-3-319-30921-7_2

$$\partial f(x) = \{l \in X : f(y) - f(x) \geq \langle l, y - x \rangle \text{ for all } y \in U\}. \tag{2.1}$$

Suppose that there exist $L > 0$, $M_0 > 0$ such that

$$C \subset B_X(0, M_0), \tag{2.2}$$

$$|f(x) - f(y)| \leq L\|x - y\| \text{ for all } x, y \in U. \tag{2.3}$$

In view of (2.1) and (2.3), for each $x \in U$,

$$\emptyset \neq \partial f(x) \subset B_X(0, L). \tag{2.4}$$

It is easy to see that the following result is true.

Lemma 2.1. *Let $z, y_0, y_1 \in X$. Then*

$$\|z - y_0\|^2 - \|z - y_1\|^2 - \|y_0 - y_1\|^2 = 2\langle z - y_1, y_1 - y_0 \rangle.$$

The next result is given in [13, 14].

Lemma 2.2. *Let D be a nonempty closed convex subset of X. Then for each $x \in X$ there is a unique point $P_D(x) \in D$ satisfying*

$$\|x - P_D(x)\| = \inf\{\|x - y\| : y \in D\}.$$

Moreover,

$$\|P_D(x) - P_D(y)\| \leq \|x - y\| \text{ for all } x, y \in X$$

and for each $x \in X$ and each $z \in D$,

$$\langle z - P_D(x), x - P_D(x) \rangle \leq 0,$$
$$\|z - P_D(x)\|^2 + \|x - P_D(x)\|^2 \leq \|z - x\|^2.$$

Lemma 2.3. *Let $A > 0$ and $n \geq 2$ be an integer. Then the minimization problem*

$$\sum_{i=1}^{n} a_i^2 \to \min$$

$$a = (a_1, \ldots, a_n) \in R^n \text{ and } \sum_{i=1}^{n} a_i = A$$

has a unique solution $a^* = (a_1^*, \ldots, a_n^*)$ where $a_i^* = n^{-1}A$, $i = 1, \ldots, n$.

Proof. Clearly, the minimization problem has a solution $a^* = (a_1^*, \ldots, a_n^*) \in R^n$. Then

$$a_n^* = A - \sum_{i=1}^{n-1} a_i^*$$

and $(a_1^*, \ldots, a_{n-1}^*)$ is a minimizer of the function

$$\phi(a_1, \ldots, a_{n-1}) := \sum_{i=1}^{n-1} a_i^2 + \left(A - \sum_{i=1}^{n-1} a_i \right)^2, \quad (a_1, \ldots, a_{n-1}) \in R^{n-1}.$$

It is clear that for all $i = 1, \ldots, n-1$,

$$0 = (\partial \phi / \partial a_i)(a_1^*, \ldots, a_{n-1}^*) = 2a_i^* - 2 \left(A - \sum_{i=1}^{n-1} a_i^* \right) = 2a_i^* - 2a_n^*.$$

Thus $a_i^* = a_n^*$ for all $i = 1, \ldots, n-1$ and $a_i^* = n^{-1} A$ for all $i = 1, \ldots, n$. Lemma 2.3 is proved.

2.2 A Convex Minimization Problem

Let $\delta \in (0, 1]$ and $\{a_k\}_{k=0}^{\infty} \subset (0, \infty)$.

Let us describe our algorithm.

Subgradient Projection Algorithm

Initialization: select an arbitrary $x_0 \in U$.

Iterative step: given a current iteration vector $x_t \in U$ calculate

$$\xi_t \in \partial f(x_t) + B_X(0, \delta)$$

and the next iteration vector $x_{t+1} \in U$ such that

$$\|x_{t+1} - P_C(x_t - a_t \xi_t)\| \leq \delta.$$

In this chapter we prove the following result.

Theorem 2.4. *Let* $\delta \in (0, 1]$, $\{a_k\}_{k=0}^{\infty} \subset (0, \infty)$ *and let*

$$x_* \in C \tag{2.5}$$

satisfies

$$f(x_*) \le f(x) \text{ for all } x \in C. \tag{2.6}$$

Assume that $\{x_t\}_{t=0}^{\infty} \subset U, \{\xi_t\}_{t=0}^{\infty} \subset X,$

$$\|x_0\| \le M_0 + 1 \tag{2.7}$$

and that for each integer $t \ge 0,$

$$\xi_t \in \partial f(x_t) + B_X(0, \delta) \tag{2.8}$$

and

$$\|x_{t+1} - P_C(x_t - a_t \xi_t)\| \le \delta. \tag{2.9}$$

Then for each natural number $T,$

$$\sum_{t=0}^{T} a_t(f(x_t) - f(x_*))$$

$$\le 2^{-1}\|x_* - x_0\|^2 + \delta(T+1)(4M_0 + 1)$$

$$+ \delta(2M_0 + 1)\sum_{t=0}^{T} a_t + 2^{-1}(L+1)^2 \sum_{t=0}^{T} a_t^2. \tag{2.10}$$

Moreover, for each natural number $T,$

$$f\left(\left(\sum_{t=0}^{T} a_t\right)^{-1} \sum_{t=0}^{T} a_t x_t\right) - f(x_*), \quad \min\{f(x_t): \ t = 0, \dots, T\} - f(x_*)$$

$$\le 2^{-1}\left(\sum_{t=0}^{T} a_t\right)^{-1} \|x_* - x_0\|^2 + \left(\sum_{t=0}^{T} a_t\right)^{-1} \delta(T+1)(4M_0 + 1)$$

$$+ \delta(2M_0 + 1) + 2^{-1}\left(\sum_{t=0}^{T} a_t\right)^{-1} (L+1)^2 \sum_{t=0}^{T} a_t^2. \tag{2.11}$$

Theorem 2.4 is proved in Sect. 2.4.

We are interested in an optimal choice of $a_t,$ $t = 0, 1, \dots.$ Let T be a natural number and $A_T = \sum_{t=0}^{T} a_t$ be given. By Theorem 2.4, in order to make the best choice of $a_t,$ $t = 0, \dots, T,$ we need to minimize the function

$$\phi(a_0, \dots, a_T) := 2^{-1}A_T^{-1}\|x_* - x_0\|^2 + A_T^{-1}\delta(T+1)(4M_0 + 1)$$

$$+\delta(2M_0 + 1) + 2^{-1}A_T^{-1}(L + 1)^2 \sum_{t=0}^{T} a_t^2$$

on the set

$$\left\{ a = (a_0, \ldots, a_T) \in R^{T+1} : a_i \geq 0, \ i = 0, \ldots, T, \ \sum_{i=0}^{T} a_i = A_T \right\}.$$

By Lemma 2.3, this function has a unique minimizer $a^* = (a_0^*, \ldots, a_T^*)$ where $a_i^* = (T + 1)^{-1}A_T, i = 0, \ldots, T$. This is the best choice of $a_t, t = 0, 1, \ldots, T$.

Theorem 2.4 implies the following result.

Theorem 2.5. *Let $\delta \in (0, 1]$, $a > 0$ and let $x_* \in C$ satisfies*

$$f(x_*) \leq f(x) \text{ for all } x \in C.$$

Assume that $\{x_t\}_{t=0}^{\infty} \subset U, \{\xi_t\}_{t=0}^{\infty} \subset X,$

$$\|x_0\| \leq M_0 + 1$$

and that for each integer $t \geq 0$,

$$\xi_t \in \partial f(x_t) + B_X(0, \delta)$$

and

$$\|x_{t+1} - P_C(x_t - a\xi_t)\| \leq \delta.$$

Then for each natural number T,

$$f\left((T + 1)^{-1}\sum_{t=0}^{T} x_t\right) - f(x_*), \ \min\{f(x_t) : t = 0, \ldots, T\} - f(x_*)$$

$$\leq 2^{-1}(T + 1)^{-1}a^{-1}(2M_0 + 1)^2 + a^{-1}\delta(4M_0 + 1)$$

$$+ \delta(2M_0 + 1) + 2^{-1}(L + 1)^2 a.$$

Now we will find the best $a > 0$. Since T can be arbitrary large, we need to find a minimizer of the function

$$\phi(a) := a^{-1}\delta(4M_0 + 1) + 2^{-1}(L + 1)^2 a, \ a \in (0, \infty).$$

Clearly, the minimizer a satisfies

$$a^{-1}\delta(4M_0 + 1) = 2^{-1}(L + 1)^2 a$$

and

$$a = (2\delta(4M_0 + 1))^{1/2}(L + 1)^{-1}$$

and the minimal value of ϕ is

$$(2\delta(4M_0 + 1))^{1/2}(L + 1). \tag{2.12}$$

Theorem 2.5 implies the following result.

Theorem 2.6. *Let $\delta \in (0, 1]$,*

$$a = (2\delta(4M_0 + 1))^{1/2}(L + 1)^{-1},$$

$x_* \in C$ *satisfies*

$$f(x_*) \leq f(x) \text{ for all } x \in C.$$

Assume that $\{x_t\}_{t=0}^{\infty} \subset U$, $\{\xi_t\}_{t=0}^{\infty} \subset X$,

$$\|x_0\| \leq M_0 + 1$$

and that for each integer $t \geq 0$,

$$\xi_t \in \partial f(x_t) + B_X(0, \delta)$$

and

$$\|x_{t+1} - P_C(x_t - a\xi_t)\| \leq \delta.$$

Then for each natural number T,

$$f\left((T + 1)^{-1}\sum_{t=0}^{T} x_t\right) - f(x_*), \ \min\{f(x_t) : \ t = 0, \ldots, T\} - f(x_*)$$

$$\leq 2^{-1}(T + 1)^{-1}(2M_0 + 1)^2(L + 1)(2\delta(4M_0 + 1))^{-1/2} + \delta(2M_0 + 1)$$

$$+ 2^{-1}(2\delta(4M_0 + 1))^{1/2}(L + 1) + \delta(4M_0 + 1)(L + 1)(2\delta(4M_0 + 1))^{-1/2}.$$

Now we can think about the best choice of T. It is clear that it should be at the same order as $\lfloor \delta^{-1} \rfloor$. Putting $T = \lfloor \delta^{-1} \rfloor$, we obtain that

$$f\left((T+1)^{-1}\sum_{t=0}^{T}x_t\right)-f(x_*),\ \min\{f(x_t):\ t=0,\ldots,T\}-f(x_*)$$

$$\leq 2^{-1}(2M_0+1)^2(L+1)(8M_0+2)^{-1/2}\delta^{1/2}+\delta(2M_0+1)$$
$$+(8M_0+2)^{1/2}(L+1)\delta^{1/2}+(4M_0+1)(L+1)(8M_0+2)^{-1/2}\delta^{1/2}.$$

Note that in the theorems above δ is the computational error produced by our computer system.

In view of the inequality above, which has the right-hand side bounded by $c_1\delta^{1/2}$ with a constant $c_1 > 0$, we conclude that after $T = \lfloor\delta^{-1}\rfloor$ iterations we obtain a point $\xi \in U$ such that

$$B_X(\xi,\delta)\cap C \neq \emptyset$$

and

$$f(\xi) \leq f(x_*)+c_1\delta^{1/2},$$

where the constant $c_1 > 0$ depends only on L and M_0.

2.3   The Main Lemma

We use the notation and definitions introduced in Sect. 2.1.

Lemma 2.7. *Let $\delta \in (0,1]$, $a > 0$ and let*

$$z \in C. \tag{2.13}$$

Assume that

$$x \in U \cap B_X(0,M_0+1), \tag{2.14}$$

$$\xi \in \partial f(x)+B_X(0,\delta) \tag{2.15}$$

and that

$$u \in U \tag{2.16}$$

satisfies

$$\|u-P_C(x-a\xi)\| \leq \delta. \tag{2.17}$$

Then

$$a(f(x) - f(z)) \leq 2^{-1}\|z - x\|^2 - 2^{-1}\|z - u\|^2$$
$$+ \delta(4M_0 + 1 + a(2M_0 + 1)) + 2^{-1}a^2(L + 1)^2.$$

Proof. In view of (2.15), there exists

$$l \in \partial f(x) \tag{2.18}$$

such that

$$\|l - \xi\| \leq \delta. \tag{2.19}$$

By Lemmas 2.1 and 2.2 and (2.13),

$$0 \leq \langle z - P_C(x - a\xi), P_C(x - a\xi) - (x - a\xi) \rangle$$
$$= \langle z - P_C(x - a\xi), P_C(x - a\xi) - x \rangle$$
$$+ \langle a\xi, z - P_C(x - a\xi) \rangle$$
$$= 2^{-1}[\|z - x\|^2 - \|z - P_C(x - a\xi)\|^2 - \|x - P_C(x - a\xi)\|^2]$$
$$+ \langle a\xi, z - x \rangle + \langle a\xi, x - P_C(x - a\xi) \rangle. \tag{2.20}$$

Clearly,

$$|\langle a\xi, x - P_C(x - a\xi) \rangle| \leq 2^{-1}(\|a\xi\|^2 + \|x - P_C(x - a\xi)\|^2). \tag{2.21}$$

It follows from (2.20) and (2.21) that

$$0 \leq 2^{-1}[\|z - x\|^2 - \|z - P_C(x - a\xi)\|^2 - \|x - P_C(x - a\xi)\|^2]$$
$$+ \langle a\xi, z - x \rangle + 2^{-1}a^2\|\xi\|^2 + 2^{-1}\|x - P_C(x - a\xi)\|^2$$
$$\leq 2^{-1}\|z - x\|^2 - 2^{-1}\|z - P_C(x - a\xi)\|^2 + 2^{-1}a^2\|\xi\|^2 + \langle a\xi, z - x \rangle. \tag{2.22}$$

Relations (2.2), (2.13), and (2.17) imply that

$$|\|z - P_C(x - a\xi)\|^2 - \|z - u\|^2|$$
$$= |\|z - P_C(x - a\xi)\| - \|z - u\||(\|z - P_C(x - a\xi)\| + \|z - u\|)$$
$$\leq \|u - P_C(x - a\xi)\|(4M_0 + 1) \leq (4M_0 + 1)\delta. \tag{2.23}$$

By (2.2), (2.13), (2.14), and (2.19),

$$\langle a\xi, z - x \rangle = \langle al, z - x \rangle + \langle a(\xi - l), z - x \rangle$$

$$\leq \langle al, z - x \rangle + a\|\xi - l\|\|z - x\|$$
$$\leq \langle al, z - x \rangle + a\delta(2M_0 + 1). \tag{2.24}$$

It follows from (2.4), (2.18), (2.19), (2.22), (2.23), and (2.24) that

$$0 \leq 2^{-1}\|z - x\|^2 - 2^{-1}\|z - P_C(x - a\xi)\|^2 + 2^{-1}a^2\|\xi\|^2 + \langle a\xi, z - x \rangle$$
$$\leq 2^{-1}\|z - x\|^2 - 2^{-1}\|z - u\|^2 + \delta(4M_0 + 1) + 2^{-1}a^2(L + 1)^2$$
$$+ \langle al, z - x \rangle + a\delta(2M_0 + 1). \tag{2.25}$$

By (2.1), (2.18), and (2.25),

$$a(f(z) - f(x)) \geq \langle al, z - x \rangle$$

and

$$a(f(x) - f(z)) \leq \langle al, x - z \rangle$$
$$\leq 2^{-1}\|z - x\|^2 - 2^{-1}\|z - u\|^2 + \delta(4M_0 + 1) + 2^{-1}a^2(L + 1)^2$$
$$+ a\delta(2M_0 + 1).$$

This completes the proof of Lemma 2.7.

2.4   Proof of Theorem 2.4

It is clear that

$$\|x_t\| \leq M_0 + 1, \ t = 0, 1, \ldots.$$

Let $t \geq 0$ be an integer. Applying Lemma 2.7 with

$$z = x_*, \ a = a_t, \ x = x_t, \ \xi = \xi_t, \ u = x_{t+1}$$

we obtain that

$$a_t(f(x_t) - f(x_*)) \leq 2^{-1}\|x_* - x_t\|^2 - 2^{-1}\|x_* - x_{t+1}\|^2$$
$$+ \delta(4M_0 + 1 + a_t(2M_0 + 1)) + 2^{-1}a_t^2(L + 1)^2. \tag{2.26}$$

By (2.26), for each natural number T,

$$\sum_{t=0}^{T} a_t(f(x_t) - f(x_*))$$

$$\leq \sum_{t=0}^{T} (2^{-1} \|x_* - x_t\|^2 - 2^{-1} \|x_* - x_{t+1}\|^2$$

$$+ \delta(4M_0 + 1) + a_t(2M_0 + 1)\delta + 2^{-1} a_t^2 (L + 1)^2)$$

$$\leq 2^{-1} \|x_* - x_0\|^2 + \delta(T + 1)(4M_0 + 1)$$

$$+ \delta(2M_0 + 1) \sum_{t=0}^{T} a_t + 2^{-1}(L + 1)^2 \sum_{t=0}^{T} a_t^2.$$

Thus (2.10) is true. Evidently, (2.10) implies (2.11). Theorem 2.4 is proved.

2.5   Subgradient Algorithm on Unbounded Sets

We use the notation and definitions introduced in Sect. 2.1. Let X be a Hilbert space with an inner product $\langle \cdot, \cdot \rangle$, D be a nonempty closed convex subset of X, V be an open convex subset of X such that

$$D \subset V, \tag{2.27}$$

and $f : V \to R^1$ be a convex function which is Lipschitz on all bounded subsets of V. Set

$$D_{\min} = \{x \in D : f(x) \leq f(y) \text{ for all } y \in D\}. \tag{2.28}$$

We suppose that

$$D_{\min} \neq \emptyset. \tag{2.29}$$

We will prove the following result.

Theorem 2.8. *Let* $\delta \in (0, 1]$, $M > 0$ *satisfy*

$$D_{\min} \cap B_X(0, M) \neq \emptyset, \tag{2.30}$$

$$M_0 \geq 4M + 4, \tag{2.31}$$

$L > 0$ *satisfy*

$$|f(v_1) - f(v_2)| \leq L \|v_1 - v_2\| \text{ for all } v_1, v_2 \in V \cap B_X(0, M_0 + 2), \tag{2.32}$$

$$0 < \tau_0 \leq \tau_1 \leq (L+1)^{-1}, \tag{2.33}$$

$$\epsilon_0 = 2\tau_0^{-1}\delta(4M_0 + 1) + 2\delta(2M_0 + 1) + 2\tau_1(L+1)^2 \tag{2.34}$$

and let

$$n_0 = \lfloor \tau_0^{-1}(2M + 2)^2 \epsilon_0^{-1} \rfloor. \tag{2.35}$$

Assume that $\{x_t\}_{t=0}^\infty \subset V$, $\{\xi_t\}_{t=0}^\infty \subset X$,

$$\{a_t\}_{t=0}^\infty \subset [\tau_0, \tau_1], \tag{2.36}$$

$$\|x_0\| \leq M \tag{2.37}$$

and that for each integer $t \geq 0$,

$$\xi_t \in \partial f(x_t) + B_X(0, \delta) \tag{2.38}$$

and

$$\|x_{t+1} - P_D(x_t - a_t\xi_t)\| \leq \delta. \tag{2.39}$$

Then there exists an integer $q \in [1, n_0 + 1]$ *such that*

$$\|x_i\| \leq 3M + 2, \quad i = 0, \ldots, q$$

and

$$f(x_q) \leq f(x) + \epsilon_0 \text{ for all } x \in D.$$

We are interested in the best choice of a_t, $t = 0, 1, \ldots$. Assume for simplicity that $\tau_1 = \tau_0$. In order to meet our goal we need to minimize the function

$$2\tau^{-1}\delta(4M_0 + 1) + 2(L+1)^2\tau, \quad \tau \in (0, \infty).$$

This function has a minimizer

$$\tau = (\delta(4M_0 + 1))^{1/2}(L+1)^{-1},$$

the minimal value of ϵ_0 is

$$2\delta(2M_0 + 1) + 4(\delta(4M_0 + 1))^{1/2}(L+1)$$

and $n_0 = \lfloor \Delta \rfloor$ where

$$\Delta = (2(\delta(4M_0 + 1))^{1/2}(L+1)^{-1})^{-1}(2M + 2)^2(2\delta(2M_0 + 1)$$

$$+ 4(\delta(4M_0 + 1))^{1/2}(L + 1)^{-1})$$
$$\leq \delta^{-1/2}(4M_0 + 1)^{-1/2}(L + 1)(2M + 2)^2(4L + 4)^{-1}(4M_0 + 1)^{-1/2}\delta^{-1/2}$$
$$= \delta^{-1}(4M_0 + 1)^{-1}(2M + 2)^2 4^{-1}.$$

Note that in the theorem above δ is the computational error produced by our computer system. In view of the inequality above, in order to obtain a good approximate solution we need $\lfloor c_1 \delta^{-1} \rfloor + 1$ iterations, where

$$c_1 = 4^{-1}(4M_0 + 1)^{-1}(2M + 1)^2.$$

As a result, we obtain a point $\xi \in V$ such that

$$B_X(\xi, \delta) \cap D \neq \emptyset$$

and

$$f(\xi) \leq \inf\{f(x) : x \in D\} + c_2 \delta^{1/2},$$

where the constant $c_2 > 0$ depends only on L and M_0.

2.6   Proof of Theorem 2.8

By (2.30) there exists

$$z \in D_{\min} \cap B_X(0, M). \tag{2.40}$$

Assume that T is a natural number and that

$$f(x_t) - f(z) > \epsilon_0, \ t = 1, \ldots, T. \tag{2.41}$$

Lemma 2.2, (2.36), (2.37), (2.39), and (2.40) imply that

$$\|x_1 - z\| \leq \|x_1 - P_D(x_0 - a_0 \xi_0)\| + \|P_D(x_0 - a_0 \xi_0) - z\|$$
$$\leq \delta + \|x_0 - z\| + a_0 \|\xi_0\| \leq 1 + 2M + \tau_1 \|\xi_0\|. \tag{2.42}$$

In view of (2.32), (2.37), and (2.38),

$$\xi_0 \in \partial f(x_0) + B_X(0, 1) \subset B_X(0, L) + 1,$$
$$\|\xi_0\| \leq L + 1. \tag{2.43}$$

It follows from (2.33), (2.40), (2.42), and (2.43) that

$$\|x_1 - z\| \leq 2M + 2, \tag{2.44}$$

$$\|x_1\| \leq 3M + 2. \tag{2.45}$$

Set

$$U = V \cap \{v \in X : \|v\| < M_0 + 2\} \tag{2.46}$$

and

$$C = D \cap B_X(0, M_0). \tag{2.47}$$

By induction we show that for every integer $t \in [1, T]$,

$$\|x_t - z\| \leq 2M + 2, \tag{2.48}$$

$$f(x_t) - f(z)$$
$$\leq (2\tau_0)^{-1}(\|z - x_t\|^2 - \|z - x_{t+1}\|^2)$$
$$+ \tau_0^{-1}\delta(4M_0 + 1) + \delta(2M_0 + 1) + 2^{-1}\tau_1(L + 1)^2. \tag{2.49}$$

In view of (2.44), (2.48) holds for $t = 1$.

Assume that an integer $t \in [1, T]$ and that (2.48) holds. It follows from (2.31), (2.40), (2.46), (2.47), and (2.48) that

$$z \in C \subset B_X(0, M_0), \tag{2.50}$$

$$x_t \in U \cap B_X(0, M_0 + 1). \tag{2.51}$$

Relation (2.39) implies that $x_{t+1} \in V$ satisfies

$$\|x_{t+1} - P_D(x_t - a_t\xi_t)\| \leq 1. \tag{2.52}$$

By (2.32), (2.38), and (2.51),

$$\xi_t \in \partial f(x_t) + B_X(0, 1) \subset B_X(0, L + 1). \tag{2.53}$$

It follows from (2.33), (2.36), (2.40), (2.48), (2.53), and Lemma 2.2 that

$$\|z - P_D(x_t - a_t\xi_t)\| \leq \|z - x_t + a_t\xi_t\|$$
$$\leq \|z - x_t\| + \|\xi_t\|a_t \leq 2M + 3,$$
$$\|P_D(x_t - a_t\xi_t)\| \leq 3M + 3. \tag{2.54}$$

In view of (2.47) and (2.54),

$$P_D(x_t - a_t\xi_t) \in C, \qquad (2.55)$$

and

$$P_D(x_t - a_t\xi_t) = P_C(x_t - a_t\xi_t). \qquad (2.56)$$

Relations (2.44), (2.52), and (2.54) imply that

$$\|x_{t+1}\| \leq 3M + 4, \ x_{t+1} \in U. \qquad (2.57)$$

By (2.32), (2.38), (2.39), (2.46), (2.47), (2.50), (2.51), (2.55), (2.56), (2.57), and Lemma 2.7 which holds with

$$x = x_t, \ a = a_t, \ \xi = \xi_t, \ u = x_{t+1},$$

we have

$$a_t(f(x_t) - f(z)) \leq 2^{-1}\|z - x_t\|^2 - 2^{-1}\|z - x_{t+1}\|^2$$
$$+ \delta(4M_0 + 1 + a_t(2M_0 + 1)) + 2^{-1}a_t^2(L + 1)^2.$$

The relation above, (2.34) and (2.36) imply that

$$f(x_t) - f(z) \leq (2\tau_0)^{-1}\|z - x_t\|^2 - (2\tau_0)^{-1}\|z - x_{t+1}\|^2$$
$$+ \tau_0^{-1}\delta(4M_0 + 1) + (2M_0 + 1)\delta + 2^{-1}\tau_1(L + 1)^2. \qquad (2.58)$$

In view of (2.41), (2.58) and the inclusion $t \in [1, T]$,

$$\|z - x_t\|^2 - \|z - x_{t+1}\|^2 \geq 0,$$
$$\|z - x_{t+1}\| \leq \|z - x_t\| \leq 2M + 2. \qquad (2.59)$$

Therefore we assumed that (2.48) is true and showed that (2.58) and (2.59) hold. Hence by induction we showed that (2.49) holds for all $t = 1, \ldots, T$ and (2.48) holds for all $t = 1, \ldots, T + 1$.

It follows from (2.49) which holds for all $t = 1, \ldots, T$, (2.41) and (2.44) that

$$T\epsilon_0 < T(\min\{f(x_t) : \ t = 1, \ldots, T\} - f(z))$$
$$\leq \sum_{t=1}^{T}(f(x_t) - f(z))$$

$$\leq (2\tau_0)^{-1} \sum_{t=1}^{T} (\|z - x_t\|^2 - \|z - x_{t+1}\|^2)$$

$$+ T\tau_0^{-1}\delta(4M_0 + 1) + T(2M_0 + 1)\delta + 2^{-1}T\tau_1(L+1)^2$$
$$\leq (2\tau_0)^{-1}(2M+2)^2 + T\tau_0^{-1}\delta(4M_0 + 1)$$
$$+ T(2M_0 + 1)\delta + 2^{-1}T\tau_1(L+1)^2.$$

Together with (2.34) and (2.35) this implies that

$$\epsilon_0 < (2\tau_0 T)^{-1}(2M+2)^2 + \tau_0^{-1}\delta(4M_0 + 1)$$
$$+ (2M_0 + 1)\delta + 2^{-1}\tau_1(L+1)^2,$$
$$2^{-1}\epsilon_0 < (2\tau_0 T)^{-1}(2M+2)^2,$$
$$T < \tau_0^{-1}(2M+2)^2\epsilon_0^{-1} \leq n_0 + 1.$$

Thus we have shown that if an integer T satisfies (2.41), then $T \leq n_0$ and

$$\|z - x_t\| \leq 2M + 2, \ t = 1, \ldots, T + 1,$$
$$\|x_t\| \leq 3M + 2, \ t = 0, \ldots, T + 1.$$

This implies that there exists an integer $q \in [1, n_0 + 1]$ such that

$$\|x_t\| \leq 3M + 2, \ t = 0, \ldots, q$$

and

$$f(x_q) - f(z) \leq \epsilon_0.$$

Theorem 2.8 is proved.

2.7 Zero-Sum Games with Two-Players

We use the notation and definitions introduced in Sect. 2.1.

Let X, Y be Hilbert spaces, C be a nonempty closed convex subset of X, D be a nonempty closed convex subset of Y, U be an open convex subset of X, and V be an open convex subset of Y such that

$$C \subset U, \ D \subset V \tag{2.60}$$

and let a function $f : U \times V \to R^1$ possess the following properties:

(i) for each $v \in V$, the function $f(\cdot, v) : U \to R^1$ is convex;
(ii) for each $u \in U$, the function $f(u, \cdot) : V \to R^1$ is concave.

Assume that a function $\phi : R^1 \to [0, \infty)$ is bounded on all bounded sets and positive numbers M_0, L satisfy

$$C \subset B_X(0, M_0), \ D \subset B_Y(0, M_0), \tag{2.61}$$

$$|f(u, v_1) - f(u, v_2)| \le L\|v_1 - v_2\| \text{ for all } u \in U \text{ and all } v_1, v_2 \in V, \tag{2.62}$$

$$|f(u_1, v) - f(u_2, v)| \le L\|u_1 - u_2\| \text{ for all } v \in V \text{ and all } u_1, u_2 \in U. \tag{2.63}$$

Let

$$x_* \in C \text{ and } y_* \in D \tag{2.64}$$

satisfy

$$f(x_*, y) \le f(x_*, y_*) \le f(x, y_*) \tag{2.65}$$

for each $x \in C$ and each $y \in D$.

In the next section we prove the following result.

Proposition 2.9. *Let T be a natural number, $\delta \in (0, 1]$, $\{a_t\}_{t=0}^T \subset (0, \infty)$ and let $\{b_t\}_{t=0}^T \subset (0, \infty)$. Assume that $\{x_t\}_{t=0}^{T+1} \subset U$, $\{y_t\}_{t=0}^{T+1} \subset V$, for each $t \in \{0, \ldots, T+1\}$,*

$$B(x_t, \delta) \cap C \ne \emptyset, \ B(y_t, \delta) \cap D \ne \emptyset, \tag{2.66}$$

for each $z \in C$ and each $t \in \{0, \ldots, T\}$,

$$a_t(f(x_t, y_t) - f(z, y_t)) \le \phi(\|z - x_t\|) - \phi(\|z - x_{t+1}\|) + b_t \tag{2.67}$$

and that for each $v \in D$ and each $t \in \{0, \ldots, T\}$,

$$a_t(f(x_t, v) - f(x_t, y_t)) \le \phi(\|v - y_t\|) - \phi(\|v - y_{t+1}\|) + b_t. \tag{2.68}$$

Let

$$\hat{x}_T = \left(\sum_{i=0}^T a_i\right)^{-1} \sum_{t=0}^T a_t x_t,$$

$$\hat{y}_T = \left(\sum_{i=0}^T a_i\right)^{-1} \sum_{t=0}^T a_t y_t. \tag{2.69}$$

Then

$$B(\hat{x}_T, \delta) \cap C \neq \emptyset, \ B(\hat{y}_T, \delta) \cap D \neq \emptyset, \tag{2.70}$$

$$\left| \left(\sum_{t=0}^{T} a_t \right)^{-1} \sum_{t=0}^{T} a_t f(x_t, y_t) - f(x_*, y_*) \right|$$

$$\leq \left(\sum_{t=0}^{T} a_t \right)^{-1} \sum_{t=0}^{T} b_t + \left(\sum_{t=0}^{T} a_t \right)^{-1} \sup\{\phi(u) : \ u \in [0, 2M_0 + 1]\}, \tag{2.71}$$

$$\left| f(\hat{x}_T, \hat{y}_T) - \left(\sum_{t=0}^{T} a_t \right)^{-1} \sum_{t=0}^{T} a_t f(x_t, y_t) \right|$$

$$\leq \left(\sum_{t=0}^{T} a_t \right)^{-1} \sum_{t=0}^{T} b_t + L\delta$$

$$+ \left(\sum_{t=0}^{T} a_t \right)^{-1} \sup\{\phi(u) : \ u \in [0, 2M_0 + 1]\}, \tag{2.72}$$

and for each $z \in C$ and each $v \in D$,

$$f(z, \hat{y}_T) \geq f(\hat{x}_T, \hat{y}_T)$$

$$- 2 \left(\sum_{t=0}^{T} a_t \right)^{-1} \sup\{\phi(s) : \ s \in [0, 2M_0 + 1]\}$$

$$- 2 \left(\sum_{t=0}^{T} a_t \right)^{-1} \sum_{t=0}^{T} b_t - L\delta, \tag{2.73}$$

$$f(\hat{x}_T, v) \leq f(\hat{x}_T, \hat{y}_T)$$

$$+ 2 \left(\sum_{t=0}^{T} a_t \right)^{-1} \sup\{\phi(s) : \ s \in [0, 2M_0 + 1]\}$$

$$+ 2 \left(\sum_{t=0}^{T} a_t \right)^{-1} \sum_{t=0}^{T} b_t + L\delta. \tag{2.74}$$

Corollary 2.10. *Suppose that all the assumptions of Proposition 2.9 hold and that*

$$\tilde{x} \in C, \ \tilde{y} \in D$$

satisfy

$$\|\hat{x}_T - \tilde{x}\| \leq \delta, \ \|\hat{y}_T - \tilde{y}\| \leq \delta. \tag{2.75}$$

Then

$$|f(\tilde{x}, \tilde{y}) - f(\hat{x}_T, \hat{y}_T)| \le 2L\delta \qquad (2.76)$$

and for each $z \in C$ and each $v \in D$,

$$f(z, \tilde{y}) \ge f(\tilde{x}, \tilde{y})$$

$$-2 \left(\sum_{t=0}^{T} a_t \right)^{-1} \sup\{\phi(s) : s \in [0, 2M_0 + 1]\}$$

$$-2 \left(\sum_{t=0}^{T} a_t \right)^{-1} \sum_{t=0}^{T} b_t - 4L\delta,$$

$$f(\tilde{x}, v) \le f(\tilde{x}, \tilde{y})$$

$$+2 \left(\sum_{t=0}^{T} a_t \right)^{-1} \sup\{\phi(s) : s \in [0, 2M_0 + 1]\}$$

$$+2 \left(\sum_{t=0}^{T} a_t \right)^{-1} \sum_{t=0}^{T} b_t + 4L\delta.$$

Proof. In view of (2.62), (2.63), and (2.75),

$$|f(\tilde{x}, \tilde{y}) - f(\hat{x}_T, \hat{y}_T)|$$
$$\le |f(\tilde{x}, \tilde{y}) - f(\tilde{x}, \hat{y}_T)| + |f(\tilde{x}, \hat{y}_T) - f(\hat{x}_T, \hat{y}_T)|$$
$$\le L\|\tilde{y} - \hat{y}_T\| + L\|\tilde{x} - \hat{x}_T\| \le 2L\delta$$

and (2.76) holds.

Let $z \in C$ and $v \in D$. Relations (2.62), (2.63), and (2.75) imply that

$$|f(z, \tilde{y}) - f(z, \hat{y}_T)| \le L\delta,$$
$$|f(\tilde{x}, v) - f(\hat{x}_T, v)| \le L\delta.$$

By the relation above, (2.73), (2.74), and (2.75),

$$f(z, \tilde{y}) \ge f(z, \hat{y}_T) - L\delta$$

$$\ge f(\hat{x}_T, \hat{y}_T) - 2 \left(\sum_{t=0}^{T} a_t \right)^{-1} \sup\{\phi(s) : s \in [0, 2M_0 + 1]\}$$

$$-2 \left(\sum_{t=0}^{T} a_t \right)^{-1} \sum_{t=0}^{T} b_t - 2L\delta$$

$$\geq f(\tilde{x}, \tilde{y}) - 2 \left(\sum_{t=0}^{T} a_t \right)^{-1} \sup\{\phi(s) : s \in [0, 2M_0 + 1]\}$$

$$- 2 \left(\sum_{t=0}^{T} a_t \right)^{-1} \sum_{t=0}^{T} b_t - 4L\delta,$$

$$f(\tilde{x}, v) \leq f(\hat{x}_T, v) + L\delta$$

$$\leq f(\hat{x}_T, \hat{y}_T) + 2 \left(\sum_{t=0}^{T} a_t \right)^{-1} \sup\{\phi(s) : s \in [0, 2M_0 + 1]\}$$

$$+ 2 \left(\sum_{t=0}^{T} a_t \right)^{-1} \sum_{t=0}^{T} b_t + 2L\delta$$

$$\leq f(\tilde{x}, \tilde{y}) + 2 \left(\sum_{t=0}^{T} a_t \right)^{-1} \sup\{\phi(s) : s \in [0, 2M_0 + 1]\}$$

$$+ 2 \left(\sum_{t=0}^{T} a_t \right)^{-1} \sum_{t=0}^{T} b_t + 4L\delta.$$

This completes the proof of Corollary 2.10.

2.8   Proof of Proposition 2.9

It is clear that (2.70) is true. Let $t \in \{0, \ldots, T\}$. By (2.65), (2.67), and (2.68),

$$a_t(f(x_t, y_t) - f(x_*, y_*))$$
$$\leq a_t(f(x_t, y_t) - f(x_*, y_t))$$
$$\leq \phi(\|x_* - x_t\|) - \phi(\|x_* - x_{t+1}\|) + b_t, \tag{2.77}$$
$$a_t(f(x_*, y_*) - f(x_t, y_t))$$
$$\leq a_t(f(x_t, y_*) - f(x_t, y_t))$$
$$\leq \phi(\|y_* - y_t\|) - \phi(\|y_* - y_{t+1}\|) + b_t. \tag{2.78}$$

In view of (2.77) and (2.78),

$$\sum_{t=0}^{T} a_t f(x_t, y_t) - \sum_{t=0}^{T} a_t f(x_*, y_*)$$

$$\le \sum_{t=0}^{T} (\phi(\|x_* - x_t\|) - \phi(\|x_* - x_{t+1}\|)) + \sum_{t=0}^{T} b_t$$

$$\le \phi(\|x_* - x_0\|) + \sum_{t=0}^{T} b_t, \tag{2.79}$$

$$\sum_{t=0}^{T} a_t f(x_*, y_*) - \sum_{t=0}^{T} a_t f(x_t, y_t)$$

$$\le \sum_{t=0}^{T} (\phi(\|y_* - y_t\|) - \phi(\|y_* - y_{t+1}\|)) + \sum_{t=0}^{T} b_t$$

$$\le \phi(\|y_* - y_0\|) + \sum_{t=0}^{T} b_t. \tag{2.80}$$

Relations (2.61), (2.64), (2.66), (2.79), and (2.80) imply that

$$\left| \left(\sum_{t=0}^{T} a_t \right)^{-1} \sum_{t=0}^{T} a_t f(x_t, y_t) - f(x_*, y_*) \right|$$

$$\le \left(\sum_{t=0}^{T} a_t \right)^{-1} \sum_{t=0}^{T} b_t + \left(\sum_{t=0}^{T} a_t \right)^{-1} \sup\{\phi(s) : s \in [0, 2M_0 + 1]\}. \tag{2.81}$$

By (2.70), there exists

$$z_T \in C \tag{2.82}$$

such that

$$\|z_T - \hat{x}_T\| \le \delta. \tag{2.83}$$

In view of (2.82), we apply (2.67) with $z = z_T$ and obtain that for all $t = 0, \ldots, T$,

$$a_t(f(x_t, y_t) - f(z_T, y_t))$$

$$\le \phi(\|z_T - x_t\|) - \phi(\|z_T - x_{t+1}\|) + b_t. \tag{2.84}$$

It follows from (2.63) and (2.83) that for all $t = 0, \ldots, T$,

$$|f(z_T, y_t) - f(\hat{x}_T, y_t)| \leq L\|z_T - \hat{x}_T\| \leq L\delta. \tag{2.85}$$

By (2.84) and (2.85), for all $t = 0, \ldots, T$,

$$
\begin{aligned}
&a_t(f(x_t, y_t) - f(\hat{x}_T, y_t)) \\
&\leq a_t(f(x_t, y_t) - f(z_T, y_t)) + a_t L\delta \\
&\leq \phi(\|z_T - x_t\|) - \phi(\|z_T - x_{t+1}\|) + b_t + a_t L\delta.
\end{aligned} \tag{2.86}
$$

Combined with (2.61), (2.66), and (2.82) this implies that

$$
\begin{aligned}
&\sum_{t=0}^{T} a_t f(x_t, y_t) - \sum_{t=0}^{T} a_t f(\hat{x}_T, y_t) \\
&\leq \sum_{t=0}^{T} (\phi(\|z_T - x_t\|) - \phi(\|z_T - x_{t+1}\|)) + \sum_{t=0}^{T} b_t + \sum_{t=0}^{T} a_t L\delta \\
&\leq \phi(\|z_T - x_0\|) + \sum_{t=0}^{T} b_t + \sum_{t=0}^{T} a_t L\delta \\
&\leq \sup\{\phi(s) : s \in [0, 2M_0 + 1]\} + \sum_{t=0}^{T} b_t + \sum_{t=0}^{T} a_t L\delta.
\end{aligned} \tag{2.87}
$$

Property (ii) and (2.69) imply that

$$
\begin{aligned}
\sum_{t=0}^{T} a_t f(\hat{x}_T, y_t) &= \left(\sum_{i=0}^{T} a_i\right) \sum_{t=0}^{T} \left(a_t \left(\sum_{i=0}^{T} a_i\right)^{-1} f(\hat{x}_T, y_t)\right) \\
&\leq \left(\sum_{t=0}^{T} a_t\right) f(\hat{x}_T, \hat{y}_T).
\end{aligned} \tag{2.88}
$$

By (2.87) and (2.88),

$$
\begin{aligned}
&\sum_{t=0}^{T} a_t f(x_t, y_t) - \sum_{t=0}^{T} a_t f(\hat{x}_T, \hat{y}_T) \\
&\leq \sum_{t=0}^{T} a_t f(x_t, y_t) - \sum_{t=0}^{T} a_t f(\hat{x}_T, y_t) \\
&\leq \sup\{\phi(s) : s \in [0, 2M_0 + 1]\} + \sum_{t=0}^{T} b_t + \sum_{t=0}^{T} a_t L\delta.
\end{aligned} \tag{2.89}
$$

By (2.70), there exists

$$h_T \in D \tag{2.90}$$

such that

$$\|h_T - \hat{y}_T\| \le \delta. \tag{2.91}$$

In view of (2.90), we apply (2.68) with $v = h_T$ and obtain that for all $t = 0, \ldots, T$,

$$a_t(f(x_t, h_T) - f(x_t, y_t))$$
$$\le \phi(\|h_T - y_t\|) - \phi(\|h_T - y_{t+1}\|) + b_t. \tag{2.92}$$

It follows from (2.62) and (2.91) that for all $t = 0, \ldots, T$,

$$|f(x_t, h_T) - f(x_t, \hat{y}_T)| \le L\|h_T - \hat{y}_T\| \le L\delta. \tag{2.93}$$

By (2.92) and (2.93), for all $t = 0, \ldots, T$,

$$a_t(f(x_t, \hat{y}_T) - f(x_t, y_t))$$
$$\le a_t(f(x_t, h_T) - f(x_t, y_t)) + a_t L\delta$$
$$\le \phi(\|h_T - y_t\|) - \phi(\|h_T - y_{t+1}\|) + b_t + a_t L\delta. \tag{2.94}$$

In view of (2.94),

$$\sum_{t=0}^{T} a_t f(x_t, \hat{y}_T) - \sum_{t=0}^{T} a_t f(x_t, y_t)$$

$$\le \sum_{t=0}^{T} (\phi(\|h_T - y_t\|) - \phi(\|h_T - y_{t+1}\|)) + \sum_{t=0}^{T} b_t + \sum_{t=0}^{T} a_t L\delta. \tag{2.95}$$

Property (i) and (2.69) imply that

$$\sum_{t=0}^{T} a_t f(x_t, \hat{y}_T) = \left(\sum_{i=0}^{T} a_i \right) \sum_{t=0}^{T} \left(a_t \left(\sum_{i=0}^{T} a_i \right)^{-1} f(x_t, \hat{y}_T) \right)$$

$$\ge \sum_{t=0}^{T} a_t f(\hat{x}_T, \hat{y}_T). \tag{2.96}$$

By (2.61), (2.66), (2.90), (2.95), and (2.96),

$$\sum_{t=0}^{T} a_t f(\hat{x}_T, \hat{y}_T) - \sum_{t=0}^{T} a_t f(x_t, y_t)$$

$$\leq \sum_{t=0}^{T} a_t f(x_t, \hat{y}_T) - \sum_{t=0}^{T} a_t f(x_t, y_t)$$

$$\leq \phi(\|h_T - y_0\|) + \sum_{t=0}^{T} b_t + \sum_{t=0}^{T} a_t L\delta$$

$$\leq \sup\{\phi(s) : \ s \in [0, 2M_0 + 1]\} + \sum_{t=0}^{T} b_t + \sum_{t=0}^{T} a_t L\delta. \tag{2.97}$$

It follows from (2.89) and (2.97) that

$$\left| \sum_{t=0}^{T} a_t f(\hat{x}_T, \hat{y}_T) - \sum_{t=0}^{T} a_t f(x_t, y_t) \right|$$

$$\leq \sup\{\phi(s) : \ s \in [0, 2M_0 + 1]\} + \sum_{t=0}^{T} b_t + \sum_{t=0}^{T} a_t L\delta.$$

This implies (2.72).

Let $z \in C$. By (2.67),

$$\sum_{t=0}^{T} a_t f(x_t, y_t) - f(z, y_t))$$

$$\leq \sum_{t=0}^{T} [\phi(\|z - x_t\|) - \phi(\|z - x_{t+1}\|)] + \sum_{t=0}^{T} b_t. \tag{2.98}$$

By property (ii) and (2.69),

$$\sum_{t=0}^{T} a_t f(z, y_t) = \left(\sum_{i=0}^{T} a_i \right) \sum_{t=0}^{T} \left(a_t \left(\sum_{i=0}^{T} a_i \right)^{-1} f(z, y_t) \right)$$

$$\leq \left(\sum_{t=0}^{T} a_t \right) f(z, \hat{y}_T). \tag{2.99}$$

In view of (2.98) and (2.99),

$$\sum_{t=0}^{T} a_t f(x_t, y_t) - \sum_{t=0}^{T} a_t f(z, \hat{y}_T)$$

$$\leq \sum_{t=0}^{T} a_t (f(x_t, y_t) - f(z, y_t))$$

$$\leq \sum_{t=0}^{T} [\phi(\|z - x_t\|) - \phi(\|z - x_{t+1}\|)] + \sum_{t=0}^{T} b_t$$

$$\leq \phi(\|z - x_0\|) + \sum_{t=0}^{T} b_t. \qquad (2.100)$$

It follows from (2.61), (2.70), and (2.72) that

$$f(z, \hat{y}_T) \geq \left(\sum_{i=0}^{T} a_i\right)^{-1} \sum_{t=0}^{T} a_t f(x_t, y_t)$$

$$- \left(\sum_{i=0}^{T} a_i\right)^{-1} \sup\{\phi(s) : s \in [0, 2M_0 + 1]\} - \left(\sum_{i=0}^{T} a_i\right)^{-1} \sum_{t=0}^{T} b_t$$

$$\geq f(\hat{x}_T, \hat{y}_T) - 2\left(\sum_{t=0}^{T} a_t\right)^{-1} \sup\{\phi(s) : s \in [0, 2M_0 + 1]\}$$

$$- 2\left(\sum_{t=0}^{T} a_t\right)^{-1} \sum_{t=0}^{T} b_t - L\delta$$

and (2.73) holds.

Let $v \in D$. By (2.68),

$$\sum_{t=0}^{T} a_t (f(x_t, v) - f(x_t, y_t))$$

$$\leq \sum_{t=0}^{T} [\phi(\|v - y_t\|) - \phi(\|v - y_{t+1}\|)] + \sum_{t=0}^{T} b_t. \qquad (2.101)$$

By property (i) and (2.69),

$$\sum_{t=0}^{T} a_t f(x_t, v) = \left(\sum_{i=0}^{T} a_i\right) \sum_{t=0}^{T} \left(a_t \left(\sum_{i=0}^{T} a_i\right)^{-1} f(x_t, v)\right)$$

$$\geq \left(\sum_{t=0}^{T} a_t\right) f(\hat{x}_T, v). \tag{2.102}$$

In view of (2.101) and (2.102),

$$\sum_{t=0}^{T} a_t f(\hat{x}_T, v) - \sum_{t=0}^{T} a_t f(x_t, y_t)$$

$$\leq \phi(\|v - y_0\|) + \sum_{t=0}^{T} b_t.$$

Together with (2.61), (2.66), and (2.72) this implies that

$$f(\hat{x}_T, v) \leq \left(\sum_{t=0}^{T} a_t\right)^{-1} \sum_{t=0}^{T} a_t f(x_t, y_t)$$

$$+ \left(\sum_{t=0}^{T} a_t\right)^{-1} \sup\{\phi(s) : s \in [0, 2M_0 + 1]\} + \left(\sum_{t=0}^{T} a_t\right)^{-1} \sum_{t=0}^{T} b_t$$

$$\leq f(\hat{x}_T, \hat{y}_T) + 2\left(\sum_{t=0}^{T} a_t\right)^{-1} \sup\{\phi(s) : s \in [0, 2M_0 + 1]\}$$

$$+ 2\left(\sum_{t=0}^{T} a_t\right)^{-1} \sum_{t=0}^{T} b_t + L\delta.$$

Therefore (2.74) holds. This completes the proof of Proposition 2.9.

2.9   Subgradient Algorithm for Zero-Sum Games

We use the notation and definitions introduced in Sect. 2.1.

Let X, Y be Hilbert spaces, C be a nonempty closed convex subset of X, D be a nonempty closed convex subset of Y, U be an open convex subset of X, and V be an open convex subset of Y such that

$$C \subset U, \ D \subset V. \tag{2.103}$$

For each concave function $g : V \to R^1$ and each $x \in V$ set

$$\partial g(x) = \{l \in Y : \langle l, y - x \rangle \geq g(y) - g(x) \text{ for all } y \in V\}. \tag{2.104}$$

Clearly, for each $x \in V$,

$$\partial g(x) = -(\partial(-g)(x)). \tag{2.105}$$

Suppose that there exist $L > 0$, $M_0 > 0$ such that

$$C \subset B_X(0, M_0), \quad D \subset B_Y(0, M_0), \tag{2.106}$$

a function $f : U \times V \to R^1$ possesses the following properties:

(i) for each $v \in V$, the function $f(\cdot, v) : U \to R^1$ is convex;
(ii) for each $u \in U$, the function $f(u, \cdot) : V \to R^1$ is concave,

for each $v \in V$,

$$|f(u_1, v) - f(u_2, v)| \le L\|u_1 - u_2\| \text{ for all } u_1, u_2 \in U \tag{2.107}$$

and that for each $u \in U$,

$$|f(u, v_1) - f(u, v_2)| \le L\|v_1 - v_2\| \text{ for all } v_1, v_2 \in V. \tag{2.108}$$

For each $(\xi, \eta) \in U \times V$, set

$$\partial_x f(\xi, \eta) = \{l \in X : f(y, \eta) - f(\xi, \eta) \ge \langle l, y - \xi \rangle \text{ for all } y \in U\}, \tag{2.109}$$

$$\partial_y f(\xi, \eta) = \{l \in Y : \langle l, y - \eta \rangle \ge f(\xi, y) - f(\xi, \eta) \text{ for all } y \in V\}. \tag{2.110}$$

In view of properties (i) and (ii) and (2.107)–(2.110), for each $\xi \in U$ and each $\eta \in V$,

$$\emptyset \ne \partial_x f(\xi, \eta) \subset B_X(0, L), \tag{2.111}$$

$$\emptyset \ne \partial_y f(\xi, \eta) \subset B_Y(0, L). \tag{2.112}$$

Let

$$x_* \in C \text{ and } y_* \in D$$

satisfy

$$f(x_*, y) \le f(x_*, y_*) \le f(x, y_*) \tag{2.113}$$

for each $x \in C$ and each $y \in D$.
    Let $\delta \in (0, 1]$ and $\{a_k\}_{k=0}^{\infty} \subset (0, \infty)$.
    Let us describe our algorithm.

Subgradient Projection Algorithm for Zero-Sum Games
Initialization: select arbitrary $x_0 \in U$ and $y_0 \in V$.
 Iterative step: given current iteration vectors $x_t \in U$ and $y_t \in V$ calculate

$$\xi_t \in \partial_x f(x_t, y_t) + B_X(0, \delta),$$

$$\eta_t \in \partial_y f(x_t, y_t) + B_Y(0, \delta)$$

and the next pair of iteration vectors $x_{t+1} \in U, y_{t+1} \in V$ such that

$$\|x_{t+1} - P_C(x_t - a_t \xi_t)\| \leq \delta,$$

$$\|y_{t+1} - P_D(y_t + a_t \eta_t)\| \leq \delta.$$

In this chapter we prove the following result.

Theorem 2.11. *Let $\delta \in (0, 1]$ and $\{a_k\}_{k=0}^{\infty} \subset (0, \infty)$. Assume that $\{x_t\}_{t=0}^{\infty} \subset U,$
$\{y_t\}_{t=0}^{\infty} \subset V, \{\xi_t\}_{t=0}^{\infty} \subset X, \{\eta_t\}_{t=0}^{\infty} \subset Y,$*

$$B_X(x_0, \delta) \cap C \neq \emptyset, \ B_Y(y_0, \delta) \cap D \neq \emptyset \tag{2.114}$$

and that for each integer $t \geq 0$,

$$\xi_t \in \partial_x f(x_t, y_t) + B_X(0, \delta), \tag{2.115}$$

$$\eta_t \in \partial_y f(x_t, y_t) + B_Y(0, \delta), \tag{2.116}$$

$$\|x_{t+1} - P_C(x_t - a_t \xi_t)\| \leq \delta \tag{2.117}$$

and that

$$\|y_{t+1} - P_D(y_t + a_t \eta_t)\| \leq \delta. \tag{2.118}$$

Let for each natural number T,

$$\hat{x}_T = \left(\sum_{i=0}^{T} a_t\right)^{-1} \sum_{t=0}^{T} a_t x_t, \ \hat{y}_T = \left(\sum_{i=0}^{T} a_t\right)^{-1} \sum_{t=0}^{T} a_t y_t. \tag{2.119}$$

Then for each natural number T,

$$\left|\left(\sum_{t=0}^{T} a_t\right)^{-1} \sum_{t=0}^{T} a_t f(x_t, y_t) - f(x_*, y_*)\right|$$

$$\leq [2^{-1}(2M_0 + 1)^2 + \delta(T + 1)(4M_0 + 1)] \left(\sum_{t=0}^{T} a_t\right)^{-1}$$

$$+ \delta(2M_0 + 1) + 2^{-1} \left(\sum_{t=0}^{T} a_t \right)^{-1} (L+1)^2 \sum_{t=0}^{T} a_t^2, \qquad (2.120)$$

$$\left| f(\hat{x}_T, \hat{y}_T) - \left(\sum_{t=0}^{T} a_t \right)^{-1} \sum_{t=0}^{T} a_t f(x_t, y_t) \right|$$

$$\leq [2^{-1}(2M_0 + 1)^2 + \delta(T+1)(4M_0 + 1)] \left(\sum_{t=0}^{T} a_t \right)^{-1}$$

$$+ \delta(2M_0 + 1) + 2^{-1} \left(\sum_{t=0}^{T} a_t \right)^{-1} (L+1)^2 \sum_{t=0}^{T} a_t^2 + L\delta, \qquad (2.121)$$

and for each natural number T, each $z \in C$, and each $u \in D$,

$$f(z, \hat{y}_T) \geq f(\hat{x}_T, \hat{y}_T)$$

$$- (2M_0 + 1)^2 \left(\sum_{t=0}^{T} a_t \right)^{-1} - 2 \left(\sum_{t=0}^{T} a_t \right)^{-1} (T+1)\delta(4M_0 + 1)$$

$$- 2\delta(2M_0 + 1) - \left(\sum_{t=0}^{T} a_t \right)^{-1} (L+1)^2 \sum_{t=0}^{T} a_t^2 - L\delta,$$

$$f(\hat{x}_T, v) \leq f(\hat{x}_T, \hat{y}_T)$$

$$+ (2M_0 + 1)^2 \left(\sum_{t=0}^{T} a_t \right)^{-1} + 2 \left(\sum_{t=0}^{T} a_t \right)^{-1} \delta(T+1)(4M_0 + 1)$$

$$+ 2\delta(2M_0 + 1) + \left(\sum_{t=0}^{T} a_t \right)^{-1} (L+1)^2 \sum_{t=0}^{T} a_t^2 + L\delta.$$

We are interested in the optimal choice of a_t, $t = 0, 1, \ldots, T$. Let T be a natural number and $A_T = \sum_{t=0}^{T} a_t$ be given. By Theorem 2.11, in order to make the best choice of a_t, $t = 0, \ldots, T$, we need to minimize the function $\sum_{t=0}^{T} a_t^2$ on the set

$$\left\{ a = (a_0, \ldots, a_T) \in R^{T+1} : a_i \geq 0, \ i = 0, \ldots, T, \ \sum_{i=0}^{T} a_i = A_T \right\}.$$

By Lemma 2.3, this function has a unique minimizer $a^* = (a_0^*, \ldots, a_T^*)$ where $a_i^* = (T+1)^{-1} A_T$, $i = 0, \ldots, T$ which is the best choice of a_t, $t = 0, 1, \ldots, T$.

Now we will find the best $a > 0$. Let T be a natural number and $a_t = a$ for all $t = 0, \ldots, T$. We need to choose a which is a minimizer of the function

$$\Psi_T(a) = ((T+1)a)^{-1}[(2M_0 + 1)^2 + 2\delta(T+1)(4M_0 + 1)]$$
$$+ 2\delta(2M_0 + 1) + a(L+1)^2$$
$$= (2M_0 + 1)^2((T+1)a)^{-1} + 2\delta(4M_0 + 1)a^{-1} + 2\delta(2M_0 + 1) + (L+1)^2 a.$$

Since T can be arbitrary large, we need to find a minimizer of the function

$$\phi(a) := 2a^{-1}\delta(4M_0 + 1) + (L+1)^2 a, \ a \in (0, \infty).$$

In Sect. 2.2 we have already shown that the minimizer is

$$a = (2\delta(4M_0 + 1))^{1/2}(L+1)^{-1}$$

and the minimal value of ϕ is

$$(8\delta(4M_0 + 1))^{1/2}(L+1).$$

Now our goal is to find the best integer $T > 0$ which gives us an appropriate value of $\Psi_T(a)$. Since in view of the inequalities above, this value is bounded from below by $c_0\delta^{1/2}$ with the constant c_0 depending on L, M_0, it is clear that in order to make the best choice of T, it should be at the same order as $\lfloor \delta^{-1} \rfloor$. For example, $T = \lfloor \delta^{-1} \rfloor$.

Note that in the theorem above δ is the computational error produced by our computer system. We obtain a good approximate solution after $T = \lfloor \delta^{-1} \rfloor$ iterations. Namely, we obtain a pair of points $\hat{x} \in U, \hat{y} \in V$ such that

$$B_X(\hat{x}, \delta) \cap C \neq \emptyset, \ B_Y(\hat{y}, \delta) \cap D \neq \emptyset$$

and for each $z \in C$ and each $v \in D$,

$$f(z, \hat{y}) \geq f(\hat{x}, \hat{y}) - c\delta^{1/2}, \ f(\hat{x}, v) \leq f(\hat{x}, \hat{y}) + c\delta^{1/2},$$

where the constant $c > 0$ depends only on L and M_0.

2.10   Proof of Theorem 2.11

By (2.106), (2.114), (2.117), and (2.118), for all integers $t \geq 0$,

$$\|x_t\| \leq M_0 + 1, \ \|y_t\| \leq M_0 + 1. \tag{2.122}$$

Let $t \geq 0$ be an integer. Applying Lemma 2.7 with

$$a = a_t, \ x = x_t, \ f = f(\cdot, y_t), \ \xi = \xi_t, \ u = x_{t+1}$$

we obtain that for each $z \in C$,

$$a_t(f(x_t, y_t) - f(z, y_t)) \leq 2^{-1}\|z - x_t\|^2 - 2^{-1}\|z - x_{t+1}\|^2$$
$$+ \delta(4M_0 + 1 + a_t(2M_0 + 1)) + 2^{-1}a_t^2(L + 1)^2. \quad (2.123)$$

Applying Lemma 2.7 with

$$a = a_t, \; x = y_t, \; f = -f(x_t, \cdot), \; \xi = -\eta_t, \; u = y_{t+1}$$

we obtain that for each $v \in D$,

$$a_t(f(x_t, v) - f(x_t, y_t)) \leq 2^{-1}\|v - y_t\|^2 - 2^{-1}\|v - y_{t+1}\|^2$$
$$+ \delta(4M_0 + 1 + a_t(2M_0 + 1)) + 2^{-1}a_t^2(L + 1)^2. \quad (2.124)$$

For all integers $t \geq 0$ set

$$b_t = \delta(4M_0 + 1 + a_t(2M_0 + 1)) + 2^{-1}a_t^2(L + 1)^2$$

and define

$$\phi(s) = 2^{-1}s^2, \; s \in R^1.$$

It is easy to see that all the assumptions of Proposition 2.9 hold and it implies Theorem 2.11.

Chapter 3
The Mirror Descent Algorithm

In this chapter we analyze the convergence of the mirror descent algorithm under the presence of computational errors. We show that the algorithms generate a good approximate solution, if computational errors are bounded from above by a small positive constant. Moreover, for a known computational error, we find out what an approximate solution can be obtained and how many iterates one needs for this.

3.1 Optimization on Bounded Sets

Let X be a Hilbert space equipped with an inner product $\langle \cdot, \cdot \rangle$ which induces a complete norm $\| \cdot \|$. For each $x \in X$ and each $r > 0$ set

$$B_X(x, r) = \{ y \in X : \|x - y\| \leq r \}.$$

For each $x \in X$ and each nonempty set $E \subset X$ set

$$d(x, E) = \inf\{ \|x - y\| : y \in E \}.$$

Let C be a nonempty closed convex subset of X, U be an open convex subset of X such that $C \subset U$ and let $f : U \to R^1$ be a convex function. Recall that for each $x \in U$,

$$\partial f(x) = \{ l \in X : f(y) - f(x) \geq \langle l, y - x \rangle \text{ for all } y \in U \}. \tag{3.1}$$

© Springer International Publishing Switzerland 2016
A.J. Zaslavski, *Numerical Optimization with Computational Errors*, Springer
Optimization and Its Applications 108, DOI 10.1007/978-3-319-30921-7_3

Suppose that there exist $L > 0$, $M_0 > 0$ such that

$$C \subset B_X(0, M_0), \tag{3.2}$$

$$|f(x) - f(y)| \leq L\|x - y\| \text{ for all } x, y \in U. \tag{3.3}$$

In view of (3.1) and (3.3), for each $x \in U$,

$$\emptyset \neq \partial f(x) \subset B_X(0, L). \tag{3.4}$$

For each nonempty set $D \subset X$ and each function $h : D \to R^1$ put

$$\inf(h, D) = \inf\{h(y) : y \in D\}$$

and

$$\operatorname{argmin}\{h(y) : y \in D\} = \{y \in D : h(y) = \inf(h; D)\}.$$

We study the convergence of the mirror descent algorithm under the presence of computational errors. This method was introduced by Nemirovsky and Yudin for solving convex optimization problems [90]. Here we use a derivation of this algorithm proposed by Beck and Teboulle [19].

Let $\delta \in (0, 1]$ and $\{a_k\}_{k=0}^\infty \subset (0, \infty)$.

We describe the inexact version of the mirror descent algorithm.

Mirror Descent Algorithm

Initialization: select an arbitrary $x_0 \in U$.

Iterative step: given a current iteration vector $x_t \in U$ calculate

$$\xi_t \in \partial f(x_t) + B_X(0, \delta),$$

define

$$g_t(x) = \langle \xi_t, x \rangle + (2a_t)^{-1}\|x - x_t\|^2, \ x \in X$$

and calculate the next iteration vector $x_{t+1} \in U$ such that

$$B_X(x_{t+1}, \delta) \cap \operatorname{argmin}\{g_t(y) : y \in C\} \neq \emptyset.$$

Note that g_t is a convex bounded from below function on X which possesses a minimizer on C.

In this chapter we prove the following result.

Theorem 3.1. Let $\delta \in (0, 1]$, $\{a_k\}_{k=0}^\infty \subset (0, \infty)$ and let

$$x_* \in C \tag{3.5}$$

satisfies

$$f(x_*) \le f(x) \text{ for all } x \in C. \tag{3.6}$$

Assume that $\{x_t\}_{t=0}^{\infty} \subset U$, $\{\xi_t\}_{t=0}^{\infty} \subset X$,

$$\|x_0\| \le M_0 + 1 \tag{3.7}$$

and that for each integer $t \ge 0$,

$$\xi_t \in \partial f(x_t) + B_X(0, \delta) \tag{3.8}$$

and

$$B_X(x_{t+1}, \delta) \cap argmin\{\langle \xi_t, v \rangle + (2a_t)^{-1} \|v - x_t\|^2 : v \in C\} \ne \emptyset. \tag{3.9}$$

Then for each natural number T,

$$\sum_{t=0}^{T} a_t(f(x_t) - f(x_*))$$

$$\le 2^{-1}(2M_0 + 1)^2 + \delta(2M_0 + L + 2) \sum_{t=0}^{T} a_t$$

$$+ \delta(T + 1)(8M_0 + 8) + 2^{-1}(L + 1)^2 \sum_{t=0}^{T} a_t^2. \tag{3.10}$$

Moreover, for each natural number T,

$$f\left(\left(\sum_{t=0}^{T} a_t\right)^{-1} \sum_{t=0}^{T} a_t x_t\right) - f(x_*), \ \min\{f(x_t) : t = 0, \dots, T\} - f(x_*)$$

$$\le 2^{-1}(2M_0 + 1)^2 \left(\sum_{t=0}^{T} a_t\right)^{-1} + \delta(2M_0 + L + 2)$$

$$+ \delta(T + 1)(8M_0 + 8) \left(\sum_{t=0}^{T} a_t\right)^{-1} + 2^{-1}(L + 1)^2 \sum_{t=0}^{T} a_t^2 \left(\sum_{t=0}^{T} a_t\right)^{-1}.$$

Theorem 3.1 is proved in Sect. 3.3.

We are interested in an optimal choice of a_t, $t = 0, 1, \dots$. Let T be a natural number and $A_T = \sum_{t=0}^{T} a_t$ be given. By Theorem 3.1, in order to make the best choice of a_t, $t = 0, \dots, T$, we need to minimize the function $\sum_{t=0}^{T} a_t^2$ on the set

$$\left\{ a = (a_0, \dots, a_T) \in R^{T+1} : a_i \ge 0, \ i = 0, \dots, T, \ \sum_{i=0}^{T} a_i = A_T \right\}.$$

By Lemma 2.3, this function has a unique minimizer $a^* = (a_0^*, \ldots, a_T^*)$ where $a_i^* = (T+1)^{-1}A_T$, $i = 0, \ldots, T$. This is the best choice of a_t, $t = 0, 1, \ldots, T$.

Let T be a natural number and $a_t = a$, $t = 0, \ldots, T$. Now we will find the best $a > 0$. By Theorem 3.1, we need to choose a which is a minimizer of the function

$$2^{-1}((T+1)a)^{-1}(2M_0 + 1)^2 + \delta(2M_0 + L + 2)$$
$$+ a^{-1}\delta(8M_0 + 8) + 2^{-1}(L+1)^2 a.$$

Since T can be arbitrary large, we need to find a minimizer of the function

$$\phi(a) := a^{-1}\delta(16M_0 + 16) + (L+1)^2 a, \quad a \in (0, \infty).$$

Clearly, the minimizer is

$$a = (16\delta(M_0 + 1))^{1/2}(L+1)^{-1}$$

and the minimal value of ϕ is

$$2(\delta(16M_0 + 16))^{1/2}(L+1).$$

Now we can think about the best choice of T. It is clear that it should be at the same order as $\lfloor \delta^{-1} \rfloor$.

Note that in the theorem above δ is the computational error produced by our computer system. In order to obtain a good approximate solution we need T iterations which is at the same order as $\lfloor \delta^{-1} \rfloor$. As a result, we obtain a point $\xi \in U$ such that

$$B_X(\xi, \delta) \cap C \neq \emptyset$$

and

$$f(\xi) \leq f(x_*) + c_1 \delta^{1/2},$$

where the constant $c_1 > 0$ depends only on L and M_0.

3.2 The Main Lemma

We use the notation and definitions introduced in Sect. 3.1.

Lemma 3.2. *Let $\delta \in (0, 1]$, $a > 0$ and let*

$$z \in C. \tag{3.11}$$

Assume that

$$x \in U \cap B_X(0, M_0 + 1), \tag{3.12}$$

$$\xi \in \partial f(x) + B_X(0, \delta), \tag{3.13}$$

$$g(v) = \langle \xi, v \rangle + (2a)^{-1} \|v - x\|^2, \ v \in X \tag{3.14}$$

and that

$$u \in U \tag{3.15}$$

satisfies

$$B_X(u, \delta) \cap \{v \in C : \ g(v) = \inf(g; C)\} \neq \emptyset. \tag{3.16}$$

Then

$$a(f(x) - f(z)) \leq \delta a(2M_0 + L + 2) + 8\delta(M_0 + 1)$$
$$+ 2^{-1} a^2 (L + 1)^2 + 2^{-1} \|z - x\|^2 - 2^{-1} \|z - u\|^2.$$

Proof. In view of (3.13), there exists

$$l \in \partial f(x) \tag{3.17}$$

such that

$$\|l - \xi\| \leq \delta. \tag{3.18}$$

Clearly, the function g is Frechet differentiable on X. We denote by $g'(v)$ its Frechet derivative at $v \in X$. It is easy to see that

$$g'(v) = \xi + a^{-1}(v - x), \ v \in X. \tag{3.19}$$

By (3.16), there exists

$$\hat{u} \in B_X(u, \delta) \cap C \tag{3.20}$$

such that

$$g(\hat{u}) = \inf(g; C). \tag{3.21}$$

It follows from (3.19) and (3.21) that

$$0 \le \langle g'(\hat{u}), v - \hat{u} \rangle = \langle \xi + a^{-1}(\hat{u} - x), v - \hat{u} \rangle. \tag{3.22}$$

By (3.2), (3.11), (3.12), (3.17), and (3.18),

$$
\begin{aligned}
a(f(x) - f(z)) &\le a\langle x - z, l \rangle \\
&= a\langle x - z, \xi \rangle + a\langle x - z, l - \xi \rangle \\
&\le a\langle x - z, \xi \rangle + a\|x - z\|\|l - \xi\| \\
&\le \delta a(2M_0 + 1) + a\langle x - z, \xi \rangle \\
&= \delta a(2M_0 + 1) + a\langle \xi, x - \hat{u} \rangle + a\langle \xi, \hat{u} - z \rangle \\
&= \delta a(2M_0 + 1) + \langle x - \hat{u}, a\xi \rangle \\
&\quad + \langle z - \hat{u}, x - \hat{u} - a\xi \rangle + \langle z - \hat{u}, \hat{u} - x \rangle.
\end{aligned} \tag{3.23}
$$

Relations (3.11) and (3.22) imply that

$$\langle z - \hat{u}, x - \hat{u} - a\xi \rangle \le 0. \tag{3.24}$$

In view of Lemma 2.1,

$$\langle z - \hat{u}, \hat{u} - x \rangle = 2^{-1}[\|z - x\|^2 - \|z - \hat{u}\|^2 - \|\hat{u} - x\|^2]. \tag{3.25}$$

It follows from (3.4), (3.13), and (3.20) that

$$
\begin{aligned}
\langle x - \hat{u}, a\xi \rangle &= \langle x - u, a\xi \rangle + \langle u - \hat{u}, a\xi \rangle \\
&\le a\delta(L + 1) + \langle x - u, a\xi \rangle \\
&\le a\delta(L + 1) + 2^{-1}\|x - u\|^2 + 2^{-1}a^2\|\xi\|^2.
\end{aligned} \tag{3.26}
$$

By (3.4), (3.13), and (3.23)–(3.26),

$$
\begin{aligned}
a(f(x) - f(z)) &\le \delta a(2M_0 + 1) + a\delta(L + 1) \\
&\quad + 2^{-1}\|x - u\|^2 + 2^{-1}a^2\|\xi\|^2 \\
&\quad + 2^{-1}\|z - x\|^2 - 2^{-1}\|z - \hat{u}\|^2 - 2^{-1}\|\hat{u} - x\|^2 \\
&\le \delta a(2M_0 + L + 2) + 2^{-1}a^2(L + 1)^2 \\
&\quad + 2^{-1}\|z - x\|^2 - 2^{-1}\|z - \hat{u}\|^2 \\
&\quad + 2^{-1}\|x - u\|^2 - 2^{-1}\|\hat{u} - x\|^2.
\end{aligned} \tag{3.27}
$$

In view of (3.2), (3.12), (3.16), and (3.20),

$$| \|x - u\|^2 - \|\hat{u} - x\|^2 |$$
$$\leq | \|x - u\| - \|\hat{u} - x\| | (\|x - u\| + \|\hat{u} - x\|)$$
$$\leq 4\|u - \hat{u}\| (M_0 + 1) \leq 4(M_0 + 1)\delta. \tag{3.28}$$

Relations (3.2), (3.11), (3.12), and (3.20) imply that

$$| \|z - \hat{u}\|^2 - \|z - u\|^2 |$$
$$\leq | \|z - \hat{u}\| - \|z - u\| | (\|z - \hat{u}\| + \|z - u\|)$$
$$\leq \|u - \hat{u}\| (4M_0 + 1) \leq (4M_0 + 1)\delta. \tag{3.29}$$

By (3.27), (3.28), and (3.29),

$$a(f(x) - f(z)) \leq \delta a(2M_0 + L + 2) + +2^{-1}a^2(L + 1)^2$$
$$+2^{-1}\|z - x\|^2 - 2^{-1}\|z - u\|^2 + 8(M_0 + 1)\delta.$$

This completes the proof of Lemma 3.2.

3.3 Proof of Theorem 3.1

It is clear that

$$\|x_t\| \leq M_0 + 1, \ t = 0, 1, \ldots.$$

Let $t \geq 0$ be an integer. Applying Lemma 3.2 with

$$z = x_*, \ a = a_t, \ x = x_t, \ \xi = \xi_t, \ u = x_{t+1}$$

we obtain that

$$a_t(f(x_t) - f(x_*)) \leq 2^{-1}\|x_* - x_t\|^2 - 2^{-1}\|x_* - x_{t+1}\|^2$$
$$+\delta(8(M_0 + 1) + a_t(2M_0 + L + 2)) + 2^{-1}a_t^2(L + 1)^2. \tag{3.30}$$

By (3.2), (3.5), (3.7), and (3.30), for each natural number T,

$$\sum_{t=0}^{T} a_t(f(x_t) - f(x_*))$$

$$\leq \sum_{t=0}^{T} (2^{-1}\|x_* - x_t\|^2 - 2^{-1}\|x_* - x_{t+1}\|^2)$$

$$+\delta(2M_0 + L + 2)\sum_{t=0}^{T} a_t + (T+1)\delta(8M_0 + 8) + 2^{-1}(L+1)^2 \sum_{t=0}^{T} a_t^2$$

$$\leq 2^{-1}(2M_0 + 1)^2 + \delta(2M_0 + L + 2)\sum_{t=0}^{T} a_t$$

$$+(T+1)\delta(8M_0 + 8) + 2^{-1}(L+1)^2 \sum_{t=0}^{T} a_t^2.$$

Thus (3.10) is true. Evidently, (3.10) implies the last relation of the statement of Theorem 3.1. This completes the proof of Theorem 3.1

3.4   Optimization on Unbounded Sets

We use the notation and definitions introduced in Sect. 3.1. Let X be a Hilbert space with an inner product $\langle \cdot, \cdot \rangle$, D be a nonempty closed convex subset of X, V be an open convex subset of X such that

$$D \subset V, \tag{3.31}$$

and $f : V \to R^1$ be a convex function which is Lipschitz on all bounded subsets of V. Set

$$D_{\min} = \{x \in D : f(x) \leq f(y) \text{ for all } y \in D\}. \tag{3.32}$$

We suppose that

$$D_{\min} \neq \emptyset.$$

We will prove the following result.

Theorem 3.3. *Let* $\delta \in (0, 1]$, $M > 1$ *satisfy*

$$D_{\min} \cap B_X(0, M) \neq \emptyset, \tag{3.33}$$

$$M_0 > 80M + 6, \tag{3.34}$$

$L > 1$ *satisfy*

$$|f(v_1) - f(v_2)| \leq L\|v_1 - v_2\| \text{ for all } v_1, v_2 \in V \cap B_X(0, M_0 + 2), \tag{3.35}$$

$$0 < \tau_0 \leq \tau_1 \leq (4L + 4)^{-1}, \tag{3.36}$$

$$\epsilon_0 = 16\tau_0^{-1}\delta(M_0 + 1) + 4\delta(2M_0 + L + 2) + \tau_1(L+1)^2 \tag{3.37}$$

and let

$$n_0 = \lfloor \tau_0^{-1}(2M+1)^2 \epsilon_0^{-1} \rfloor + 1. \tag{3.38}$$

Assume that $\{x_t\}_{t=0}^\infty \subset V$, $\{\xi_t\}_{t=0}^\infty \subset X$,

$$\{a_t\}_{t=0}^\infty \subset [\tau_0, \tau_1],$$

$$\|x_0\| \le M \tag{3.39}$$

and that for each integer $t \ge 0$,

$$\xi_t \in \partial f(x_t) + B_X(0, \delta) \tag{3.40}$$

and

$$B_X(x_{t+1}, \delta) \cap argmin\{\langle \xi_t, v \rangle + (2a_t)^{-1} \|v - x_t\|^2 : v \in C\} \ne \emptyset. \tag{3.41}$$

Then there exists an integer $q \in [1, n_0]$ *such that*

$$f(x_q) \le \inf(f; D) + \epsilon_0,$$

$$\|x_q\| \le 15M + 1.$$

We are interested in the best choice of a_t, $t = 0, 1, \ldots$. Assume for simplicity that $\tau_1 = \tau_0$. In order to meet our goal we need to minimize ϵ_0 which obtains its minimal value when

$$\tau_0 = (4\delta(M_0 + 1))^{1/2}(L+1)^{-1}$$

and the minimal value of ϵ_0 is

$$4\delta(2M_0 + L + 2) + 4(4\delta(M_0 + 1))^{1/2}(L+1),$$

Thus ϵ_0 is at the same order as $\lfloor \delta^{1/2} \rfloor$. By (3.38) and the inequalities above, n_0 is at the same order as $\lfloor \delta^{-1} \rfloor$.

3.5 Proof of Theorem 3.3

By (3.33) there exists

$$z \in D_{min} \cap B_X(0, M). \tag{3.42}$$

Assume that T is a natural number and that

$$f(x_t) - f(z) > \epsilon_0, \quad t = 1, \ldots, T. \tag{3.43}$$

In view of (3.41), there exists

$$\eta \in B_X(x_1, \delta) \cap \mathrm{argmin}\{\langle \xi_0, v \rangle + (2a_0)^{-1}\|v - x_0\|^2 : v \in D\}. \tag{3.44}$$

Relations (3.42) and (3.44) imply that

$$\langle \xi_0, \eta \rangle + (2a_0)^{-1}\|\eta - x_0\|^2$$
$$\leq \langle \xi_0, z \rangle + (2a_0)^{-1}\|z - x_0\|^2. \tag{3.45}$$

It follows from (3.34), (3.35), (3.39), and (3.40) that

$$\|\xi_0\| \leq L + 1. \tag{3.46}$$

In view of (3.36),

$$a_0^{-1} \geq \tau_1^{-1} \geq 4(L + 1). \tag{3.47}$$

By (3.35), (3.39), (3.40), (3.42), and (3.45)–(3.47),

$$(L + 1)M + (2a_0)^{-1}(2M + 1)^2$$
$$\geq \langle \xi_0, z \rangle + (2a_0)^{-1}\|z - x_0\|^2$$
$$\geq (2a_0)^{-1}\|\eta - x_0\|^2 + \langle \xi_0, \eta - x_0 \rangle + \langle \xi_0, x_0 \rangle$$
$$\geq (2a_0)^{-1}\|\eta - x_0\|^2 - (L + 1)\|\eta - x_0\| - (L + 1)M.$$

Together with (3.36) this implies that

$$M + (2M + 1)^2 \geq \|\eta - x_0\|^2 - 2^{-1}\|\eta - x_0\|,$$
$$(\|\eta - x_0\| - 4^{-1})^2 \leq (4M + 1)^2,$$
$$\|\eta - x_0\| \leq 8M.$$

Together with (3.39) and (3.44) this implies that

$$\|\eta\| \leq 9M, \ \|x_1\| \leq 9M + 1,$$
$$\|\eta - z\| \leq 10M,$$
$$\|x_1 - z\| \leq 10M + 1. \tag{3.48}$$

By induction we show that for every integer $t \in [1, T]$,

$$\|x_t - z\| \leq 14M + 1, \tag{3.49}$$
$$f(x_t) - f(z) \leq \delta(2M_0 + L + 2) + 8\tau_0^{-1}\delta(M_0 + 1) + 2^{-1}\tau_1(L + 1)^2$$
$$+ (2\tau_0)^{-1}(\|z - x_t\|^2 - \|z - x_{t+1}\|^2). \tag{3.50}$$

Set

$$U = V \cap \{v \in X : \|v\| < M_0 + 2\} \tag{3.51}$$

and

$$C = D \cap B_X(0, M_0). \tag{3.52}$$

In view of (3.48), (3.49) holds for $t = 1$.

Assume that an integer $t \in [1, T]$ and that (3.49) holds. It follows from (3.34), (3.42) and (3.52) that

$$z \in C \subset B_X(0, M_0). \tag{3.53}$$

In view of (3.34), (3.42), and (3.49),

$$x_t \in U \cap B_X(0, M_0 + 1). \tag{3.54}$$

Relations (3.35), (3.40), and (3.54) imply that

$$\xi_t \in \partial f(x_t) + B_X(0, \delta) \subset B_X(0, L + 1). \tag{3.55}$$

In view of (3.41), there exists

$$h \in B_X(x_{t+1}, \delta) \cap \operatorname{argmin}\{\langle \xi_t, v \rangle + (2a_t)^{-1}\|v - x_t\|^2 : v \in D\}. \tag{3.56}$$

By (3.42) and (3.56),

$$\begin{aligned} &\langle \xi_t, h \rangle + (2a_t)^{-1}\|h - x_t\|^2 \\ &\leq \langle \xi_t, z \rangle + (2a_t)^{-1}\|z - x_t\|^2. \end{aligned} \tag{3.57}$$

In view of (3.57),

$$\begin{aligned} &\langle \xi_t, z \rangle + (2a_t)^{-1}\|z - x_t\|^2 \\ &\geq (2a_t)^{-1}\|h - x_t\|^2 + \langle \xi_t, h - x_t \rangle + \langle \xi_t, x_t \rangle. \end{aligned}$$

It follows from the inequality above, (3.34), (3.36), (3.42), (3.49), and (3.55) that

$$\begin{aligned} &(L+1)M + (2a_t)^{-1}(14M + 1)^2 \\ &\geq (2a_t)^{-1}\|h - x_t\|^2 - (L+1)\|h - x_t\| - (L+1)(15M + 1) \\ &\geq (2a_t)^{-1}(\|h - x_t\|^2 - \|h - x_t\|) - (L+1)(15M + 1) \\ &\geq (2a_t)^{-1}(\|h - x_t\| - 1)^2 - (2a_t)^{-1} - (L+1)(15M + 1), \\ &\quad (2a_t)^{-1} + (L+1)(16M + 1) + (2a_t)^{-1}(14M + 1)^2 \end{aligned}$$

$$\geq (2a_t)^{-1}(\|h - x_t\| - 1)^2,$$

$$4(14M + 1)^2 \geq 2 + 16M + (14M + 1)^2 \geq (\|h - x_t\| - 1)^2,$$

$$\|h - x_t\| \leq 28M + 4,$$

$$\|h\| \leq 44M + 5 < M_0. \tag{3.58}$$

By (3.52), (3.56), and (3.58),

$$h \in C. \tag{3.59}$$

Relations (3.52), (3.56), and (3.59) imply that

$$h \in \arg\min\{\langle \xi_t, v \rangle + (2a_t)^{-1}\|v - x_t\|^2 : v \in C\}$$

and

$$h \in B_X(x_{t+1}, \delta) \cap \arg\min\{\langle \xi_t, v \rangle + (2a_t)^{-1}\|v - x_t\|^2 : v \in C\}. \tag{3.60}$$

It follows from (3.31), (3.35), (3.51), (3.52)–(3.55), (3.60), and Lemma 3.2 which holds with

$$x = x_t, \ a = a_t, \ \xi = \xi_t, \ u = h$$

that

$$a_t(f(x_t) - f(z)) \leq \delta a_t(2M_0 + L + 2)$$
$$+ 8\delta(M_0 + 1) + 2^{-1}a_t^2(L + 1)^2$$
$$+ 2^{-1}\|z - x_t\|^2 - 2^{-1}\|z - x_{t+1}\|^2$$

Together with the inclusion $a_t \in [\tau_0, \tau_1]$ this implies that

$$f(x_t) - f(z) \leq \tau_0^{-1}8\delta(M_0 + 1)$$
$$+ (2M_0 + L + 2)\delta + 2^{-1}\tau_1(L + 1)^2$$
$$+ (2\tau_0)^{-1}\|z - x_t\|^2 - (2\tau_0)^{-1}\|z - x_{t+1}\|^2 \tag{3.61}$$

and (3.50) holds.

In view of (3.37), (3.43), (3.49), and (3.61),

$$\|z - x_t\|^2 - \|z - x_{t+1}\|^2 \geq 0,$$

$$\|z - x_{t+1}\| \leq \|z - x_t\| \leq 14M + 1.$$

Hence by induction we showed that (3.50) holds for all $t = 1, \ldots, T$ and (3.49) holds for all $t = 1, \ldots, T + 1$.

It follows from (3.37)–(3.39), (3.42), (3.43), and (3.50) that

$$T\epsilon_0 < T(\min\{f(x_t) : t = 1, \ldots, T\} - f(z))$$

$$\leq \sum_{t=1}^{T}(f(x_t) - f(z))$$

$$\leq (2\tau_0)^{-1}\sum_{t=1}^{T}(\|z - x_t\|^2 - \|z - x_{t+1}\|^2)$$

$$+8T\tau_0^{-1}\delta(M_0 + 1) + T(2M_0 + L + 2)\delta + 2^{-1}T\tau_1(L + 1)^2$$

$$\leq (2\tau_0)^{-1}(2M + 1)^2 + 8T\tau_0^{-1}\delta(M_0 + 1)$$

$$+T(2M_0 + L + 2)\delta + 2^{-1}T\tau_1(L + 1)^2,$$

$$\epsilon_0 < (2\tau_0 T)^{-1}(2M + 1)^2 + 8\tau_0^{-1}\delta(M_0 + 1)$$

$$+(2M_0 + L + 2)\delta + 2^{-1}\tau_1(L + 1)^2,$$

$$2^{-1}\epsilon_0 < (2\tau_0 T)^{-1}(2M + 1)^2$$

and

$$T < \tau_0^{-1}(2M + 1)^2\epsilon_0^{-1} < n_0.$$

Thus we have shown that if an integer $T \geq 1$ satisfies

$$f(x_t) - f(z) > \epsilon_0, \quad t = 1, \ldots, T,$$

then $T < n_0$ and (3.49) holds for all $t = 1, \ldots, T + 1$. This implies that there exists an integer $t \in \{1, \ldots, n_0\}$ such that

$$f(x_t) - f(z) \leq \epsilon_0,$$

$$\|x_t\| \leq 15M + 1.$$

Theorem 3.3 is proved.

3.6 Zero-Sum Games

We use the notation and definitions introduced in Sect. 3.1.

Let X, Y be Hilbert spaces, C be a nonempty closed convex subset of X, D be a nonempty closed convex subset of Y, U be an open convex subset of X, and V be an open convex subset of Y such that

$$C \subset U, \ D \subset V. \tag{3.62}$$

Suppose that there exist $L > 0$, $M_0 > 0$ such that

$$C \subset B_X(0, M_0), \quad D \subset B_Y(0, M_0), \tag{3.63}$$

a function $f : U \times V \to R^1$ possesses the following properties:

(i) for each $v \in V$, the function $f(\cdot, v) : U \to R^1$ is convex;
(ii) for each $u \in U$, the function $f(u, \cdot) : V \to R^1$ is concave,

and that for each $u \in U$,

$$|f(u, v_1) - f(u, v_2)| \le L\|v_1 - v_2\| \text{ for all } v_1, v_2 \in V, \tag{3.64}$$

for each $v \in V$,

$$|f(u_1, v) - f(u_2, v)| \le L\|u_1 - u_2\| \text{ for all } u_1, u_2 \in U. \tag{3.65}$$

Recall that for each $(\xi, \eta) \in U \times V$,

$$\partial_x f(\xi, \eta) = \{l \in X : f(y, \eta) - f(\xi, \eta) \ge \langle l, y - \xi \rangle \text{ for all } y \in U\},$$
$$\partial_y f(\xi, \eta) = \{l \in Y : \langle l, y - \eta \rangle \ge f(\xi, y) - f(\xi, \eta) \text{ for all } y \in V\}.$$

In view of properties (i) and (ii) and (3.63)–(3.65), for each $\xi \in U$ and each $\eta \in V$,

$$\emptyset \ne \partial_x f(\xi, \eta) \subset B_X(0, L),$$
$$\emptyset \ne \partial_y f(\xi, \eta) \subset B_Y(0, L).$$

Let

$$x_* \in C \text{ and } y_* \in D \tag{3.66}$$

satisfy

$$f(x_*, y) \le f(x_*, y_*) \le f(x, y_*) \tag{3.67}$$

for each $x \in C$ and each $y \in D$.
  Let $\delta \in (0, 1]$ and $\{a_k\}_{k=0}^{\infty} \subset (0, \infty)$.
  Let us describe our algorithm.

Mirror Descent Algorithm for Zero-Sum Games
  Initialization: select arbitrary $x_0 \in U$ and $y_0 \in V$.
  Iterative step: given current iteration vectors $x_t \in U$ and $y_t \in V$ calculate

$$\xi_t \in \partial_x f(x_t, y_t) + B_X(0, \delta),$$
$$\eta_t \in \partial_y f(x_t, y_t) + B_Y(0, \delta)$$

and the next pair of iteration vectors $x_{t+1} \in U$, $y_{t+1} \in V$ such that

$$B_X(x_{t+1}, \delta) \cap \mathrm{argmin}\{\langle \xi_t, v \rangle + (2a_t)^{-1}\|v - x_t\|^2 : v \in C\} \neq \emptyset,$$

$$B_Y(y_{t+1}, \delta) \cap \mathrm{argmin}\{\langle -\eta_t, u \rangle + (2a_t)^{-1}\|u - y_t\|^2 : u \in D\} \neq \emptyset.$$

In this chapter we prove the following result.

Theorem 3.4. *Let* $\delta \in (0, 1]$ *and* $\{a_k\}_{k=0}^{\infty} \subset (0, \infty)$. *Assume that* $\{x_t\}_{t=0}^{\infty} \subset U$, $\{y_t\}_{t=0}^{\infty} \subset V$, $\{\xi_t\}_{t=0}^{\infty} \subset X$, $\{\eta_t\}_{t=0}^{\infty} \subset Y$,

$$B_X(x_0, \delta) \cap C \neq \emptyset, \quad B_Y(y_0, \delta) \cap D \neq \emptyset$$

and that for each integer $t \geq 0$,

$$\xi_t \in \partial_x f(x_t, y_t) + B_X(0, \delta),$$
$$\eta_t \in \partial_y f(x_t, y_t) + B_Y(0, \delta),$$
$$B_X(x_{t+1}, \delta) \cap argmin\{\langle \xi_t, v \rangle + (2a_t)^{-1}\|v - x_t\|^2 : v \in C\} \neq \emptyset,$$
$$B_Y(y_{t+1}, \delta) \cap argmin\{\langle -\eta_t, u \rangle + (2a_t)^{-1}\|u - y_t\|^2 : u \in D\} \neq \emptyset.$$

Let for each natural number T,

$$\hat{x}_T = \left(\sum_{i=0}^{T} a_t \right)^{-1} \sum_{t=0}^{T} a_t x_t, \quad \hat{y}_T = \left(\sum_{i=0}^{T} a_t \right)^{-1} \sum_{t=0}^{T} a_t y_t.$$

Then for each natural number T,

$$B_X(\hat{x}_T, \delta) \cap C \neq \emptyset, \quad B_Y(\hat{y}_T, \delta) \cap D \neq \emptyset, \tag{3.68}$$

$$\left| \left(\sum_{t=0}^{T} a_t \right)^{-1} \sum_{t=0}^{T} a_t f(x_t, y_t) - f(x_*, y_*) \right|$$

$$\leq [2^{-1}(2M_0 + 1)^2 + 8\delta(T + 1)(M_0 + 1)] \left(\sum_{t=0}^{T} a_t \right)^{-1}$$

$$+ \delta(2M_0 + L + 2) + 2^{-1} \left(\sum_{t=0}^{T} a_t \right)^{-1} (L + 1)^2 \sum_{t=0}^{T} a_t^2,$$

$$\left| f(\hat{x}_T, \hat{y}_T) - \left(\sum_{t=0}^{T} a_t \right)^{-1} \sum_{t=0}^{T} a_t f(x_t, y_t) \right|$$

$$\leq [2^{-1}(2M_0 + 1)^2 + 8\delta(T + 1)(M_0 + 1)] \left(\sum_{t=0}^{T} a_t \right)^{-1}$$

$$+ \delta(2M_0 + 2L + 2) + 2^{-1} \left(\sum_{t=0}^{T} a_t \right)^{-1} (L + 1)^2 \sum_{t=0}^{T} a_t^2,$$

and for each natural number T, each $z \in C$, and each $v \in D$,

$$f(\hat{x}_T, v) \leq f(\hat{x}_T, \hat{y}_T)$$

$$+ (2M_0 + 1)^2 \left(\sum_{t=0}^{T} a_t \right)^{-1} + \left(\sum_{t=0}^{T} a_t \right)^{-1} 16\delta(T + 1)(M_0 + 1)$$

$$+ 2\delta(2M_0 + 2L + 2) + \left(\sum_{t=0}^{T} a_t \right)^{-1} (L + 1)^2 \sum_{t=0}^{T} a_t^2,$$

$$f(z, \hat{y}_T) \geq f(\hat{x}_T, \hat{y}_T)$$

$$- (2M_0 + 1)^2 \left(\sum_{t=0}^{T} a_t \right)^{-1} - 16 \left(\sum_{t=0}^{T} a_t \right)^{-1} (T + 1)\delta(M_0 + 1)$$

$$- 2\delta(2M_0 + 2L + 2) - \left(\sum_{t=0}^{T} a_t \right)^{-1} (L + 1)^2 \sum_{t=0}^{T} a_t^2.$$

Proof. Evidently, (3.68) holds. It is not difficult to see that

$$\|x_t\| \leq M_0 + 1, \ \|y_t\| \leq M_0 + 1, \ t = 0, 1, \dots.$$

Let $t \geq 0$ be an integer. Applying Lemma 3.2 with

$$a = a_t, \ x = x_t, \ f = f(\cdot, y_t), \ \xi = \xi_t, \ u = x_{t+1}$$

we obtain that for each $z \in C$,

$$a_t(f(x_t, y_t) - f(z, y_t)) \leq 2^{-1}\|z - x_t\|^2 - 2^{-1}\|z - x_{t+1}\|^2$$
$$+ \delta(8M_0 + 8 + a_t(2M_0 + L + 2)) + 2^{-1}a_t^2(L + 1)^2.$$

Applying Lemma 3.2 with

$$a = a_t, \ x = y_t, \ f = -f(x_t, \cdot), \ \xi = -\eta_t, \ u = y_{t+1}$$

we obtain that for each $v \in D$,

$$a_t(f(x_t, v) - f(x_t, y_t)) \leq 2^{-1}\|v - y_t\|^2 - 2^{-1}\|v - y_{t+1}\|^2$$
$$+\delta(8M_0 + 8 + a_t(2M_0 + L + 2)) + 2^{-1}a_t^2(L + 1)^2.$$

For all integers $t \geq 0$ set

$$b_t = \delta(8M_0 + 8 + a_t(2M_0 + L + 2)) + 2^{-1}a_t^2(L + 1)^2$$

and define

$$\phi(s) = 2^{-1}s^2, \; s \in R^1.$$

It is easy to see that all the assumptions of Proposition 2.9 hold and it implies Theorem 3.4.

We are interested in the optimal choice of a_t, $t = 0, 1, \ldots$. Let T be a natural number and $A_T = \sum_{t=0}^{T} a_t$ be given. By Theorem 3.4, in order to make the best choice of a_t, $t = 0, \ldots, T$, we need to minimize the function $\sum_{t=0}^{T} a_t^2$ on the set

$$\left\{ a = (a_0, \ldots, a_T) \in R^{T+1} : \; a_i \geq 0, \; i = 0, \ldots, T, \; \sum_{i=0}^{T} a_i = A_T \right\}.$$

By Lemma 2.3, this function has a unique minimizer $a^* = (a_0^*, \ldots, a_T^*)$ where $a_i^* = (T + 1)^{-1}A_T$, $i = 0, \ldots, T$ which is the best choice of a_t, $t = 0, 1, \ldots, T$.

Now we will find the best $a > 0$. Let T be a natural number and $a_t = a$ for all $t = 0, \ldots, T$. We need to choice a which is a minimizer of the function

$$\Psi_T(a) = ((T + 1)a)^{-1}(2M_0 + 1)^2 + 2\delta(2M_0 + L + 2)$$
$$+16\delta(T + 1)(M_0 + 1)(a(T + 1))^{-1} + a(L + 1)^2$$
$$= (2M_0 + 1)^2((T + 1)a)^{-1} + 2\delta(2M_0 + L + 2) + 16\delta(M_0 + 1)a^{-1}$$
$$+(L + 1)^2a.$$

Since T can be arbitrary large, we need to find a minimizer of the function

$$\phi(a) := 16a^{-1}\delta(M_0 + 1) + (L + 1)^2a, \; a \in (0, \infty).$$

This function has a minimizer

$$a = 4(\delta(M_0 + 1))^{1/2}(L + 1)^{-1}$$

and the minimal value of ϕ is

$$8(\delta(M_0 + 1))^{1/2}(L + 1).$$

Now our goal is to find the best $T > 0$ which gives us an appropriate value of $\Psi_T(a)$. Since in view of the inequalities above, this value is bounded from below by $c_0 \delta^{1/2}$ with the constant c_0 depending on L, M_0 it is clear that in order to make the best choice of T, it should be at the same order as $\lfloor \delta^{-1} \rfloor$. For example, $T = \lfloor \delta^{-1} \rfloor$.

Note that in the theorem above δ is the computational error produced by our computer system. We obtain a good approximate solution after $T = \lfloor \delta^{-1} \rfloor$ iterations. Namely, we obtain a pair of points $\hat{x} \in U, \hat{y} \in V$ such that

$$B_X(\hat{x}, \delta) \cap C \neq \emptyset, \; B_Y(\hat{y}, \delta) \cap D \neq \emptyset$$

and for each $z \in C$ and each $v \in D$,

$$f(z, \hat{y}) \geq f(\hat{x}, \hat{y}) - c\delta^{1/2}, \; f(\hat{x}, v) \leq f(\hat{x}, \hat{y}) + c\delta^{1/2},$$

where the constant $c > 0$ depends only on L and M_0.

Chapter 4
Gradient Algorithm with a Smooth Objective Function

In this chapter we analyze the convergence of a projected gradient algorithm with a smooth objective function under the presence of computational errors. We show that the algorithm generates a good approximate solution, if computational errors are bounded from above by a small positive constant. Moreover, for a known computational error, we find out what an approximate solution can be obtained and how many iterates one needs for this.

4.1 Optimization on Bounded Sets

Let X be a Hilbert space equipped with an inner product $\langle \cdot, \cdot \rangle$ which induces a complete norm $\| \cdot \|$. For each $x \in X$ and each $r > 0$ set

$$B_X(x, r) = \{y \in X : \|x - y\| \le r\}.$$

For each $x \in X$ and each nonempty set $E \subset X$ put

$$d(x, E) = \inf\{\|x - y\| : y \in E\}.$$

Let C be a nonempty closed convex subset of X, U be an open convex subset of X such that $C \subset U$ and let $f : U \to R^1$ be a convex continuous function.

We suppose that the function f is Frechet differentiable at every point $x \in U$ and for every $x \in U$ we denote by $f'(x) \in X$ the Frechet derivative of f at x. It is clear that for any $x \in U$ and any $h \in X$

$$\langle f'(x), h \rangle = \lim_{t \to 0} t^{-1}(f(x + th) - f(x)). \tag{4.1}$$

© Springer International Publishing Switzerland 2016

A.J. Zaslavski, *Numerical Optimization with Computational Errors*, Springer Optimization and Its Applications 108, DOI 10.1007/978-3-319-30921-7_4

For each nonempty set D and each function $g : D \to R^1$ set

$$\inf(g; D) = \inf\{g(y) : y \in D\}, \tag{4.2}$$

$$\operatorname{argmin}\{g(z) : z \in D\} = \{z \in D : g(z) = \inf(g; D)\}. \tag{4.3}$$

We suppose that the mapping $f' : U \to X$ is Lipschitz on all bounded subsets of U. It is well known (see Lemma 2.2) that for each nonempty closed convex set $D \subset X$ and each $x \in X$ there exists a unique point $P_D(x) \in D$ such that

$$\|x - P_D(x)\| = \inf\{\|x - y\| : y \in D\}.$$

In this chapter we study the behavior of a projected gradient algorithm with a smooth objective function which is used for solving convex constrained minimization problems [91, 92, 98].

In the sequel we use the following proposition [91, 92, 98] which is proved in Sect. 4.2.

Proposition 4.1. *Assume that* $x, u \in U$, $L > 0$ *and that for each* $v_1, v_2 \in \{tx + (1 - t)u : t \in [0, 1]\}$,

$$\|f'(v_1) - f'(v_2)\| \leq L\|v_1 - v_2\|.$$

Then

$$f(u) \leq f(x) + \langle f'(x), u - x \rangle + 2^{-1}L\|u - x\|^2.$$

Suppose that there exist $L > 1$, $M_0 > 0$ such that

$$C \subset B_X(0, M_0), \tag{4.4}$$

$$|f(v_1) - f(v_2)| \leq L\|v_1 - v_2\| \text{ for all } v_1, v_2 \in U, \tag{4.5}$$

$$\|f'(v_1) - f'(v_2)\| \leq L\|v_1 - v_2\| \text{ for all } v_1, v_2 \in U. \tag{4.6}$$

Let $\delta \in (0, 1]$. We describe below our algorithm.

Gradient Algorithm

Initialization: select an arbitrary $x_0 \in U \cap B_X(0, M_0)$.

Iterative step: given a current iteration vector $x_t \in U$ calculate

$$\xi_t \in f'(x_t) + B_X(0, \delta)$$

and calculate the next iteration vector $x_{t+1} \in U$ such that

$$\|x_{t+1} - P_C(x_t - L^{-1}\xi_t)\| \leq \delta.$$

In this chapter we prove the following result.

Theorem 4.2. *Let $\delta \in (0, 1]$ and let*

$$x_0 \in U \cap B_X(0, M_0). \tag{4.7}$$

Assume that $\{x_t\}_{t=1}^{\infty} \subset U$, $\{\xi_t\}_{t=0}^{\infty} \subset X$ and that for each integer $t \geq 0$,

$$\|\xi_t - f'(x_t)\| \leq \delta \tag{4.8}$$

and

$$\|x_{t+1} - P_C(x_t - L^{-1}\xi_t)\| \leq \delta. \tag{4.9}$$

Then for each natural number T,

$$
\begin{aligned}
f(x_{T+1}) &- \inf(f; C) \\
&\leq (2T)^{-1}L(2M_0 + 1)^2 + L\delta(8M_0 + 8)(T + 1)
\end{aligned}
\tag{4.10}
$$

and

$$
\min\{f(x_t): \ t = 2, \ldots, T + 1\} - \inf(f; C), \ f\left(\sum_{t=2}^{T+1} T^{-1}x_t\right) - \inf(f; C)
$$

$$
\leq (2T)^{-1}L(2M_0 + 1)^2 + L\delta(8M_0 + 8). \tag{4.11}
$$

We are interested in an optimal choice of T. If we choose T in order to minimize the right-hand side of (4.11) we obtain that T should be at the same order as δ^{-1}. In this case the right-hand side of (4.11) is at the same order as δ. For example, if $T = \lfloor \delta^{-1} \rfloor + 1 \geq \delta^{-1}$, then the right-hand side of (4.11) does not exceed $2L\delta(4M_0 + 4 + (2M_0 + 1)^2)$.

4.2   An Auxiliary Result and the Proof of Proposition 4.1

Proposition 4.3. *Let D be a nonempty closed convex subset of X, $x \in X$ and $y \in D$. Assume that for each $z \in D$,*

$$\langle z - y, x - y \rangle \leq 0. \tag{4.12}$$

Then $y = P_D(x)$.

Proof. Let $z \in D$. By (4.12),

$$
\begin{aligned}
\langle z - x, z - x \rangle &= \langle z - y + (y - x), z - y + (y - x) \rangle \\
&= \langle y - x, y - x \rangle + 2\langle z - y, y - x \rangle + \langle z - y, z - y \rangle \\
&\geq \langle y - x, y - x \rangle + \langle z - y, z - y \rangle \\
&\geq \|y - x\|^2 + \|z - y\|^2.
\end{aligned}
$$

Thus $y = P_D(x)$. Proposition 4.3 is proved.

Proof of Proposition 4.1. For each $t \in [0, 1]$ set

$$
\phi(t) = f(x + t(u - x)). \tag{4.13}
$$

Clearly, ϕ is a differentiable function and for each $t \in [0, 1]$,

$$
\phi'(t) = \langle f'(x + t(u - x)), u - x \rangle. \tag{4.14}
$$

By (4.13), (4.14), and the proposition assumptions,

$$
\begin{aligned}
f(u) - f(x) &= \phi(1) - \phi(0) \\
&= \int_0^1 \phi'(t)dt = \int_0^1 \langle f'(x + t(u - x)), u - x \rangle dt \\
&= \int_0^1 \langle f'(x), u - x \rangle dt + \int_0^1 \langle f'(x + t(u - x)) - f'(x), u - x \rangle dt \\
&\leq \langle f'(x), u - x \rangle + \int_0^1 Lt\|u - x\|^2 dt \\
&= \langle f'(x), u - x \rangle + L\|u - x\|^2 \int_0^1 t dt \\
&= \langle f'(x), u - x \rangle + L\|u - x\|^2/2.
\end{aligned}
$$

Proposition 4.1 is proved.

4.3   The Main Lemma

We use the notation, definitions, and assumptions introduced in Sect. 4.1.

Lemma 4.4. *Let $\delta \in (0, 1]$,*

$$
u \in B_X(0, M_0 + 1) \cap U, \tag{4.15}
$$

$\xi \in X$ *satisfy*

$$\|\xi - f'(u)\| \leq \delta \tag{4.16}$$

and let $v \in U$ *satisfy*

$$\|v - P_C(u - L^{-1}\xi)\| \leq \delta. \tag{4.17}$$

Then for each $x \in U$ *satisfying*

$$B(x, \delta) \cap C \neq \emptyset \tag{4.18}$$

the following inequalities hold:

$$f(x) - f(v) \geq 2^{-1}L\|x - v\|^2 - 2^{-1}L\|x - u\|^2 - \delta L(8M_0 + 8), \tag{4.19}$$

$$f(x) - f(v) \geq 2^{-1}L\|u - v\|^2 + L\langle v - u, u - x\rangle - \delta L(8M_0 + 12). \tag{4.20}$$

Proof. For each $x \in U$ define

$$g(x) = f(u) + \langle f'(u), x - u\rangle + 2^{-1}L\|x - u\|^2. \tag{4.21}$$

Clearly, $g : U \to R^1$ is a convex Frechet differentiable function, for each $x \in U$,

$$g'(x) = f'(u) + L(x - u), \tag{4.22}$$

$$\lim_{\|x\| \to \infty} g(x) = \infty \tag{4.23}$$

and there exists

$$x_0 \in C \tag{4.24}$$

such that

$$g(x_0) \leq g(x) \text{ for all } x \in C. \tag{4.25}$$

By (4.24) and (4.25), for all $z \in C$,

$$\langle g'(x_0), z - x_0\rangle \geq 0. \tag{4.26}$$

In view of (4.22) and (4.26),

$$\langle L^{-1}f'(u) + x_0 - u, z - x_0\rangle \geq 0 \text{ for all } z \in C. \tag{4.27}$$

Proposition 4.3, (4.24), and (4.27) imply that

$$x_0 = P_C(u - L^{-1}f'(u)). \tag{4.28}$$

It follows from (4.16), (4.28), and Lemma 2.2 that

$$\|v - x_0\|$$
$$\leq \|v - P_C(u - L^{-1}\xi)\| + \|P_C(u - L^{-1}\xi) - P_C(u - L^{-1}f'(u))\|$$
$$\leq \delta + L^{-1}\|\xi - f'(u)\| \leq \delta(1 + L^{-1}). \tag{4.29}$$

In view of (4.4) and (4.24),

$$\|x_0\| \leq M_0. \tag{4.30}$$

Relations (4.4) and (4.17) imply that

$$\|v\| \leq M_0 + 1. \tag{4.31}$$

By (4.6), (4.15), (4.21), and Proposition 4.1, for all $x \in U$,

$$f(x) \leq f(u) + \langle f'(u), x - u \rangle + 2^{-1}L\|u - x\|^2 = g(x). \tag{4.32}$$

Let

$$x \in U \tag{4.33}$$

satisfy

$$B(x, \delta) \cap C \neq \emptyset. \tag{4.34}$$

It follows from (4.5) and (4.29) that

$$|f(x_0) - f(v)| \leq L\|v - x_0\| \leq \delta(L + 1). \tag{4.35}$$

In view of (4.24) and (4.32),

$$g(x_0) \geq f(x_0). \tag{4.36}$$

By (4.21), (4.36) and convexity of f,

$$f(x) - f(x_0) \geq f(x) - g(x_0)$$
$$= f(x) - f(u) - \langle f'(u), x_0 - u \rangle - 2^{-1}L\|u - x_0\|^2$$
$$\geq f(u) + \langle f'(u), x - u \rangle - f(u) - \langle f'(u), x_0 - u \rangle - 2^{-1}L\|u - x_0\|^2$$
$$= \langle f'(u), x - x_0 \rangle - 2^{-1}L\|u - x_0\|^2. \tag{4.37}$$

Relation (4.34) implies that there exists

$$x_1 \in C \tag{4.38}$$

such that

$$\|x_1 - x\| \leq \delta. \tag{4.39}$$

By (4.5), (4.33), (4.38), and (4.39),

$$|f(x_1) - f(x)| \leq L\delta. \tag{4.40}$$

It follows from (4.26) (with $z = x_1$) and (4.38) that

$$0 \leq \langle g'(x_0), x_1 - x_0 \rangle = \langle g'(x_0), x_1 - x \rangle + \langle g'(x_0), x - x_0 \rangle. \tag{4.41}$$

By (4.4), (4.5), (4.15), (4.22), (4.24), (4.38), and (4.39),

$$\langle g'(x_0), x_1 - x \rangle = \langle f'(u), x_1 - x \rangle + L\langle x_0 - u, x_1 - x \rangle$$
$$\leq L\delta + L\delta(2M_0 + 1). \tag{4.42}$$

In view of (4.22) and (4.24),

$$\langle g'(x_0), x - x_0 \rangle = \langle f'(u) + L(x_0 - u), x - x_0 \rangle. \tag{4.43}$$

Relations (4.41) and (4.43) imply that

$$\langle f'(u), x - x_0 \rangle = \langle g'(x_0), x - x_0 \rangle - L\langle x_0 - u, x - x_0 \rangle$$
$$\geq -\langle g'(x_0), x_1 - x \rangle - L\langle x_0 - u, x - x_0 \rangle$$
$$\geq -L\langle x_0 - u, x - x_0 \rangle - L\delta(2M_0 + 2). \tag{4.44}$$

It follows from (4.37) and (4.44) that

$$f(x) - f(x_0) \geq \langle f'(u), x - x_0 \rangle - 2^{-1}L\|x_0 - u\|^2$$
$$\geq -L\delta(2M_0 + 2) - L\langle x_0 - u, x - x_0 \rangle - 2^{-1}L\|x_0 - u\|^2. \tag{4.45}$$

In view of (4.45) and Lemma 2.1,

$$f(x) - f(x_0) \geq -L\delta(2M_0 + 2) - 2^{-1}L\|x_0 - u\|^2$$
$$-2^{-1}L[\|x - u\|^2 - \|x - x_0\|^2 - \|u - x_0\|^2]$$
$$= 2^{-1}L\|x - x_0\|^2 - 2^{-1}L\|x - u\|^2 - L\delta(2M_0 + 2). \tag{4.46}$$

By (4.4), (4.24), (4.34), and (4.35),

$$f(x) - f(v) \geq f(x) - f(x_0) - \delta(L + 1). \tag{4.47}$$

It follows from (4.15), (4.17), (4.24), and (4.29) that

$$| \|x - x_0\|^2 - \|x - v\|^2 |$$
$$= | \|x - x_0\| - \|x - v\| | (\|x - x_0\| + \|x - v\|) \leq \delta(8M_0 + 8) \quad (4.48)$$

and

$$| \|u - x_0\|^2 - \|u - v\|^2 |$$
$$= | \|u - x_0\| - \|u - v\| | (\|u - x_0\| + \|u - v\|) \leq \delta(8M_0 + 8). \quad (4.49)$$

In view of (4.45),

$$f(x) - f(x_0) \geq -L\delta(2M_0 + 2) + 2^{-1}L\|x_0 - u\|^2$$
$$-L\langle x_0 - u, x_0 - u \rangle - L\langle x_0 - u, x - x_0 \rangle$$
$$\geq -L\delta(2M_0 + 2) + 2^{-1}L\|x_0 - u\|^2$$
$$-L\langle x_0 - u, x - u \rangle. \quad (4.50)$$

By (4.35), (4.46), and (4.48),

$$f(x) - f(v) \geq f(x) - f(x_0) - \delta(L + 1)$$
$$\geq 2^{-1}L\|x - x_0\|^2 - 2^{-1}L\|x - u\|^2 - L\delta(2M_0 + 4)$$
$$\geq 2^{-1}L\|x - v\|^2 - 2^{-1}L\|x - u\|^2 - 2L\delta(4M_0 + 4)$$

and (4.19) holds. It follows from (4.15), (4.29), (4.34), (4.35), (4.49), and (4.50) that

$$f(x) - f(v) \geq f(x) - f(x_0) - \delta(L + 1)$$
$$\geq -L\delta(2M_0 + 4) + 2^{-1}L\|x_0 - u\|^2 - L\langle x_0 - u, x - u \rangle$$
$$\geq -L\delta(2M_0 + 4) + 2^{-1}L\|u - v\|^2 - 4L\delta(M_0 + 1)$$
$$-L\langle v - u, x - u \rangle - L\delta(2M_0 + 4)$$
$$\geq 2^{-1}L\|u - v\|^2 - L\langle v - u, x - u \rangle - L\delta(8M_0 + 12)$$

and (4.20) holds. Lemma 4.4 is proved.

4.4   Proof of Theorem 4.2

Clearly, the function f has a minimizer on the set C. Fix

$$z \in C \quad (4.51)$$

such that

$$f(z) = \inf(f; C). \tag{4.52}$$

It is easy to see that

$$\|x_t\| \le M_0 + 1, \ t = 0, 1, \dots. \tag{4.53}$$

Let T be a natural number and $t \ge 0$ be an integer. Applying Lemma 4.4 with

$$u = x_t, \ \xi = \xi_t, \ v = x_{t+1}, \ x = z$$

we obtain that

$$f(z) - f(x_{t+1}) \ge 2^{-1}L\|z - x_{t+1}\|^2 - 2^{-1}L\|z - x_t\|^2 - \delta L(8M_0 + 8).$$

This implies that

$$Tf(z) - \sum_{t=1}^{T} f(x_{t+1})$$

$$\ge \sum_{t=1}^{T} (2^{-1}L\|z - x_{t+1}\|^2 - 2^{-1}L\|z - x_t\|^2) - \delta LT(8M_0 + 8)$$

$$\ge 2^{-1}L(\|z - x_{T+1}\|^2 - \|z - x_1\|^2) - \delta LT(8M_0 + 8). \tag{4.54}$$

Let $t \ge 0$ be an integer. Applying Lemma 4.4 with

$$x = x_{t+1}, \ u = x_{t+1}, \ \xi = \xi_{t+1}, \ v = x_{t+2},$$

we obtain that

$$f(x_{t+1}) - f(x_{t+2}) \ge 2^{-1}L\|x_{t+2} - x_{t+1}\|^2 - \delta L(8M_0 + 8)$$

and

$$t(f(x_{t+1}) - f(x_{t+2})) \ge 2^{-1}Lt\|x_{t+2} - x_{t+1}\|^2 - \delta Lt(8M_0 + 8).$$

We can write the relation above as

$$tf(x_{t+1}) - (t + 1)f(x_{t+2}) + f(x_{t+2})$$
$$\ge 2^{-1}Lt\|x_{t+2} - x_{t+1}\|^2 - \delta Lt(8M_0 + 8). \tag{4.55}$$

Summing (4.55) with $t = 0, \dots, T - 1$ we obtain that

$$-Tf(x_{T+1}) + \sum_{t=0}^{T-1} f(x_{t+2})$$

$$= \sum_{t=0}^{T-1} [tf(x_{t+1}) - (t+1)f(x_{t+2}) + f(x_{t+2})]$$

$$\geq \sum_{t=0}^{T-1} [2^{-1} Lt \|x_{t+2} - x_{t+1}\|^2] - \delta L(8M_0 + 8) \sum_{t=0}^{T-1} t. \qquad (4.56)$$

By (4.54) and (4.56),

$$T(f(z) - f(x_{T+1}))$$
$$\geq -2^{-1} L \|z - x_1\|^2 - L\delta T(8M_0 + 8) - L\delta(4M_0 + 4)T(T-1)$$

and in view of (4.51) and (4.53),

$$f(x_{T+1}) - f(z)$$
$$\leq (2T)^{-1} L(2M_0 + 1)^2 + L\delta(8M_0 + 8)(T + 1).$$

In view of (4.54)

$$T(\min\{f(x_t) : t = 2, \ldots, T+1\} - f(z)), \ T\left(f\left(\sum_{t=2}^{T+1} T^{-1} x_t\right) - f(z)\right)$$

$$\leq \sum_{t=1}^{T} f(x_{t+1}) - Tf(z)$$

$$\leq 2^{-1} L(2M_0 + 1)^2 + L\delta T(8M_0 + 8).$$

This completes the proof of Theorem 4.2.

4.5 Optimization on Unbounded Sets

We use the notation and definitions introduced in Sect. 4.1. Let X be a Hilbert space with an inner product $\langle \cdot, \cdot \rangle$ which induces a complete norm $\| \cdot \|$.

Let D be a nonempty closed convex subset of X, V be an open convex subset of X such that

$$D \subset V$$

and $f : V \to R^1$ be a convex Fréchet differentiable function which is Lipschitz on all bounded subsets of V. Set

$$D_{\min} = \{x \in D : f(x) \le f(y) \text{ for all } y \in D\}. \tag{4.57}$$

We suppose that

$$D_{\min} \ne \emptyset. \tag{4.58}$$

We will prove the following result.

Theorem 4.5. *Let $\delta \in (0, 1]$, $M > 0$ satisfy*

$$D_{\min} \cap B_X(0, M) \ne \emptyset, \tag{4.59}$$

$M_0 \ge 4M + 8, L \ge 1$ *satisfy*

$$|f(v_1) - f(v_2)| \le L\|v_1 - v_2\| \text{ for all } v_1, v_2 \in V \cap B_X(0, M_0 + 2), \tag{4.60}$$

$$\|f'(v_1) - f'(v_2)\| \le L\|v_1 - v_2\| \text{ for all } v_1, v_2 \in V \cap B_X(0, M_0 + 2), \tag{4.61}$$

$$\epsilon_0 = 4L\delta(2M_0 + 3) \tag{4.62}$$

and let

$$n_0 = \lfloor (4\delta)^{-1}(2M + 1)^2 \rfloor + 1. \tag{4.63}$$

Assume that $\{x_t\}_{t=0}^{\infty} \subset V$, $\{\xi_t\}_{t=0}^{\infty} \subset X$,

$$\|x_0\| \le M \tag{4.64}$$

and that for each integer $t \ge 0$,

$$\|\xi_t - f'(x_t)\| \le \delta \tag{4.65}$$

and

$$\|x_{t+1} - P_D(x_t - L^{-1}\xi_t)\| \le \delta. \tag{4.66}$$

Then there exists an integer $q \in [1, n_0 + 1]$ such that

$$f(x_q) \le \inf(f; D) + \epsilon_0,$$
$$\|x_i\| \le 3M + 3, \ i = 0, \ldots, q.$$

Proof. By (4.59) there exists

$$z \in D_{\min} \cap B_X(0, M). \tag{4.67}$$

By (4.60), (4.64)–(4.67), and Lemma 2.2,

$$\|x_1 - z\| \le \|x_1 - P_D(x_0 - L^{-1}\xi_0)\| + \|P_D(x_0 - L^{-1}\xi_0) - z\|$$
$$\le \delta + \|x_0 - z\| + L^{-1}\|\xi_0\|$$
$$\le 1 + 2M + L^{-1}(L + 1) \le 2M + 3. \tag{4.68}$$

In view (4.67) and (4.68),

$$\|x_1\| \le 3M + 3. \tag{4.69}$$

Assume that an integer $T \ge 0$ and that for all $t = 1, \ldots, T + 1$,

$$f(x_t) - f(z) > \epsilon_0. \tag{4.70}$$

Set

$$U = V \cap \{v \in X : \|v\| < M_0 + 2\} \tag{4.71}$$

and

$$C = D \cap B_X(0, M_0). \tag{4.72}$$

Assume that an integer $t \in [0, T]$ and that

$$\|x_t - z\| \le 2M + 3. \tag{4.73}$$

(In view of (4.64), (4.67), and (4.68), our assumption is true for $t = 0$.) By (4.67) and (4.72),

$$z \in C \subset B_X(0, M_0). \tag{4.74}$$

Relations (4.67), (4.71), and (4.73) imply that

$$x_t \in U \cap B_X(0, M_0 + 1). \tag{4.75}$$

It follows from (4.60), (4.65), and (4.75) that

$$\xi_t \in f'(x_t) + B_X(0, 1) \subset B_X(0, L + 1). \tag{4.76}$$

By (4.67), (4.73), (4.76), and Lemma 2.2,

$$\|z - P_D(x_t - L^{-1}\xi_t)\| \le \|z - x_t + L^{-1}\xi_t\|$$
$$\le \|z - x_t\| + L^{-1}\|\xi_t\| \le 2M + 5. \tag{4.77}$$

In view of (4.67) and (4.77),

$$\|P_D(x_t - L^{-1}\xi_t)\| \le 3M + 5. \tag{4.78}$$

Relations (4.72) and (4.78) imply that

$$P_D(x_t - L^{-1}\xi_t) \in C, \tag{4.79}$$
$$P_D(x_t - L^{-1}\xi_t) = P_C(x_t - L^{-1}\xi_t). \tag{4.80}$$

It follows from (4.66), (4.71), and (4.78) that

$$\|x_{t+1}\| \le 3M + 6, \ x_{t+1} \in U. \tag{4.81}$$

By (4.65), (4.66), (4.79), (4.80), (4.81), and Lemma 4.4 applied with

$$u = x_t, \ \xi = \xi_t, \ v = x_{t+1}, \ x = z$$

we obtain that

$$\begin{aligned}
f(z) - f(x_{t+1}) \\
\ge 2^{-1}L\|z - x_{t+1}\|^2 - 2^{-1}L\|z - x_t\|^2 - L\delta(8M_0 + 8).
\end{aligned} \tag{4.82}$$

By (4.62), (4.70), and (4.82),

$$\begin{aligned}
4L\delta(2M_0 + 3) = \epsilon_0 < f(x_{t+1}) - f(z) \\
\le 2^{-1}L(\|z - x_t\|^2 - \|z - x_{t+1}\|^2) + L\delta(8M_0 + 8).
\end{aligned} \tag{4.83}$$

In view of (4.73) and (4.83),

$$\|z - x_{t+1}\| \le \|z - x_t\| \le 2M + 3.$$

Thus by induction we showed that for all $t = 0, \ldots, T + 1$

$$\|z - x_t\| \le 2M + 3, \ \|x_t\| \le 3M + 3$$

and that (4.83) holds for all $t = 0, \ldots, T$.

It follows from (4.63), (4.64), (4.67), and (4.83) that

$$\begin{aligned}
4(1 + T)L\delta(2M_0 + 3) \\
\le (1 + T)(\min\{f(x_t) : \ t = 1, \ldots, T + 1\} - f(z))
\end{aligned}$$

$$\leq \sum_{t=0}^{T}(f(x_{t+1}) - f(z))$$

$$\leq 2^{-1}L\sum_{t=0}^{T}(\|z - x_t\|^2 - \|z - x_{t+1}\|^2) + (T+1)L\delta(8M_0 + 8),$$

$$2L\delta(1+T) \leq 2^{-1}L\|z - x_0\|^2 \leq 2^{-1}L(2M+1)^2$$

and

$$T < (2M+1)^2(4\delta)^{-1} \leq n_0.$$

Thus we assumed that an integer $T \geq 0$ satisfies

$$f(x_t) - f(z) > \epsilon_0, \ t = 1, \ldots, T+1$$

and showed that $T \leq n_0 - 1$ and

$$\|x_t\| \leq 3M + 3, \ t = 0, \ldots, T+1.$$

This implies that there exists a natural number $q \leq n_0 + 1$ such that

$$f(x_q) - f(z) \leq \epsilon_0 \leq 4L\delta(2M_0 + 3)$$
$$\|x_t\| \leq 3M + 3, \ t = 0, \ldots, q.$$

Theorem 4.2 is proved.

Note that in the theorem above δ is the computational error produced by our computer system. We obtain a good approximate solution after $\lfloor(4\delta)^{-1}(2M+1)^2\rfloor + 2$ iterations. Namely, we obtain a point $x \in X$ such that

$$B_X(x, \delta) \cap D \neq \emptyset$$

and

$$f(x) \leq \inf(f; D) + 4L\delta(2M_0 + 3).$$

Chapter 5
An Extension of the Gradient Algorithm

In this chapter we analyze the convergence of a gradient type algorithm, under the presence of computational errors, which was introduced by Beck and Teboulle [20] for solving linear inverse problems arising in signal/image processing. We show that the algorithm generates a good approximate solution, if computational errors are bounded from above by a small positive constant. Moreover, for a known computational error, we find out what an approximate solution can be obtained and how many iterates one needs for this.

5.1 Preliminaries and the Main Result

Let X be a Hilbert space equipped with an inner product $\langle \cdot, \cdot \rangle$ which induces a complete norm $\| \cdot \|$. For each $x \in X$ and each $r > 0$ set

$$B_X(x, r) = \{y \in X : \ \|x - y\| \leq r\}.$$

Suppose that $f : X \to R^1$ is a convex Fréchet differentiable function on X and for every $x \in X$ denote by $f'(x) \in X$ the Fréchet derivative of f at x. It is clear that for any $x \in X$ and any $h \in X$

$$\langle f'(x), h \rangle = \lim_{t \to 0} t^{-1}(f(x + th) - f(x)).$$

For each function $\phi : X \to R^1$ set

$$\inf(\phi) = \inf\{\phi(y) : \ y \in X\},$$
$$\operatorname{argmin}(\phi) = \operatorname{argmin}\{\phi(z) : \ z \in X\} := \{z \in X : \ \phi(z) = \inf(\phi)\}.$$

We suppose that the mapping $f' : X \to X$ is Lipschitz on all bounded subsets of X.

© Springer International Publishing Switzerland 2016
A.J. Zaslavski, *Numerical Optimization with Computational Errors*, Springer Optimization and Its Applications 108, DOI 10.1007/978-3-319-30921-7_5

Let $g : X \rightarrow R^1$ be a convex continuous function which is Lipschitz on all bounded subsets of X. Define

$$F(x) = f(x) + g(x), \ x \in X.$$

We suppose that

$$\operatorname{argmin}(F) \neq \emptyset \tag{5.1}$$

and that there exists $c_* \in R^1$ such that

$$g(x) \geq c_* \text{ for all } x \in X. \tag{5.2}$$

For each $u \in X$, each $\xi \in X$ and each $L > 0$ define a convex function

$$G_{u,\xi}^{(L)}(w) = f(u) + \langle \xi, w - u \rangle + 2^{-1} L \|w - u\|^2 + g(w), \ w \in X \tag{5.3}$$

which has a minimizer.

In this chapter we analyze the gradient type algorithm, which was introduced by Beck and Teboulle in [20] for solving linear inverse problems, and prove the following result.

Theorem 5.1. *Let $\delta \in (0, 1], M \geq 1$ satisfy*

$$\operatorname{argmin}(F) \cap B_X(0, M) \neq \emptyset, \tag{5.4}$$

$L > 1$ satisfy

$$|f(w_1) - f(w_2)| \leq L\|w_1 - w_2\| \text{ for all } w_1, w_2 \in B_X(0, 3M + 2) \tag{5.5}$$

and

$$\|f'(w_1) - f'(w_2)\| \leq L\|w_1 - w_2\| \text{ for all } w_1, w_2 \in X, \tag{5.6}$$

$M_1 \geq 3M$ satisfy

$$|f(w)|, \ F(w) \leq M_1 \text{ for all } w \in B_X(0, 3M + 2), \tag{5.7}$$

$$M_2 = (8(M_1 + |c_*| + 1 + (L + 1)^2))^{1/2} + 3M + 2, \tag{5.8}$$

$L_0 > 1$ satisfy

$$|g(w_1) - g(w_2)| \leq L_0\|w_1 - w_2\| \text{ for all } w_1, w_2 \in B_X(0, M_2), \tag{5.9}$$

$$\epsilon_0 = 2\delta((M_2 + 3M + 2)(2L + 3) + L_0) \tag{5.10}$$

and let

$$n_0 = \lfloor 4LM^2\epsilon_0^{-1} \rfloor. \tag{5.11}$$

Assume that $\{x_t\}_{t=0}^{\infty} \subset X$, $\{\xi_t\}_{t=0}^{\infty} \subset X$,

$$\|x_0\| \leq M \tag{5.12}$$

and that for each integer $t \geq 0$,

$$\|\xi_t - f'(x_t)\| \leq \delta \tag{5.13}$$

and

$$B_X(x_{t+1}, \delta) \cap argmin(G^{(L)}_{x_t,\xi_t}) \neq \emptyset. \tag{5.14}$$

Then there exists an integer $q \in [0, n_0 + 2]$ *such that*

$$\|x_i\| \leq M_2, \ i = 0, \ldots, q$$

and

$$F(x_q) \leq \inf(F) + \epsilon_0.$$

Note that in the theorem above δ is the computational error produced by our computer system. We obtain a good approximate solution after $\lfloor c_1\delta^{-1} \rfloor$ iterations [see (5.10) and (5.11)], where $c_1 > 0$ is a constant which depends only on L, L_0, M, M_2. As a result we obtain a point $x \in X$ such that

$$F(x) \leq \inf(F) + c_2\delta,$$

where $c_2 > 0$ is a constant which depends only on L, L_0, M, M_2.

5.2 Auxiliary Results

Lemma 5.2 ([20]). *Let* $u, \xi \in X$ *and* $L > 0$. *Then the function* $G^{(L)}_{u,\xi}$ *has a point of minimum and* $z \in X$ *is a minimizer of* $G^{(L)}_{u,\xi}$ *if and only if*

$$0 \in \xi + L(z - u) + \partial g(z).$$

Proof. By (5.2) and (5.3),

$$\lim_{\|w\| \to \infty} G^{(L)}_{u,\xi}(w) = \infty.$$

This implies that the function $G_{u,\xi}^{(L)}$ has a minimizer. Clearly, z is a minimizer of $G_{u,\xi}^{(L)}$ if and only if

$$0 \in \partial G_{u,\xi}^{(L)}(z) = \xi + L(z - u) + \partial g(z).$$

Lemma 5.2 is proved.

Lemma 5.3. *Let* $M_0 \geq 1, L > 1$ *satisfy*

$$|f(w_1) - f(w_2)| \leq L\|w_1 - w_2\| \ for \ all \ w_1, w_2 \in B_X(0, M_0 + 2)$$

and

$$\|f'(w_1) - f'(w_2)\| \leq L\|w_1 - w_2\| \ for \ all \ w_1, w_2 \in X, \qquad (5.15)$$

$M_1 \geq M_0$ *satisfy*

$$|f(w)|, \ F(w) \leq M_1 \ for \ all \ w \in B_X(0, M_0 + 2). \qquad (5.16)$$

Assume that

$$u \in B_X(0, M_0 + 1), \qquad (5.17)$$

$\xi \in X$ *satisfies*

$$\|\xi - f'(u)\| \leq 1 \qquad (5.18)$$

and that $v \in X$ *satisfies*

$$B_X(v, 1) \cap \{z \in X : \ G_{u,\xi}^{(L)}(z) \leq \inf(G_{u,\xi}^{(L)}) + 1\} \neq \emptyset. \qquad (5.19)$$

Then

$$\|v\| \leq (8(M_1 + |c_*| + (L + 1)^2 + 1))^{1/2} + M_0 + 2.$$

Proof. In view of (5.19), there exists

$$\hat{v} \in B_X(v, 1) \qquad (5.20)$$

such that

$$G_{u,\xi}^{(L)}(\hat{v}) \leq \inf(G_{u,\xi}^{(L)}) + 1. \qquad (5.21)$$

By (5.3), (5.16), (5.17), and (5.21),

$$f(u) + \langle \xi, \hat{v} - u \rangle + 2^{-1}L\|\hat{v} - u\|^2 + g(\hat{v})$$
$$= G_{u,\xi}^{(L)}(\hat{v}) \leq G_{u,\xi}^{(L)}(u) + 1 = F(u) + 1 \leq M_1 + 1. \qquad (5.22)$$

It follows from (5.2), (5.16), (5.17), and (5.22) that

$$\langle \xi, \hat{v} - u \rangle + 2^{-1}L\|\hat{v} - u\|^2 \leq 2M_1 + 1 + |c_*|. \qquad (5.23)$$

It is clear that

$$2L^{-1}|\langle \xi, \hat{v} - u \rangle| \leq L^{-1}(4^{-1}\|\hat{v} - u\|^2 + 4\|\xi\|^2). \qquad (5.24)$$

Since the function f is Lipschitz on $B_X(0, M_0 + 2)$ relations (5.17) and (5.18) imply that

$$\|\xi\| \leq \|f'(u)\| + 1 \leq L + 1. \qquad (5.25)$$

By (5.23)–(5.25),

$$2L^{-1}(2M_1 + 1 + |c_*|)$$
$$\geq \|\hat{v} - u\|^2 - 2L^{-1}|\langle \xi, \hat{v} - u \rangle|$$
$$\geq \|\hat{v} - u\|^2 - 4^{-1}\|\hat{v} - u\|^2 - 4\|\xi\|^2$$
$$\geq 2^{-1}\|\hat{v} - u\|^2 - 4(L + 1)^2.$$

This implies that

$$\|\hat{v} - u\|^2 \leq 4(M_1 + 1 + |c_*|) + 8(L + 1)^2$$

and

$$\|\hat{v} - u\| \leq (4(M_1 + 1 + |c_*|) + 8(L + 1)^2)^{1/2}.$$

Together with (5.17) and (5.20) this implies that

$$\|v\| \leq \|\hat{v}\| + 1 \leq \|\hat{v} - u\| + \|u\| + 1$$
$$\leq (4(M_1 + 1 + |c_*|) + 8(L + 1)^2)^{1/2} + M_0 + 2.$$

Lemma 5.3 is proved.

5.3   The Main Lemma

Lemma 5.4. *Let $\delta \in (0, 1]$, $M_0 \geq 1$, $L > 1$ satisfy*

$$|f(w_1) - f(w_2)| \leq L\|w_1 - w_2\| \text{ for all } w_1, w_2 \in B_X(0, M_0 + 2) \tag{5.26}$$

and

$$\|f'(w_1) - f'(w_2)\| \leq L\|w_1 - w_2\| \text{ for all } w_1, w_2 \in X, \tag{5.27}$$

$M_1 \geq M_0$ *satisfy*

$$|f(w)|, \ F(w) \leq M_1 \text{ for all } w \in B_X(0, M_0 + 2), \tag{5.28}$$

$$M_2 = (8(M_1 + |c_*| + (L + 1)^2 + 1))^{1/2} + M_0 + 2 \tag{5.29}$$

and let $L_0 > 1$ satisfy

$$|g(w_1) - g(w_2)| \leq L_0\|w_1 - w_2\| \text{ for all } w_1, w_2 \in B_X(0, M_2). \tag{5.30}$$

Assume that

$$u \in B_X(0, M_0 + 1), \tag{5.31}$$

$\xi \in X$ *satisfies*

$$\|\xi - f'(u)\| \leq \delta \tag{5.32}$$

and that $v \in X$ satisfies

$$B_X(v, \delta) \cap argmin(G_{u,\xi}^{(L)}) \neq \emptyset. \tag{5.33}$$

Then for each $x \in B_X(0, M_0 + 1)$,

$$F(x) - F(v) \geq 2^{-1}L\|v - x\|^2 - 2^{-1}L\|u - x\|^2$$
$$-\delta((M_2 + M_0 + 2)(2L + 3) + L_0)$$

Proof. By (5.33) there exists $\hat{v} \in X$ such that

$$\hat{v} \in argmin(G_{u,\xi}^{(L)}) \tag{5.34}$$

and

$$\|v - \hat{v}\| \leq \delta. \tag{5.35}$$

In view of the assumptions of the lemma, Lemma 5.3 and (5.34),

$$\|v\|, \|\hat{v}\| \leq M_2. \tag{5.36}$$

Let

$$x \in B_X(0, M_0 + 1). \tag{5.37}$$

Clearly,

$$F(x) = f(x) + g(x), \; F(v) = f(v) + g(v). \tag{5.38}$$

Proposition 4.1 and (5.31) imply that

$$g(v) + f(v) \leq f(u) + \langle f'(u), v - u \rangle + 2^{-1}L\|v - u\|^2 + g(v). \tag{5.39}$$

By (5.3), (5.31), (5.32), (5.36), (5.38), and (5.39),

$$\begin{aligned}
F(x) - F(v) &\geq [f(x) + g(x)] - [f(v) + g(v)] \\
&\geq f(x) + g(x) - [f(u) + \langle f'(u), v - u \rangle + 2^{-1}L\|v - u\|^2 + g(v)] \\
&= [f(x) + g(x)] - [f(u) + \langle \xi, v - u \rangle + 2^{-1}L\|v - u\|^2 + g(v)] \\
&\quad + \langle \xi - f'(u), v - u \rangle \\
&\geq [f(x) + g(x)] - G_{u,\xi}^{(L)}(v) - \|\xi - f'(u)\|\|v - u\| \\
&\geq [f(x) + g(x)] - G_{u,\xi}^{(L)}(v) - \delta(M_2 + M_0 + 1). \tag{5.40}
\end{aligned}$$

It follows from (5.3) that

$$\begin{aligned}
G_{u,\xi}^{(L)}(v) &- G_{u,\xi}^{(L)}(\hat{v}) \\
&= \langle \xi, v - u \rangle + 2^{-1}L\|v - u\|^2 + g(v) \\
&\quad - [\langle \xi, \hat{v} - u \rangle + 2^{-1}L\|\hat{v} - u\|^2 + g(\hat{v})] \\
&= \langle \xi, v - \hat{v} \rangle + 2^{-1}L[\|v - u\|^2 - \|\hat{v} - u\|^2] + g(v) - g(\hat{v}). \tag{5.41}
\end{aligned}$$

Relation (5.27) and (5.32) imply that

$$\|\xi\| \leq \|f'(u)\| + 1 \leq L + 1. \tag{5.42}$$

In view of (5.35) and (5.42),

$$|\langle \xi, v - \hat{v} \rangle| \leq (L + 1)\delta. \tag{5.43}$$

By (5.31), (5.35), and (5.36),

$$|\|v-u\|^2 - \|\hat{v}-u\|^2| \le |\|v-u\| - \|\hat{v}-u\||(\|v-u\| + \|\hat{v}-u\|)$$
$$\le \|v-\hat{v}\|(2M_2 + 2M_0 + 2) \le \delta(2M_2 + 2M_0 + 2).$$
(5.44)

In view of (5.30), (5.35), and (5.36),

$$|g(v) - g(\hat{v})| \le L_0\|v - \hat{v}\| \le L_0\delta.$$
(5.45)

It follows from (5.41) and (5.43)–(5.45) that

$$|G_{u,\xi}^{(L)}(v) - G_{u,\xi}^{(L)}(\hat{v})|$$
$$\le (L+1)\delta + 2^{-1}L\delta(2M_0 + 2M_2 + 2) + L_0\delta$$
$$\le \delta(L + 1 + L_0 + L(M_0 + M_2 + 1)).$$
(5.46)

Relations (5.40) and (5.46) imply that

$$F(x) - F(v)$$
$$\ge f(x) + g(x) - G_{u,\xi}^{(L)}(v) - \delta(M_2 + M_0 + 1)$$
$$\ge f(x) + g(x) - G_{u,\xi}^{(L)}(\hat{v}) - \delta(L_0 + L + 1 + (L+1)M_2 + M_0 + 1)).$$
(5.47)

By the convexity of f, (5.31), (5.32), and (5.37),

$$f(x) \ge f(u) + \langle f'(u), x - u \rangle$$
$$\ge f(u) + \langle \xi, x - u \rangle - |\langle f'(u) - \xi, x - u \rangle|$$
$$\ge f(u) + \langle \xi, x - u \rangle - \|f'(u) - \xi\|\|x - u\|$$
$$\ge f(u) + \langle \xi, x - u \rangle - \delta(2M_0 + 2).$$
(5.48)

Lemma 5.2 and (5.34) imply that there exists

$$l \in \partial g(\hat{v})$$
(5.49)

such that

$$\xi + L(\hat{v} - u) + l = 0.$$
(5.50)

In view of (5.49) and the convexity of g,

$$g(x) \ge g(\hat{v}) + \langle l, x - \hat{v} \rangle.$$
(5.51)

It follows from (5.48) and (5.51) that

$$f(x) + g(x)$$
$$\geq f(u) + \langle \xi, x - u \rangle - \delta(2M_0 + 2) + g(\hat{v}) + \langle l, x - \hat{v} \rangle. \quad (5.52)$$

In view of (5.3),

$$G_{u,\xi}^{(L)}(\hat{v}) = f(u) + \langle \xi, \hat{v} - u \rangle + 2^{-1}L\|\hat{v} - u\|^2 + g(\hat{v}). \quad (5.53)$$

By (5.50), (5.52), and (5.53),

$$f(x) + g(x) - G_{u,\xi}^{(L)}(\hat{v})$$
$$\geq \langle \xi, x - \hat{v} \rangle + \langle l, x - \hat{v} \rangle - 2^{-1}L\|\hat{v} - u\|^2 - \delta(2M_0 + 2)$$
$$= -\delta(2M_0 + 2) + \langle \xi + l, x - \hat{v} \rangle - 2^{-1}L\|\hat{v} - u\|^2$$
$$= -2^{-1}L\|\hat{v} - u\|^2 - L\langle \hat{v} - u, x - \hat{v} \rangle - \delta(2M_0 + 2)$$
$$= 2^{-1}L\|\hat{v} - u\|^2 + L\langle \hat{v} - u, u - x \rangle - \delta(2M_0 + 2). \quad (5.54)$$

In view of (5.35)–(5.37),

$$|\|\hat{v} - x\|^2 - \|v - x\|^2|$$
$$\leq |\|\hat{v} - x\| - \|v - x\||(\|\hat{v} - x\| + \|v - x\|)$$
$$\leq \|\hat{v} - v\|(2M_2 + 2M_0 + 2) \leq \delta(2M_2 + 2M_0 + 2). \quad (5.55)$$

Lemma 2.1 implies that

$$\langle \hat{v} - u, u - x \rangle = 2^{-1}[\|\hat{v} - x\|^2 - \|\hat{v} - u\|^2 - \|u - x\|^2]. \quad (5.56)$$

By (5.54) and (5.56),

$$f(x) + g(x) - G_{u,\xi}^{(L)}(\hat{v})$$
$$\geq 2^{-1}L\|\hat{v} - u\|^2 + 2^{-1}L\|\hat{v} - x\|^2$$
$$\quad -2^{-1}L\|\hat{v} - u\|^2 - 2^{-1}L\|u - x\|^2 - \delta(2M_0 + 2)$$
$$\geq 2^{-1}L\|\hat{v} - x\|^2 - 2^{-1}L\|u - x\|^2 - \delta(2M_0 + 2)$$
$$\geq 2^{-1}L\|v - x\|^2 - 2^{-1}L\|u - x\|^2 - 2^{-1}L\delta(2M_2 + 2M_0 + 2) - \delta(2M_0 + 2).$$
$$(5.57)$$

It follows from (5.47) and (5.57) that

$$F(x) - F(v)$$
$$\geq 2^{-1}L\|v - x\|^2 - 2^{-1}L\|u - x\|^2 - \delta(L(M_2 + M_0 + 1)$$
$$+2M_0 + 2 + L_0 + (L + 1)(M_2 + M_0 + 2)).$$

Lemma 5.4 is proved.

5.4   Proof of Theorem 5.1

By (5.4), there exists

$$z \in \operatorname{argmin}(F) \cap B_X(0, M). \tag{5.58}$$

In view of (5.12) and (5.58),

$$\|x_0 - z\| \leq 2M. \tag{5.59}$$

If $f(x_0) \leq f(z) + \epsilon_0$, then the assertion of the theorem holds. Let

$$f(x_0) > f(z) + \epsilon_0.$$

If $f(x_1) \leq f(z) + \epsilon_0$, then in view of Lemma 5.3, $\|x_1\| \leq M_2$ and the assertion of the theorem holds. Let

$$f(x_1) > f(z) + \epsilon_0.$$

Assume that $T \geq 0$ is an integer and that for all integers $t = 0, \ldots, T$,

$$F(x_{t+1}) - F(z) > \epsilon_0. \tag{5.60}$$

We show that for all $t \in \{0, \ldots, T\}$,

$$\|x_t - z\| \leq 2M \tag{5.61}$$

and

$$F(z) - F(x_{t+1})$$
$$\geq 2^{-1}L\|z - x_{t+1}\|^2 - 2^{-1}L\|z - x_t\|^2 - \delta((M_2 + M_0 + 2)(2L + 3) + L_0).$$
$$\tag{5.62}$$

In view of (5.59), (5.61) is true for $t = 0$.

Assume that $t \in \{0, \ldots, T\}$ and (5.61) holds. Relations (5.58) and (5.61) imply that

$$\|x_t\| \leq 3M. \tag{5.63}$$

Set

$$M_0 = 3M \tag{5.64}$$

and

$$M_3 = (M_2 + M_0 + 2)(2L + 3) + L_0. \tag{5.65}$$

By (5.5)–(5.9), (5.13), (5.14), (5.58), (5.63), (5.64), and Lemma 5.4 applied with

$$x = z, \ u = x_t, \ \xi = \xi_t, \ v = x_{t+1},$$

we have

$$F(z) - F(x_{t+1}) \geq 2^{-1}L\|z - x_{t+1}\|^2 - 2^{-1}L\|z - x_t\|^2 - \delta M_3. \tag{5.66}$$

It follows from (5.60) and (5.66) that

$$\epsilon_0 < F(x_{t+1}) - F(z)$$
$$\leq 2^{-1}L\|z - x_t\|^2 - 2^{-1}L\|z - x_{t+1}\|^2 + \delta M_3. \tag{5.67}$$

In view of (5.10), (5.65) and (5.67),

$$2M \geq \|z - x_t\| \geq \|z - x_{t+1}\|.$$

Thus we have shown by induction that (5.62) holds for all $t = 0, \ldots, T$ and that (5.61) holds for all $t = 0, \ldots, T + 1$.

By (5.60), (5.62) and (5.65),

$$T\epsilon_0 < \sum_{t=0}^{T}(F(x_{t+1}) - F(z))$$

$$\leq \sum_{t=0}^{T}[2^{-1}L\|z - x_t\|^2 - 2^{-1}L\|z - x_{t+1}\|^2] + T\delta M_3$$

$$\leq 2^{-1}L\|z - x_0\|^2 + T\delta M_3. \tag{5.68}$$

It follows from (5.10), (5.59), (5.65), and (5.68) that

$$T\epsilon_0/2 \leq 2^{-1}L(4M^2),$$
$$T \leq 4LM^2\epsilon_0^{-1} < n_0 + 1.$$

Thus we have shown that if $T \geq 0$ is an integer and (5.60) holds for all $t = 0, \ldots, T$, then (5.61) holds for all $t = 0, \ldots, T + 1$ and $T < n_0 + 1$. This implies that there exists an integer $q \in \{1, \ldots, n_0 + 2\}$ such that

$$\|x_i\| \leq 3M, \ i = 0, \ldots, q,$$
$$F(x_q) \leq F(z) + \epsilon_0.$$

Theorem 5.1 is proved.

Chapter 6
Weiszfeld's Method

In this chapter we analyze the behavior of Weiszfeld's method for solving the Fermat–Weber location problem. We show that the algorithm generates a good approximate solution, if computational errors are bounded from above by a small positive constant. Moreover, for a known computational error, we find out what an approximate solution can be obtained and how many iterates one needs for this.

6.1 The Description of the Problem

Let X be a Hilbert space equipped with an inner product $\langle \cdot, \cdot \rangle$ which induces a complete norm $\| \cdot \|$.

If $x \in X$ and h is a real-valued function defined in a neighborhood of x which is Frechet differentiable at x, then its Frechet derivative at x is denoted by $h'(x) \in X$.

For each $x \in X$ and each $r > 0$ set

$$B_X(x, r) = \{y \in X : \|x - y\| \le r\}.$$

Let $a \in X$. The function $g(x) = \|x - a\|$, $x \in X$ is convex. For every $x \in X \setminus \{a\}$, g is Frechet differentiable at x and

$$g'(x) = \|x - a\|^{-1}(x - a).$$

It is easy to see that

$$\partial g(a) = B_X(0, 1).$$

Recall that the definition of the subdifferential is given in Sect. 2.1.

© Springer International Publishing Switzerland 2016
A.J. Zaslavski, *Numerical Optimization with Computational Errors*, Springer Optimization and Its Applications 108, DOI 10.1007/978-3-319-30921-7_6

In this chapter we assume that X is the finite-dimensional Euclidean space R^n, for each pair $x = (x_1, \ldots, x_n)$, $y = (y_1, \ldots, y_n) \in R^n$,

$$\langle x, y \rangle = \sum_{i=1}^{n} x_i y_i,$$

m is a natural number,

$$\omega_i > 0, \ i = 1, \ldots, m$$

and that

$$A = \{a_i \in R^n : \ i = 1, \ldots, m\},$$

where $a_{i_1} \neq a_{i_2}$ for all $i_1, i_2 \in \{1, \ldots, m\}$ satisfying $i_1 \neq i_2$.
    Set

$$f(x) = \sum_{i=1}^{m} \omega_i \|x - a_i\| \text{ for all } x \in R^n \tag{6.1}$$

and

$$\inf(f) = \inf\{f(z) : \ z \in R^n\}.$$

We say that the vectors $a_i, \ldots, a_m$ are collinear if there exist $y, b \in R^n$ and $t_1, \ldots, t_m \in R^1$ such that $a_i = y + t_i b, \ i = 1, \ldots, m$. We suppose that the vectors $a_i, \ldots, a_m$ are not collinear.
    In this chapter we study the Fermat–Weber location problem

$$f(x) \to \min, \ x \in R^n$$

by using Weiszfeld's method [110] which was recently revisited in [18]. This problem is often called the Fermat–Torricelli problem named after the mathematicians originally formulated (Fermat) and solved (Torricelli) it in the case of three points; Weber (as well as Steiner) considered its extension for finitely many points. For a full treatment of this problem with a modified proof of the Weiszfeld algorithm by using the subdifferential theory of convex analysis (as well for generalized versions of the Fermat–Torricelli and related problems) with no presence of computational errors see [86].
    Since the function f is continuous and satisfies a growth condition, this problem has a solution which is denoted by $x^* \in R^n$. Thus

$$f(x^*) = \inf(f). \tag{6.2}$$

In view of Theorem 2.1 of [18] this solution is unique but in our study we do not use this fact.

If $x^* \notin A$, then

$$f'(x^*) = \sum_{i=1}^{m} \omega_i \|x^* - a_i\|^{-1}(x^* - a_i) = 0. \tag{6.3}$$

If $x^* = a_i$ with some $i \in \{1, \ldots, m\}$, then

$$\left\| \sum_{j=1, j\neq i}^{m} \omega_j \|x^* - a_j\|^{-1}(x^* - a_j) \right\| \leq \omega_i. \tag{6.4}$$

6.2 Preliminaries

For each $x \in R^n \setminus A$ set

$$T(x) = \left(\sum_{i=1}^{m} \omega_i \|x - a_i\|^{-1} \right)^{-1} \sum_{i=1}^{m} (\|x - a_i\|^{-1}\omega_i a_i). \tag{6.5}$$

Let $y \in R^n \setminus A$ satisfy

$$T(y) = y.$$

This equality is equivalent to the relation

$$y = \left(\sum_{i=1}^{m} \omega_i \|y - a_i\|^{-1} \right)^{-1} \sum_{i=1}^{m} (\|y - a_i\|^{-1}\omega_i a_i)$$

which in its turn is equivalent to the equality

$$\sum_{i=1}^{m} (\omega_i \|y - a_i\|^{-1}(y - a_i)) = 0.$$

It is easy to see that the last equality is equivalent to the relation

$$f'(y) = 0.$$

Thus for every $y \in R^n \setminus A$,

$$T(y) = y \text{ if and only if } f'(y) = 0. \tag{6.6}$$

For each $x \in R^n$ and every $y \in R^n \setminus A$ set

$$h(x, y) = \sum_{i=1}^{m} \omega_i \|y - a_i\|^{-1} \|x - a_i\|^2. \tag{6.7}$$

Let $y \in R^n \setminus A$ and consider the function

$$s = h(\cdot, y) : R^n \to R^1$$

which is strictly convex and possesses a unique minimizer x satisfying the relation

$$0 = s'(x) = 2 \sum_{i=1}^{m} \omega_i \|y - a_i\|^{-1} (x - a_i)$$

which is equivalent to the equality

$$T(y) = x.$$

This implies that

$$h(T(y), y) \leq h(x, y) \text{ for all } x \in R^n. \tag{6.8}$$

Lemma 6.1 ([18]).

  (i) *For every $y \in R^n \setminus A$,*

$$h(T(y), y) \leq h(x, y) \text{ for all } x \in R^n.$$

  (ii) *For every $y \in R^n \setminus A$,*

$$h(y, y) = f(y).$$

(iii) *For every $x \in R^n$ and every $y \in R^n \setminus A$,*

$$h(x, y) \geq 2f(x) - f(y).$$

Proof. Assertion (i) was already proved [see (6.8)]. Assertion (ii) is evident. Let us prove assertion (iii).

Let $x \in R^n$ and $y \in R^n \setminus A$. Clearly, for each $a \in R^1$ and each $b > 0$,

$$a^2 b^{-1} \geq 2a - b.$$

Therefore for all $i = 1, \ldots, m$,

$$\|x - a_i\|^2 \|y - a_i\|^{-1} \geq 2\|x - a_i\| - \|y - a_i\|.$$

This implies that

$$h(x, y) = \sum_{i=1}^{m} \omega_i \|y - a_i\|^{-1} \|x - a_i\|^2$$

$$\geq 2 \sum_{i=1}^{m} \omega_i \|x - a_i\| - \sum_{i=1}^{m} \omega_i \|y - a_i\| = 2f(x) - f(y).$$

Lemma 6.1 is proved.

Lemma 6.2 ([18]). *For every $y \in R^n \setminus A$,*

$$f(T(y)) \leq f(y)$$

and the equality holds if and only if $T(y) = y$.

Proof. Let $y \in R^n \setminus A$. In view of Lemma 6.1 (i), (6.8) holds. By the strict convexity of the function $x \to h(x, y)$, $x \in R^n$, $T(y)$ is its unique minimizer, by Lemma 6.2 (ii),

$$h(T(y), y) \leq h(y, y) = f(y)$$

and if $T(y) \neq y$, then this implies that

$$h(T(y), y) < h(y, y) = f(y).$$

Together with Lemma 6.1 (iii) this implies that

$$2f(T(y)) - f(y) \leq h(T(y), y) \leq f(y)$$

and if $T(y) \neq y$, then

$$2f(T(y)) - f(y) \leq h(T(y), y) < h(y, y) = f(y).$$

This completes the proof of Lemma 6.2.

For every $x \in R^n \setminus A$ set

$$L(x) = \sum_{i=1}^{m} \omega_i \|x - a_i\|^{-1}. \tag{6.9}$$

For $j = 1, \ldots, m$ set

$$L(a_j) = \sum_{i=1, i \neq j}^{m} \omega_i \|a_j - a_i\|^{-1}. \tag{6.10}$$

Clearly, for each $x \in R^n \setminus A$,

$$T(x) = x - L(x)^{-1} f'(x). \tag{6.11}$$

Lemma 6.3. *Let $y \in R^n \setminus A$. Then*

$$f(T(y)) \leq f(y) + \langle f'(y), T(y) - y \rangle + 2^{-1} L(y) \|T(y) - y\|^2.$$

Proof. Clearly, the function $x \to h(x, y)$, $x \in R^n$ is quadratic. Therefore its second-order Taylor expansion around y is exact and can be written as

$$h(x, y) = h(y, y) + \langle \partial_x h(y, y), x - y \rangle + L(y) \|x - y\|^2.$$

Combined with the relations

$$h(y, y) = f(y) \text{ and } \partial_x h(y, y) = 2 f'(y)$$

this implies that

$$h(x, y) = f(y) + 2 \langle f'(y), x - y \rangle + L(y) \|x - y\|^2.$$

For $x = T(y)$ the relation above implies that

$$h(T(y), y) = f(y) + 2 \langle f'(y), T(y) - y \rangle + L(y) \|T(y) - y\|^2.$$

Together with Lemma 6.1 (iii) this implies that

$$2f(T(y)) - f(y) \leq h(T(y), y)$$
$$= f(y) + 2 \langle f'(y), T(y) - y \rangle + L(y) \|T(y) - y\|^2,$$
$$2f(T(y)) \leq 2f(y) + 2 \langle f'(y), T(y) - y \rangle + L(y) \|T(y) - y\|^2$$

and

$$f(T(y)) \leq f(y) + \langle f'(y), T(y) - y \rangle + 2^{-1} L(y) \|T(y) - y\|^2.$$

Lemma 6.3 is proved.

6.3 The Basic Lemma

Set

$$\tilde{M} = \max\{\|a_i\| : i = 1, \ldots, m\}. \tag{6.12}$$

By (6.1) and (6.12),

$$f(0) \leq \sum_{i=1}^{m} \omega_i \tilde{M}. \tag{6.13}$$

We show that $\|x^*\| \leq 2\tilde{M}$. Since x^* is the minimizer of f it follows from (6.1) and (6.13) that

$$\sum_{i=1}^{m} \omega_i \tilde{M} \geq f(0) \geq f(x^*) = \sum_{i=1}^{m} \omega_i \|x^* - a_i\|. \tag{6.14}$$

There exists $j_0 \in \{1, \ldots, m\}$ such that

$$\|x^* - a_j\| \leq \|x^* - a_i\|, \quad i = 1, \ldots, m. \tag{6.15}$$

By (6.12), (6.14), and (6.15),

$$\sum_{i=1}^{m} \omega_i \tilde{M} \geq \sum_{i=1}^{m} \omega_i \|x^* - a_j\|,$$

$$\|x^* - a_j\| \leq \tilde{M}$$

and

$$\|x^*\| \leq 2\tilde{M}. \tag{6.16}$$

Lemma 6.4. *Let $M \geq \tilde{M}$ and $y \in R^n \setminus A$ satisfy*

$$\|y\| \leq M. \tag{6.17}$$

Then

$$\|T(y)\| \leq 3M.$$

Proof. In view of Lemma 6.2,

$$f(T(y)) \leq f(y). \tag{6.18}$$

It follows from (6.1), (6.12), and (6.17) that

$$f(y) = \sum_{i=1}^{m} \omega_i \|y - a_i\| \le 2M \sum_{i=1}^{m} \omega_i. \tag{6.19}$$

By (6.1), (6.18), and (6.19),

$$\sum_{i=1}^{m} \omega_i \|T(y) - a_i\| = f(T(y)) \le f(y) \le 2M \sum_{i=1}^{m} \omega_i. \tag{6.20}$$

There exists $j \in \{1, \ldots, m\}$ such that

$$\|T(y) - a_j\| \le \|T(y) - a_i\|, \ i = 1, \ldots, m. \tag{6.21}$$

Relations (6.20) and (6.21) imply that

$$\sum_{i=1}^{m} \omega_i \|T(y) - a_j\| \le 2M \sum_{i=1}^{m} \omega_i,$$

$$\|T(y) - a_j\| \le 2M.$$

Together with (6.12) this implies that $\|T(y)\| \le 3M$. Lemma 6.4 is proved.

Lemma 6.5 (The Basic Lemma). *Let $M \ge \tilde{M}$, $\delta \in (0, 1]$, $y \in R^n \setminus A$ satisfy*

$$\|y\| \le M, \tag{6.22}$$

$x \in R^n$ *satisfy*

$$\|T(y) - x\| \le \delta \tag{6.23}$$

and let $z \in R^n$ satisfy

$$\|z\| \le M. \tag{6.24}$$

Then

$$f(x) - f(z) \le 2^{-1} L(y)(\|z - y\|^2 - \|z - x\|^2) + \delta \left(8M + 1 + \sum_{i=1}^{m} \omega_i \right).$$

Proof. Relations (6.1) and (6.23) imply that

$$|f(x) - f(T(y))| \le \|x - T(y)\| \sum_{i=1}^{m} \omega_i \le \delta \sum_{i=1}^{m} \omega_i. \tag{6.25}$$

In view of Lemma 6.3,

$$f(T(y)) \leq f(y) + \langle f'(y), T(y) - y \rangle + 2^{-1}L(y)\|T(y) - y\|^2. \tag{6.26}$$

Since the function f is convex we have

$$f(y) \leq f(z) + \langle f'(y), y - z \rangle. \tag{6.27}$$

By (6.26) and (6.27),

$$\begin{aligned}
f(T(y)) &\leq f(z) + \langle f'(y), y - z \rangle \\
&\quad + \langle f'(y), T(y) - y \rangle + 2^{-1}L(y)\|T(y) - y\|^2 \\
&= f(z) + \langle f'(y), T(y) - z \rangle + 2^{-1}L(y)\|T(y) - y\|^2.
\end{aligned} \tag{6.28}$$

It follows from (6.25) and (6.28) that

$$f(x) \leq f(z) + \langle f'(y), T(y) - z \rangle + 2^{-1}L(y)\|T(y) - y\|^2 + \delta \sum_{i=1}^{m} \omega_i. \tag{6.29}$$

In view of (6.11),

$$f'(y) = L(y)(y - T(y)). \tag{6.30}$$

By (6.29) and (6.30),

$$\begin{aligned}
f(x) \leq f(z) + L(y)\langle y - T(y), T(y) - z \rangle \\
+ 2^{-1}L(y)\|T(y) - y\|^2 + \delta \sum_{i=1}^{m} \omega_i.
\end{aligned} \tag{6.31}$$

Lemma 2.1 implies that

$$\langle y - T(y), T(y) - z \rangle = 2^{-1}[\|z - y\|^2 - \|z - T(y)\|^2 - \|y - T(y)\|^2]. \tag{6.32}$$

It follows from (6.31) and (6.32) that

$$f(x) \leq f(z) + 2^{-1}L(y)(\|z - y\|^2 - \|z - T(y)\|^2) + \delta \sum_{i=1}^{m} \omega_i. \tag{6.33}$$

Lemma 6.4 and (6.22) imply that

$$\|T(y)\| \leq 3M. \tag{6.34}$$

By (6.23), (6.24), and (6.34),

$$
\begin{aligned}
|\|z - T(y)\|^2 &- \|z - x\|^2| \\
&= |\|z - T(y)\| - \|z - x\||(\|z - T(y)\| + \|z - x\|) \\
&\le \|x - T(y)\|(8M + 1) \le \delta(8M + 1).
\end{aligned}
\tag{6.35}
$$

It follows from (6.33) and (6.35) that

$$
f(x) - f(z) \le 2^{-1}L(y)(\|z - y\|^2 - \|z - x\|^2) + \delta\left(8M + 1 + \sum_{i=1}^{m}\omega_i\right).
$$

Lemma 6.5 is proved.

6.4 The Main Result

Let $\delta \in (0, 1]$ and a positive number $\epsilon_0 \ge \delta$. Choose $p \in \{1, \ldots, m\}$ such that

$$
f(a_p) \le f(a_i), \quad i = 1, \ldots, m.
\tag{6.36}
$$

For each $j = 1, \ldots, m$ set

$$
r_j = \sum_{i=1, i \ne j}^{m} \omega_i \|a_i - a_j\|^{-1}(a_j - a_i).
\tag{6.37}
$$

In order to solve our minimization problem we need to calculate

$$
r_p = \sum_{i=1, i \ne p}^{m} \omega_i \|a_i - a_p\|^{-1}(a_p - a_i).
$$

Since our computer system produces computational errors we can obtain only a vector $\hat{r}_p \in R^n$ such that $\|\hat{r}_p - r_p\| \le \epsilon_0$.

Proposition 6.6. *Assume that* $\hat{r}_p \in R^n$ *satisfies*

$$
\|\hat{r}_p - r_p\| \le \epsilon_0
\tag{6.38}
$$

and

$$
\|\hat{r}_p\| \le \omega_p + 2\epsilon_0.
\tag{6.39}
$$

Then

$$f(a_p) \le \inf(f) + 9\tilde{M}\epsilon_0.$$

(Note that $\tilde{M}$ was defined by (6.12).)

Proof. By (6.38) and (6.39),

$$\|r_p\| \le \|\hat{r}_p\| + \epsilon_0 \le \omega_p + 3\epsilon_0. \qquad (6.40)$$

It follows from (6.1), (6.37), and (6.40) that there exists

$$l \in \partial f(a_p) \qquad (6.41)$$

such that

$$\|l\| \le 3\epsilon_0. \qquad (6.42)$$

By the convexity of f, (6.2), (6.12), (6.16), (6.41), and (6.42),

$$f(x^*) \ge f(a_p) + \langle l, x^* - a_p \rangle$$
$$\ge f(a_p) - \|l\|(\|x^*\| + \|a_p\|) \ge f(a_p) - 9\tilde{M}\epsilon_0$$

and

$$f(a_p) \le f(x^*) + 9\tilde{M}\epsilon_0.$$

Proposition 6.6 is proved.

In view of Proposition 6.6, if $\hat{r}_p$ satisfies (6.38) and (6.39), then a_p is an approximate solution of our minimization problem. It is easy to see that the following proposition holds.

Proposition 6.7. *Assume that $\hat{r}_p \in R^n$ satisfies $\|\hat{r}_p - r_p\| \le \epsilon_0$ and $\|\hat{r}_p\| > \omega_p + 2\epsilon_0$. Then $\|r_p\| > \omega_p + \epsilon_0$ and a_p is not a minimizer of f.*

Lemma 6.8 ([18]). *Let $\|r_p\| > \omega_p$. Then*

$$f(a_p) - f(a_p - (\|r_p\| - \omega_p)L(a_p)^{-1}\|r_p\|^{-1}r_p)$$
$$\ge (\|r_p\| - \omega_p)^2(2L(a_p))^{-1}.$$

Proposition 6.9. *Assume that*

$$\|r_p\| > \omega_p, \qquad (6.43)$$

$d_p \in R^n$ *satisfies*

$$\|d_p + \|r_p\|^{-1} r_p\| \le \delta, \tag{6.44}$$

$t_p \ge 0$ *satisfies*

$$|t_p - L(a_p)^{-1}(\|r_p\| - \omega_p)| \le \delta \tag{6.45}$$

and that $x_0 \in R^n$ *satisfies*

$$\|x_0 - a_p - t_p d_p\| \le \delta. \tag{6.46}$$

Then

$$\|d_p\| \le 1 + \delta, \tag{6.47}$$

$$t_p \le L(a_p)^{-1}(\|r_p\| - \omega_p) + \delta, \tag{6.48}$$

$$\|x_0\| \le \tilde{M} + 2(2\delta + L(a_p)^{-1}(\|r_p\| - \omega_p)), \tag{6.49}$$

$$\|x_0 - a_p + (\|r_p\| - \omega_p)L(a_p)^{-1}\|r_p\|^{-1} r_p\|$$
$$\le \delta(3 + (\|r_p\| - \omega_p)L(a_p)^{-1}), \tag{6.50}$$

$$|f(x_0) - f(a_p - (\|r_p\| - \omega_p)L(a_p)^{-1}\|r_p\|^{-1} r_p)\|$$

$$\le \delta(3 + (\|r_p\| - \omega_p)L(a_p)^{-1}) \sum_{i=1}^{m} \omega_i \tag{6.51}$$

and

$$f(a_p) - f(x_0)$$

$$\ge (\|r_p\| - \omega_p)^2 (2L(a_p))^{-1} - \delta(3 + (\|r_p\| - \omega_p)L(a_p)^{-1}) \sum_{i=1}^{m} \omega_i. \tag{6.52}$$

Proof. In view of (6.44), (6.47) is true. Inequality (6.45) implies (6.48). By (6.12) and (6.46)–(6.48),

$$\|x_0\| \le \delta + \|a_p\| + t_p\|d_p\|$$
$$\le \delta + \tilde{M} + 2t_p \le \tilde{M} + 2(2\delta + L(a_p)^{-1}(\|r_p\| - \omega_p))$$

and (6.49) is true. It follows from (6.44)–(6.47) that

$$\|x_0 - a_p + (\|r_p\| - \omega_p)L(a_p)^{-1}\|r_p\|^{-1} r_p\|$$
$$\le \delta + \|t_p d_p + (\|r_p\| - \omega_p)L(a_p)^{-1}\|r_p\|^{-1} r_p\|$$

$$\leq \delta + \|t_p d_p - (\|r_p\| - \omega_p)L(a_p)^{-1}d_p\|$$
$$+\|(\|r_p\| - \omega_p)L(a_p)^{-1}(d_p + \|r_p\|^{-1}r_p)\|$$
$$\leq \delta + \delta(1 + \delta) + (\|r_p\| - \omega_p)L(a_p)^{-1}\delta$$
$$\leq \delta(3 + (\|r_p\| - \omega_p)L(a_p)^{-1})$$

and (6.50) holds. Relations (6.1) and (6.50) imply (6.51). Relation (6.52) follows from (6.51) and Lemma 6.8. Proposition 6.9 is proved.

The next theorem which is proved in Sect. 6.5, is our main result.

Theorem 6.10. *Let*

$$\|r_p\| > \omega_p, \tag{6.53}$$

$$M_0 = 3\tilde{M} + 4 + 2(\|r_p\| - \omega_p)L(a_p)^{-1}, \tag{6.54}$$

a positive number δ satisfy

$$\delta < 12^{-1}(\|r_p\| - \omega_p)\left(\sum_{i=1}^{m}\omega_i\right)^{-1} \tag{6.55}$$

and

$$2\delta\left(8M_0 + 1 + \sum_{i=1}^{m}\omega_i\right) < (\|r_p\| - \omega_p)^2(16L(a_p))^{-1}, \tag{6.56}$$

$$\epsilon_0 = 4\delta\left(16M_0 + 1 + \sum_{i=1}^{m}\omega_i\right)[144L(a_p)^2(\|r_p\| - \omega_p)^{-4}M_0^2$$

$$+1]((\sum_{i=1}^{m}\omega_i)^2 + 1) \tag{6.57}$$

and

$$n_0 = \lfloor \delta^{-1}\left(8M_0 + 1 + \sum_{i=1}^{m}\omega_i\right)^{-1}(\|r_p\| - \omega_p)^2(8L(a_p))^{-1}\rfloor + 1. \tag{6.58}$$

Assume that $t_p \geq 0$, $d_p \in R^n$ and $x_0 \in R^n$ satisfy (6.44)–(6.46), $\{x_i\}_{i=1}^{\infty} \subset R^n$ and that for each integer $i \geq 0$ satisfying $x_i \notin A$,

$$\|T(x_i) - x_{i+1}\| \leq \delta. \tag{6.59}$$

Then

$$x_0 \notin A$$

and there exists $j \in \{0, \ldots, n_0\}$ such that

$$x_i \notin A, \ i \in \{0, \ldots, j\} \setminus \{j\},$$

$$f(x_j) \leq \inf(f) + \epsilon_0.$$

Note that in the theorem above δ is the computational error produced by our computer system. In order to obtain a good approximate solution we need $\lfloor c_1 \delta^{-1} \rfloor$ iterations [see (6.58)], where $c_1 > 0$ is a constant depending only on M_0, $\sum_{i=1}^{m} \omega_i$, $\|r_p\| - \omega_p$ and $L(a_p)$. As a result, we obtain a point $x \in R^n$ such that

$$f(x) \leq \inf(f) + c_2 \delta$$

[see (6.57)], where the constant $c_2 > 0$ depends only on M_0, $\sum_{i=1}^{m} \omega_i$, $\|r_p\| - \omega_p$ and $L(a_p)$.

6.5 Proof of Theorem 6.10

Proposition 6.9, (6.44)–(6.46), (6.55), and (6.56) imply that

$$f(x_0) \leq f(a_p)$$

$$- (\|r_p\| - \omega_p)^2 (2L(a_p))^{-1} + \delta(3 + (\|r_p\| - \omega_p)L(a_p)^{-1}) \sum_{i=1}^{m} \omega_i$$

$$\leq f(a_p) - (\|r_p\| - \omega_p)^2 (4L(a_p))^{-1}. \tag{6.60}$$

By (6.36) and (6.60),

$$x_0 \notin A. \tag{6.61}$$

If

$$f(x_0) \leq \inf(f) + \epsilon_0 \text{ or } f(x_1) \leq \inf(f) + \epsilon_0,$$

then in view of (6.61) the assertion of the theorem holds with $j = 0$ or $j = 1$, respectively. Consider the case with

$$f(x_0) > \inf(f) + \epsilon_0 \text{ and } f(x_1) > \inf(f) + \epsilon_0. \tag{6.62}$$

Assume that $k \in [0, n_0]$ is an integer,

$$x_i \notin A, \ i = 0, \ldots, k \tag{6.63}$$

and

$$f(x_i) > \inf(f) + \epsilon_0, \ i = 0, \ldots, k + 1. \tag{6.64}$$

(Note that in view of (6.61) and (6.62), relations (6.63) and (6.64) hold for $k = 0$.) For all integers $i \geq 0$, set

$$\Delta_i = i\delta \left(8M_0 + 1 + \sum_{j=1}^{m} \omega_j \right). \tag{6.65}$$

By (6.56), (6.58), (6.60), and (6.64), for all $i = 0, \ldots, n_0$,

$$\Delta_i \leq n_0 \delta (8M_0 + 1 + \sum_{j=1}^{m} \omega_j)$$

$$\leq \delta (8M_0 + 1 + \sum_{j=1}^{m} \omega_j) + (\|r_p\| - \omega_p)^2 (8L(a_p))^{-1}$$

$$\leq (\|r_p\| - \omega_p)^2 (8L(a_p))^{-1} + (\|r_p\| - \omega_p)^2 (16L(a_p))^{-1}$$

$$\leq 4^{-1} \cdot 3(f(a_p) - f(x_0)). \tag{6.66}$$

Remind (see Sect. 6.1) that $x^* \in R^n$ satisfies

$$f(x^*) = \inf(f). \tag{6.67}$$

We show that for all $j = 0, \ldots, k + 1$,

$$f(x_j) \leq f(x_0) + \Delta_j, \tag{6.68}$$

$$\|x_j - x^*\| \leq M_0. \tag{6.69}$$

In view of (6.16),

$$\|x^*\| \leq 2\tilde{M}. \tag{6.70}$$

Proposition 6.9, (6.44)–(6.46), and (6.49) imply that

$$\|x_0\| \leq \tilde{M} + 2(2 + L(a_p)^{-1}(\|r_p\| - \omega_p)). \tag{6.71}$$

By (6.54), (6.70), and (6.71),

$$\|x_0 - x^*\| \le M_0.$$

Thus (6.68) and (6.69) hold for $j = 0$.

Assume that an integer $j \in \{0, \ldots, k\}$ and (6.68) and (6.69) hold. By (6.36), (6.60), (6.66), (6.68), and the relation $k \le n_0$,

$$f(x_j) \le f(x_0) + \Delta_{n_0}$$
$$\le f(x_0) + 4^{-1} \cdot 3(f(a_p) - f(x_0)) < f(a_p)$$

and

$$x_j \notin A. \tag{6.72}$$

Let $i \in \{1, \ldots, m\}$ and

$$v_i \in \partial f(a_i). \tag{6.73}$$

In view of (6.1),

$$\partial f(a_i) = \sum_{q=1, q \ne i}^{m} \omega_q \|a_i - a_q\|^{-1} (a_i - a_q) + \omega_i B_{R^n}(0, 1). \tag{6.74}$$

Relations (6.73) and (6.74) imply that

$$\|v_i\| \le \sum_{q=1, q \ne i}^{m} \omega_q + \omega_i = \sum_{q=1}^{m} \omega_q. \tag{6.75}$$

It follows from (6.68), (6.73), and (6.75) that

$$f(a_i) - f(x_0)$$
$$\le f(a_i) - f(x_j) + \Delta_j$$
$$\le \langle v_i, a_i - x_j \rangle + \Delta_j$$
$$\le \|v_i\| \|a_i - x_j\| + \Delta_j$$
$$\le \|a_i - x_j\| \sum_{q=1}^{m} \omega_q + \Delta_j. \tag{6.76}$$

By (6.36) and (6.76),

$$f(a_p) - f(x_0) \le \|a_i - x_j\| \sum_{q=1}^{m} \omega_q + \Delta_j, \ i = 1, \ldots, m. \tag{6.77}$$

In view of (6.66) and (6.77), for all $i = 1, \ldots, m$,

$$\|a_i - x_j\| \geq 4^{-1}(f(a_p) - f(x_0))\left(\sum_{q=1}^{m} \omega_q\right)^{-1}$$

and

$$\|a_i - x_j\|^{-1} \leq 4(f(a_p) - f(x_0))^{-1}\sum_{q=1}^{m} \omega_q. \tag{6.78}$$

It follows from (6.9), (6.72), and (6.78) that

$$L(x_j) = \sum_{i=1}^{m}(\omega_i\|x_j - a_i\|^{-1}) \leq 4\left(\sum_{i=1}^{m}\omega_i\right)^2 (f(a_p) - f(x_0))^{-1}. \tag{6.79}$$

Lemma 6.2, (6.1), (6.59), (6.64), (6.68), and (6.72) imply that

$$f(x_{j+1}) \leq f(T(x_j)) + \|x_{j+1} - T(x_j)\|\sum_{i=1}^{m}\omega_i$$

$$\leq f(x_j) + \delta\sum_{i=1}^{m}\omega_i \leq f(x_0) + \Delta_{j+1}. \tag{6.80}$$

It follows from (6.54), (6.59), (6.64), (6.67), (6.69), (6.70), (6.72), (6.79), and Lemma 6.5 applied with $M = 2M_0$, $z = x^*$, $y = x_j$, and $x = x_{j+1}$ that

$$\epsilon_0 < f(x_{j+1}) - f(x^*)$$

$$\leq 2\left(\sum_{i=1}^{m}\omega_i\right)^2 (f(a_p) - f(x_0))^{-1}(\|x^* - x_j\|^2 - \|x^* - x_{j+1}\|^2)$$

$$+ \delta\left(16M_0 + 1 + \sum_{i=1}^{m}\omega_i\right). \tag{6.81}$$

By (6.57), (6.60), and (6.81),

$$\|x^* - x_j\| \geq \|x^* - x_{j+1}\|.$$

Therefore in view of the relation above, (6.80) and (6.81), we showed by induction that (6.68) and (6.69) hold for $j = 0, \ldots, k+1$ and that (6.81) holds for $j = 0, \ldots, k$. It follows from (6.58), (6.60), (6.64), (6.68) and the relation $k \leq n_0$ that

$$f(x_{k+1}) \leq f(x_0) + \Delta_{k+1}$$

$$\leq f(x_0) + (n_0 + 1)\delta \left(8M_0 + 1 + \sum_{i=1}^{m} \omega_i \right)$$

$$< f(x_0) + (\|r_p\| - \omega_p)^2 (8L(a_p))^{-1} + 2\delta \left(8M_0 + 1 + \sum_{i=1}^{m} \omega_i \right) < f(a_p)$$

and

$$x_{k+1} \notin A. \tag{6.82}$$

By (6.81) which holds for all $j = 0, \ldots, k$,

$$(k+1)\epsilon_0 < \sum_{j=0}^{k} (f(x_{j+1}) - f(x^*))$$

$$\leq 2 \left(\sum_{i=1}^{m} \omega_i \right)^2 (f(a_p) - f(x_0))^{-1} \sum_{j=0}^{k} (\|x^* - x_j\|^2 - \|x^* - x_{j+1}\|^2)$$

$$+ (k+1)\delta \left(16M_0 + 1 + \sum_{i=1}^{m} \omega_i \right).$$

Together with (6.57), (6.60), and (6.69) this implies that

$$2^{-1}(k+1)\epsilon_0 \leq 2 \left(\sum_{i=1}^{m} \omega_i \right)^2 4L(a_p)(\|r_p\| - \omega_p)^{-2} \|x_0 - x^*\|^2$$

$$\leq 8 \left(\sum_{i=1}^{m} \omega_i \right)^2 L(a_p)(\|r_p\| - \omega_p)^{-2} M_0^2.$$

Combined with (6.56) and (6.57) this implies that

$$k+1 \leq 16 \left(\sum_{i=1}^{m} \omega_i \right)^2 L(a_p)(\|r_p\| - \omega_p)^{-2} M_0^2 \epsilon_0^{-1}$$

$$\leq 16 \left(\sum_{i=1}^{m} \omega_i \right)^2 L(a_p)(\|r_p\| - \omega_p)^{-2} M_0^2$$

$$\times (4\delta)^{-1} \left(16M_0 + 1 + \sum_{i=1}^{m} \omega_i \right)^{-1} L(a_p)^{-2} (\|r_p\| - \omega_p)^4 M_0^{-2} 144^{-1} \left(\sum_{i=1}^{m} \omega_i \right)^{-2}$$

$$= 36^{-1}L(a_p)^{-1}(\|r_p\| - \omega_p)^2\delta^{-1}\left(16M_0 + 1 + \sum_{i=1}^{m}\omega_i\right)^{-1} \leq 2^{-1}n_0.$$

Thus we assumed that an integer $k \in [0, n_0]$ satisfies (6.63) and (6.64) and showed that

$$x_{k+1} \notin A$$

[see (6.82)] and that $k + 1 \leq 2^{-1}n_0$. (Note that in view of (6.56) and (6.58), $n_0 \geq 5$.) This implies that there exists an integer $k \in [0, n_0/2]$ such that (6.63), (6.64) hold and

$$f(x_{k+1}) \leq \inf(f) + \epsilon_0.$$

Theorem 6.10 is proved.

$$\leq \log \sum_{j} \left[\tilde{a}_j^2 \log^{\alpha} j \right] \left(\log n + 1 + \left(\sum_{j} a_j \right)^2 \right) < \varepsilon^2$$

Thus we assumed that an integer $N \in \mathbb{N}$ and statistics (6.02) and (6.04) and showed that

$$\Delta_N < \frac{\varepsilon^2}{2}.$$

Hence it is implied that $\Delta_N < \frac{\varepsilon^2}{2}$. Note that in view of (6.56) and (6.55), $\Delta_N = \Delta_{N+1}$. This implies that there exists an number $k \in [0, N-2]$ such that in $[6.51)-(6.64)]$ hold and

$$\Delta(x, 1) \leq \inf_n f(x) - \sigma_n^2$$

The theorem is proved.

Chapter 7
The Extragradient Method for Convex Optimization

In this chapter we study convergence of the extragradient method for constrained convex minimization problems in a Hilbert space. Our goal is to obtain an ϵ-approximate solution of the problem in the presence of computational errors, where ϵ is a given positive number. We show that the extragradient method generates a good approximate solution, if the sequence of computational errors is bounded from above by a constant.

7.1 Preliminaries and the Main Results

Let $(X, \langle \cdot, \cdot \rangle)$ be a Hilbert space with an inner product $\langle \cdot, \cdot \rangle$ which induces a complete norm $\| \cdot \|$.

For each $x \in X$ and each nonempty set $A \subset X$ put

$$d(x, A) = \inf\{\|x - y\| : y \in A\}.$$

For each $x \in X$ and each $r > 0$ set

$$B(x, r) = \{y \in X : \|x - y\| \leq r\}.$$

Let C be a nonempty closed convex subset of X.

Assume that $f : X \to R^1$ is a convex continuous function which is bounded from below on C.

Recall that for each $x \in X$,

$$\partial f(x) = \{u \in X : f(y) - f(x) \geq \langle u, y - x \rangle \text{ for all } y \in X\}.$$

© Springer International Publishing Switzerland 2016

A.J. Zaslavski, *Numerical Optimization with Computational Errors*, Springer Optimization and Its Applications 108, DOI 10.1007/978-3-319-30921-7_7

We consider the minimization problem

$$f(x) \to \min, \; x \in C.$$

Denote

$$C_{\min} = \{x \in C : f(x) \leq f(y) \text{ for all } y \in C\}. \tag{7.1}$$

We assume that $C_{\min} \neq \emptyset$.

We suppose that f is Gâteaux differential at any point $x \in X$ and for $x \in X$ denote by $f'(x) \in X$ the Gâteaux derivative of f at x. This implies that for any $x \in X$ and any $h \in X$

$$\langle f'(x), h \rangle = \lim_{t \to 0} t^{-1}[f(x + th) - f(x)]. \tag{7.2}$$

We suppose that the mapping $f' : X \to X$ is Lipschitz on all the bounded subsets of X.

Set

$$\inf(f; C) = \inf\{f(z) : \; z \in C\}. \tag{7.3}$$

We study the minimization problem with the objective function f, over the set C, using the extragradient method introduced in Korpelevich [75]. By Lemma 2.2, for each nonempty closed convex set $D \subset X$ and for each $x \in X$, there is a unique point $P_D(x) \in D$ satisfying

$$\|x - P_D(x)\| = \inf\{\|x - y\| : \; y \in D\},$$

$$\|P_D(x) - P_D(y)\| \leq \|x - y\| \text{ for all } x, y \in X$$

and

$$\langle z - P_D(x), x - P_D(x) \rangle \leq 0$$

for each $x \in X$ and each $z \in D$.

The following theorem is our first main result of this chapter.

Theorem 7.1. *Let* $M_0 > 0$, $M_1 > 0$, $L > 0$, $\epsilon \in (0, 1)$,

$$B(0, M_0) \cap C_{\min} \neq \emptyset, \tag{7.4}$$

$$f'(B(0, 3M_0)) \subset B(0, M_1), \tag{7.5}$$

$$\|f'(z_1) - f'(z_2)\| \leq L\|z_1 - z_2\|$$

$$\text{for all } z_1, z_1 \in B(0, 3M_0 + M_1 + 1), \tag{7.6}$$

$$0 < \alpha_* < \alpha^* \leq 1, \; \alpha^* L \leq 1, \tag{7.7}$$

an integer k satisfy

$$k > 4M_0^2 \alpha_*^{-1} \epsilon^{-1} \tag{7.8}$$

and let a positive number δ satisfy

$$\delta < 4^{-1}(2M_0 + 1)^{-1}\alpha_*\epsilon. \tag{7.9}$$

Assume that

$$\{\alpha_i\}_{i=0}^\infty \subset [\alpha_*, \alpha^*], \ \{x_i\}_{i=0}^\infty \subset X, \ \{y_i\}_{i=0}^\infty \subset X, \tag{7.10}$$

$$\|x_0\| \le M_0 \tag{7.11}$$

and that for each integer i ≥ 0,

$$\|y_i - P_C(x_i - \alpha_i f'(x_i))\| \le \delta, \tag{7.12}$$

$$\|x_{i+1} - P_C(x_i - \alpha_i f'(y_i))\| \le \delta. \tag{7.13}$$

Then there is an integer j ∈ [0, k] such that

$$\|x_i\| \le 3M_0, \ i = 0, \dots, j,$$

$$f(P_C(x_j - \alpha_j f'(x_j))) \le \inf(f; C) + \epsilon.$$

In Theorem 7.1 our goal is to obtain a point $\xi \in C$ such that

$$f(\xi) \le \inf(f; C) + \epsilon,$$

where $\epsilon > 0$ is given. In order to meet this goal, the computational errors, produced by our computer system, should not exceed $c_1\epsilon$, where $c_1 > 0$ is a constant depending only on M_0, α_* [see (7.9)]. The number of iterations is $\lfloor c_2\epsilon^{-1} \rfloor$, where $c_2 > 0$ is a constant depending only on M_0, α_*.

It is easy to see that the following proposition holds.

Proposition 7.2. *If* $\lim_{x \in C, \|x\| \to \infty} f(x) = \infty$ *and the space X is finite-dimensional, then for each* $\epsilon > 0$ *there exists* $\gamma > 0$ *such that if* $x \in C$ *satisfies* $f(x) \le \inf(f; C) + \gamma$, *then* $d(x, C_{\min}) \le \epsilon$.

The following theorem is our second main result of this chapter.

Theorem 7.3. *Let*

$$\lim_{x \in C,\, \|x\| \to \infty} f(x) = \infty \tag{7.14}$$

and let the following property hold:
(C) for each $\epsilon > 0$ there exists $\gamma > 0$ such that if $x \in C$ satisfies $f(x) \le$ $\inf(f; C) + \gamma$, then $d(x, C_{\min}) \le \epsilon/2$.
Let $\epsilon \in (0, 1)$,

$$\gamma \in (0, (\epsilon/4)^2) \tag{7.15}$$

be as guaranteed by property (C), $M_0 > 1$, $M_1 > 0$, $L > 0$,

$$C_{\min} \subset B(0, M_0 - 1), \tag{7.16}$$

$$f'(B(0, 3M_0)) \subset B(0, M_1), \tag{7.17}$$

$$\|f'(z_1) - f'(z_2)\| \le L\|z_1 - z_2\|$$

$$\text{for all } z_1, z_1 \in B(0, 3M_0 + M_1 + 1), \tag{7.18}$$

$$0 < \alpha_* < \alpha^* \le 1, \ \alpha^* L < 1, \tag{7.19}$$

an integer k satisfy

$$k > 8M_0^2 \gamma^{-1} \min\{\alpha_*, 1 - (\alpha^*)^2 L^2\}^{-1} \tag{7.20}$$

and let a positive number δ satisfy

$$16\delta(8M_0 + 8) \le \gamma \min\{\alpha_*, 1 - (\alpha^*)^2 L^2\}. \tag{7.21}$$

Assume that

$$\{\alpha_i\}_{i=0}^{\infty} \subset [\alpha_*, \alpha^*], \ \{x_i\}_{i=0}^{\infty} \subset X, \ \{y_i\}_{i=0}^{\infty} \subset X, \tag{7.22}$$

$$\|x_0\| \le M_0 \tag{7.23}$$

and that for each integer $i \ge 0$,

$$\|y_i - P_C(x_i - \alpha_i f'(x_i))\| \le \delta, \tag{7.24}$$

$$\|x_{i+1} - P_C(x_i - \alpha_i f'(y_i))\| \le \delta. \tag{7.25}$$

Then $d(x_i, C_{\min}) < \epsilon$ for all integers $i \ge k$.

The chapter is organized as follows. Section 7.2 contains auxiliary results. Theorem 7.1 is proved in Sect. 7.3 while Theorem 7.3 is proved in Sect. 7.4.

The results of this chapter were obtained in [126].

7.2 Auxiliary Results

We use the assumptions, notation, and definitions introduced in Sect. 7.1.

Lemma 7.4. *Let*

$$u_* \in C_{\min}, \ u \in X, \ \alpha > 0, \tag{7.26}$$

$$v = P_C(u - \alpha f'(u)), \ \bar{u} = P_C(u - \alpha f'(v)). \tag{7.27}$$

Then

$$\|\bar{u} - u_*\|^2 \leq \|u - u_*\|^2 - \|u - v\|^2 - \|v - \bar{u}\|^2$$
$$+2\alpha[f(u_*) - f(v)] + 2\alpha\|\bar{u} - v\| \|f'(u) - f'(v)\|.$$

Proof. It is easy to see that

$$\langle f'(v), x - v \rangle \leq f(x) - f(v) \text{ for all } x \in X. \tag{7.28}$$

In view of (7.28),

$$\langle u_* - \bar{u}, f'(v) \rangle - \langle v - \bar{u}, f'(v) \rangle \leq f(u_*) - f(v). \tag{7.29}$$

It follows from (7.27) and Lemma 2.2 that

$$\langle \bar{u} - v, (u - \alpha f'(u)) - v \rangle \leq 0. \tag{7.30}$$

Relation (7.30) implies that

$$\langle \bar{u} - v, (u - \alpha f'(v)) - v \rangle \leq \alpha \langle \bar{u} - v, f'(u) - f'(v) \rangle. \tag{7.31}$$

Set

$$z = u - \alpha f'(v). \tag{7.32}$$

It follows from (7.27) and (7.32) that

$$\|\bar{u} - u_*\|^2 = \|z - u_* + P_C(z) - z\|^2$$
$$= \|z - u_*\|^2 + \|z - P_C(z)\|^2 + 2\langle P_C(z) - z, z - u_* \rangle. \tag{7.33}$$

Relation (7.26) and Lemma 2.2 imply that

$$2\|z - P_C(z)\|^2 + 2\langle P_C(z) - z, z - u_* \rangle$$
$$= 2\langle z - P_C(z), u_* - P_C(z) \rangle \leq 0. \tag{7.34}$$

It follows from (7.33), (7.34), (7.32), (7.27), and (7.28) that

$$
\begin{aligned}
\|\bar{u} - u_*\|^2 &\leq \|z - u_*\|^2 - \|z - P_C(z)\|^2 \\
&= \|u - \alpha f'(v) - u_*\|^2 - \|u - \alpha f'(v) - \bar{u}\|^2 \\
&= \|u - u_*\|^2 - \|u - \bar{u}\|^2 + 2\alpha \langle u_* - \bar{u}, f'(v) \rangle \\
&\leq \|u - u_*\|^2 - \|u - \bar{u}\|^2 \\
&\quad + 2\alpha \langle v - \bar{u}, f'(v) \rangle + 2\alpha [f(u_*) - f(v)].
\end{aligned}
\tag{7.35}
$$

In view of (7.31) and (7.35),

$$
\begin{aligned}
\|\bar{u} - u_*\|^2 &\leq \|u - u_*\|^2 \\
&\quad + 2\alpha \langle v - \bar{u}, f'(v) \rangle + 2\alpha [f(u_*) - f(v)] \\
&\quad - \langle u - v + v - \bar{u}, u - v + v - \bar{u} \rangle \\
&= \|u - u_*\|^2 + 2\alpha \langle v - \bar{u}, f'(v) \rangle \\
&\quad + 2\alpha [f(u_*) - f(v)] - \|u - v\|^2 \\
&\quad - \|v - \bar{u}\|^2 - 2 \langle u - v, v - \bar{u} \rangle \\
&= \|u - u_*\|^2 - \|u - v\|^2 - \|v - \bar{u}\|^2 \\
&\quad + 2\alpha [f(u_*) - f(v)] + 2 \langle v - \bar{u}, \alpha f'(v) - u + v \rangle \\
&\leq \|u - u_*\|^2 - \|u - v\|^2 - \|v - \bar{u}\|^2 \\
&\quad + 2\alpha [f(u_*) - f(v)] + 2\alpha \langle \bar{u} - v, f'(u) - f'(v) \rangle \\
&\leq \|u - u_*\|^2 - \|u - v\|^2 - \|v - \bar{u}\|^2 \\
&\quad + 2\alpha [f(u_*) - f(v)] + 2\alpha \|\bar{u} - v\| \|f'(u) - f'(v)\|.
\end{aligned}
$$

This completes the proof of Lemma 7.4.

Lemma 7.5. *Let*

$$
u_* \in C_{\min}, \ M_0 > 0, \ M_1 > 0, \ L > 0, \ \alpha \in (0, 1],
\tag{7.36}
$$

$$
f'(B(u_*, M_0)) \subset B(0, M_1),
\tag{7.37}
$$

$$
\|f'(z_1) - f'(z_2)\| \leq L \|z_1 - z_2\|
$$
$$
\textit{for all } z_1, z_1 \in B(u_*, M_0 + M_1),
\tag{7.38}
$$

$$
\alpha L \leq 1
\tag{7.39}
$$

and let

$$
u \in B(u_*, M_0), \ v = P_C(u - \alpha f'(u)), \ \bar{u} = P_C(u - \alpha f'(v)).
\tag{7.40}
$$

Then

$$\|\bar{u} - u_*\|^2 \leq \|u - u_*\|^2$$
$$+2\alpha[f(u_*) - f(v)] - \|u - v\|^2(1 - \alpha^2 L^2).$$

Proof. Lemma 7.4 and (7.36) imply that

$$\|\bar{u} - u_*\|^2 \leq \|u - u_*\|^2 - \|u - v\|^2 - \|v - \bar{u}\|^2$$
$$+2\alpha[f(u_*) - f(v)] + 2\alpha\|\bar{u} - v\|\|f'(u) - f'(v)\|. \qquad (7.41)$$

In view of (7.40) and (7.37),

$$\|f'(u)\| \leq M_1. \qquad (7.42)$$

It follows from (7.40), (7.36), Lemma 2.2, and (7.42) that

$$\|v - u_*\| \leq \|u - \alpha f'(u) - u_*\| \leq \|u - u_*\| + \alpha\|f'(u)\|$$
$$\leq M_0 + \alpha M_1. \qquad (7.43)$$

In view of (7.40), (7.43), (7.36), and (7.39),

$$\|f'(u) - f'(v)\| \leq L\|u - v\|. \qquad (7.44)$$

By (7.41) and (7.44),

$$\|\bar{u} - u_*\|^2 \leq \|u - u_*\|^2 - \|u - v\|^2 - \|v - \bar{u}\|^2$$
$$+2\alpha[f(u_*) - f(v)] + 2\alpha\|\bar{u} - v\|\|u - v\|L$$
$$\leq \|u - u_*\|^2 - \|u - v\|^2 - \|v - \bar{u}\|^2$$
$$+2\alpha[f(u_*) - f(v)] + \alpha^2 L^2\|u - v\|^2 + \|\bar{u} - v\|^2$$
$$\leq \|u - u_*\|^2 + 2\alpha[f(u_*) - f(v)] - \|u - v\|^2(1 - \alpha^2 L^2). \quad (7.45)$$

By (7.45),

$$\|\bar{u} - u_*\|^2 \leq \|u - u_*\|^2 + 2\alpha[f(u_*) - f(v)] - \|u - v\|^2(1 - \alpha^2 L^2).$$

This completes the proof of Lemma 7.5.

Lemma 7.6. *Let*

$$u_* \in C_{\min}, \ M_0 > 0, \ M_1 > 0, L > 0, \alpha \in (0, 1], \ \delta \in (0, 1), \qquad (7.46)$$

$$f'(B(u_*, M_0)) \subset B(0, M_1), \qquad (7.47)$$

$$\|f'(z_1) - f'(z_2)\| \le L\|z_1 - z_2\|$$

$$\text{for all } z_1, z_2 \in B(u_*, M_0 + M_1 + 1), \tag{7.48}$$

$$\alpha L \le 1. \tag{7.49}$$

Assume that

$$x \in B(u_*, M_0), \ y \in X, \tag{7.50}$$

$$\|y - P_C(x - \alpha f'(x))\| \le \delta, \tag{7.51}$$

$$\tilde{x} \in X, \ \|\tilde{x} - P_C(x - \alpha f'(y))\| \le \delta. \tag{7.52}$$

Then

$$\|\tilde{x} - u_*\|^2 \le 4\delta(M_0 + 1) + \|x - u_*\|^2$$
$$+ 2\alpha[f(u_*) - f(P_C(x - \alpha f'(x)))]$$
$$- \|x - P_C(x - \alpha f'(x))\|^2 (1 - \alpha^2 L^2).$$

Proof. Put

$$v = P_C(x - \alpha f'(x)), \ z = P_C(x - \alpha f'(v)). \tag{7.53}$$

Lemma 7.5, (7.46), (7.47), (7.48), (7.49), (7.50), and (7.53) imply that

$$\|z - u_*\|^2 \le \|x - u_*\|^2 + 2\alpha[f(u_*) - f(v)] - \|x - v\|^2 (1 - \alpha^2 L^2). \tag{7.54}$$

It is clear that

$$\|\tilde{x} - u_*\|^2 = \|\tilde{x} - z + z - u_*\|^2$$
$$= \|\tilde{x} - z\|^2 + 2\langle \tilde{x} - z, z - u_* \rangle + \|z - u_*\|^2$$
$$\le \|\tilde{x} - z\|^2 + 2\|\tilde{x} - z\|\|z - u_*\| + \|z - u_*\|^2. \tag{7.55}$$

In view of (7.51) and (7.53),

$$\|v - y\| \le \delta. \tag{7.56}$$

Put

$$\tilde{z} = P_C(x - \alpha f'(y)). \tag{7.57}$$

By (7.53), (7.46), Lemma 2.2, (7.50), and (7.47),

$$\|u_* - v\| \le \|u_* - x\| + \alpha\|f'(x)\| \le M_0 + \alpha M_1. \tag{7.58}$$

Relations (7.58), (7.56), and (7.46) imply that

$$\|u_* - y\| \le \|u_* - v\| + \|v - y\| \le M_0 + \alpha M_1 + 1. \tag{7.59}$$

It follows from (7.52), (7.57), (7.53), Lemma 2.2, (7.58), (7.59), (7.46), (7.48), (7.56), and (7.49) that

$$
\begin{aligned}
\|\tilde{x} - z\| &\le \|\tilde{x} - \tilde{z}\| + \|\tilde{z} - z\| \\
&\le \delta + \|\tilde{z} - z\| \le \delta + \alpha \|f'(y) - f'(v)\| \\
&\le \delta + \alpha L \|v - y\| \le \delta + \alpha L \delta \\
&= \delta(1 + \alpha L) \le 2\delta.
\end{aligned}
\tag{7.60}
$$

In view of (7.55), (7.60), (7.54), (7.46), (7.50), and (7.53),

$$
\begin{aligned}
\|\tilde{x} - u_*\|^2 &\le 4\delta^2 + \|x - u_*\|^2 + 2\alpha[f(u_*) - f(v)] \\
&\quad - \|x - v\|^2(1 - \alpha^2 L^2) + 4\delta\|x - u_*\| \\
&\le 4\delta(M_0 + 1) + \|x - u_*\|^2 \\
&\quad + 2\alpha[f(u_*) - f(P_C(x - \alpha f'(x)))] - \|x - v\|^2(1 - \alpha^2 L^2) \\
&= 4\delta(M_0 + 1) + \|x - u_*\|^2 + 2\alpha[f(u_*) - f(P_C(x - \alpha f'(x)))] \\
&\quad - \|x - P_C(x - \alpha f'(x))\|^2(1 - \alpha^2 L^2).
\end{aligned}
$$

This completes the proof of Lemma 7.6.

7.3   Proof of Theorem 7.1

In view of (7.4), there exists a point

$$u_* \in C_{\min} \cap B(0, M_0). \tag{7.61}$$

It follows from (7.11) and (7.61) that

$$\|x_0 - u_*\| \le 2M_0. \tag{7.62}$$

Assume that $i \ge 0$ is an integer and that

$$x_i \in B(u_*, 2M_0). \tag{7.63}$$

(It is clear that in view of (7.62), inclusion (7.63) is valid for $i = 0$.) It follows from (7.61), (7.63), (7.7), (7.11), (7.5), (7.6), (7.12), (7.13), and Lemma 7.6 applied with $x = x_i, y = y_i \ \tilde{x} = x_{i+1}, \alpha = \alpha_i$ that

$$\|x_{i+1} - u_*\|^2 \leq 4\delta(2M_0 + 1)$$
$$+ \|x_i - u_*\|^2 + 2\alpha_i[f(u_*) - f(P_C(x_i - \alpha_i f'(x_i)))]. \quad (7.64)$$

Thus we have shown that the following property holds:
   (P1) If an integer $i \geq 0$ satisfies (7.63), then inequality (7.64) is valid.
   We claim that there exists an integer $i \in [0, k]$ for which

$$f(P_C(x_i - \alpha_i f'(x_i))) \leq f(u_*) + \epsilon. \quad (7.65)$$

Assume the contrary. Then relations (7.9) and (7.10) imply that for each integer $i \in [0, k]$,

$$2\alpha_i[f(u_*) - f(P_C(x_i - \alpha_i f'(x_i)))] + 4\delta(2M_0 + 1)$$
$$\leq 2\alpha_i(-\epsilon) + 4\delta(2M_0 + 1)$$
$$\leq -2\alpha_*\epsilon + 4\delta(2M_0 + 1) \leq -\alpha_*\epsilon. \quad (7.66)$$

It follows from (7.62), (7.66), and property (P1) that for each integer $i \in [0, k]$,

$$\|x_{i+1} - u_*\|^2 \leq \|x_i - u_*\|^2 - \alpha_*\epsilon. \quad (7.67)$$

In view of (7.63) and (7.67),

$$4M_0^2 \geq \|x_0 - u_*\|^2 - \|x_k - u_*\|^2$$
$$= \sum_{i=0}^{k-1}[\|x_i - u_*\|^2 - \|x_{i+1} - u_*\|^2] \geq k\alpha_*\epsilon$$

and

$$k \leq 4M_0^2\alpha_*^{-1}\epsilon^{-1}.$$

This contradicts (7.8). The contradiction we have reached proves that there exists an integer $j \in [0, k]$ such that

$$f(P_C(x_j - \alpha_j f'(x_j))) \leq f(u_*) + \epsilon.$$

We may assume without loss of generality that for each integer $i \geq 0$ satisfying $i < j$,

$$f(P_C(x_i - \alpha_i f'(x_i))) > f(u_*) + \epsilon. \quad (7.68)$$

It follows from (7.68), (7.9), and (7.10) that for any integer $i \geq 0$ satisfying $i < j$, inequality (7.66) is valid. Combined with (7.62), property (P1), and (7.61) this implies that for each integer i satisfying $0 \leq i \leq j$, we have

$$\|x_i - u_*\| \leq 2M_0$$

and $\|x_i\| \leq 3M_0$. This completes the proof of Theorem 7.1.

7.4 Proof of Theorem 7.3

Let

$$u_* \text{ be an arbitrary element of } C_{\min}. \tag{7.69}$$

In view of (7.69) and (7.16),

$$\|u_*\| \leq M_0 - 1. \tag{7.70}$$

Relations (7.70) and (7.23) imply that

$$\|x_i - u_*\| \leq 2M_0 - 1. \tag{7.71}$$

Assume that $i \geq 0$ is an integer such that

$$x_i \in B(u_*, 2M_0). \tag{7.72}$$

(It is clear that in view of (7.71) inclusion (7.72) is valid for $i = 0$.) It follows from (7.69), (7.9), (7.17)–(7.19), (7.72), (7.70), (7.24), and Lemma 7.6 applied with $x = x_i, y = y_i, \tilde{x} = x_{i+1}, \alpha = \alpha_i$ that

$$\|x_{i+1} - u_*\|^2 \leq 4\delta(2M_0 + 1)$$
$$+ \|x_i - u_*\|^2 + 2\alpha_i[f(u_*) - f(P_C(x_i - \alpha_i f'(x_i)))]$$
$$- \|x_i - P_C(x_i - \alpha_i f'(x_i))\|^2(1 - \alpha_i^2 L^2)$$

and by (7.22),

$$\|x_i - u_*\|^2 - \|x_{i+1} - u_*\|^2$$
$$\geq 2\alpha_*[f(P_C(x_i - \alpha_i f'(x_i))) - f(u_*)]$$
$$+ \|x_i - P_C(x_i - \alpha_i f'(x_i))\|^2(1 - \alpha_i^2 L^2) - 4\delta(2M_0 + 1). \tag{7.73}$$

Thus we have shown that the following property holds:

(P2)   If an integer $i \geq 0$ satisfies (7.72), then inequality (7.73) is valid.

Assume that an integer $i \geq 0$ satisfies (7.72) and that

$$\max\{f(P_C(x_i - \alpha_i f'(x_i))) - f(u_*), \ \|x_i - P_C(x_i - \alpha_i f'(x_i))\|^2\} \leq \gamma. \qquad (7.74)$$

It follows from (7.74), property (C), and (7.15) that

$$d(P_C(x_i - \alpha_i f'(x_i)), C_{\min}) \leq \epsilon/2,$$
$$\|x_i - P_C(x_i - \alpha_i f'(x_i))\| \leq \gamma^{1/2} < \epsilon/4$$

and

$$d(x_i, C_{\min}) \leq 3\epsilon/4.$$

Thus we have shown that the following property holds:

(P3)   If an integer $i \geq 0$ satisfies (7.72) and (7.74), then

$$d(x_i, C_{\min}) \leq 3\epsilon/4.$$

Assume that an integer $i \geq 0$ satisfies inclusion (7.72) and that

$$\max\{f(P_C(x_i - \alpha_i f'(x_i))) - f(u_*), \ \|x_i - P_C(x_i - \alpha_i f'(x_i))\|^2\} > \gamma. \qquad (7.75)$$

It follows from (7.72), property (P2), (7.73), (7.75), (7.19), (7.21), and (7.22) that

$$\|x_i - u_*\|^2 - \|x_{i+1} - u_*\|^2$$
$$\geq \gamma \min\{\alpha_*, 1 - (\alpha^*)^2 L^2\} - 4\delta(2M_0 + 1)$$
$$\geq 2^{-1}\gamma \min\{\alpha_*, \ 1 - (\alpha^*)^2 L^2\}.$$

Thus we have shown that the following property holds:

(P4)   If an integer $i \geq 0$ satisfies (7.72) and (7.75), then

$$\|x_i - u_*\|^2 - \|x_{i+1} - u_*\|^2$$
$$\geq 2^{-1}\gamma \min\{\alpha_*, \ 1 - (\alpha^*)^2 L^2\}$$

and since u_* is an arbitrary point of $C_{\min}$, we have

$$d(x_{i+1}, C_{\min})^2 \leq d(x_i, C_{\min})^2 - (\gamma/2) \min\{\alpha_*, \ 1 - (\alpha^*)^2 L^2)\}.$$

We claim that there exists an integer $i \in [0, k]$ such that (7.74) is valid.

Assume the contrary. Then (7.75) holds for each integer $i \in [0, k]$. Combined with (7.71) and property (P4) this implies that

$$4M_0^2 \geq \|x_0 - u_*\|^2 - \|x_k - u_*\|^2$$

$$= \sum_{i=0}^{k-1} [\|x_i - u_*\|^2 - \|x_{i+1} - u_*\|^2]$$

$$\geq k(\gamma/2) \min\{\alpha_*, \ 1 - (\alpha^*)^2 L^2\}$$

and

$$k \leq 8M_0^2 \gamma^{-1} \min\{\alpha_*, \ 1 - (\alpha^*)^2 L^2\}^{-1}.$$

This contradicts (7.20). The contradiction we have reached proves that there exists an integer $j \in [0, k]$ such that (7.74) is valid with $i = j$.

We may assume that for all integers $i \geq 0$ satisfying $i < j$ Eq. (7.75) holds. It follows from property (P4) and (7.71) that

$$x_j \in B(u_*, 2M_0). \tag{7.76}$$

Property (P3), (7.76), and (7.74) with $i = j$ imply that

$$d(x_j, C_{\min}) \leq 3\epsilon/4. \tag{7.77}$$

Assume that an integer $i \geq j$ and that

$$d(x_i, C_{\min}) < \epsilon. \tag{7.78}$$

There are two cases: (7.74) is valid; (7.75) is valid. Assume that (7.74) is true. In view of property (P3), (7.78), and (7.16),

$$d(x_i, C_{\min}) < 3\epsilon/4.$$

Since u_* is an arbitrary point of the set $C_{\min}$ we may assume without loss of generality that

$$\|x_i - u_*\| < (4/5)\epsilon. \tag{7.79}$$

It follows from (7.79), (7.15), property (P2), (7.73), (7.19), and (7.21) that

$$\|x_{i+1} - u_*\| \leq \|x_i - u_*\| + 2(8\delta(M_0 + 1))^{1/2} < (4/5)\epsilon + \epsilon/5$$

and

$$d(x_{i+1}, C_{\min}) < \epsilon.$$

Assume that (7.75) holds. Property (P4), (7.78), and (7.16) imply that

$$d(x_{i+1}, C_{\min})^2 \leq d(x_i, C_{\min})^2 - (\gamma/2) \min\{\alpha_*, 1 - (\alpha^*)^2 L^2\}$$

and

$$d(x_{i+1}, C_{\min}) < \epsilon. \tag{7.80}$$

Thus (7.80) holds in both cases.

We have shown that if an integer $i \geq j$ and (7.78) holds, then (7.80) is true. Therefore,

$$d(x_i, C_{\min}) < \epsilon \text{ for all integers } i \geq k.$$

This completes the proof of Theorem 7.3.

Chapter 8
A Projected Subgradient Method for Nonsmooth Problems

In this chapter we study the convergence of the projected subgradient method for a class of constrained optimization problems in a Hilbert space. For this class of problems, an objective function is assumed to be convex but a set of admissible points is not necessarily convex. Our goal is to obtain an ϵ-approximate solution in the presence of computational errors, where ϵ is a given positive number.

8.1 Preliminaries and Main Results

Let $(X, \langle \cdot, \cdot \rangle)$ be a Hilbert space with an inner product $\langle \cdot, \cdot \rangle$ which induces a complete norm $\| \cdot \|$.

For each $x \in X$ and each nonempty set $A \subset X$ put

$$d(x, A) = \inf\{\|x - y\| : y \in A\}.$$

For each $x \in X$ and each $r > 0$ set

$$B(x, r) = \{y \in X : \|x - y\| \le r\}.$$

Assume that $f : X \to R^1$ is a convex continuous function which is Lipschitz on all bounded subsets of X.

For each point $x \in X$ and each positive number ϵ let

$$\partial f(x) = \{l \in X : f(y) - f(x) \ge \langle l, y - x \rangle \text{ for all } y \in X\} \tag{8.1}$$

© Springer International Publishing Switzerland 2016
A.J. Zaslavski, *Numerical Optimization with Computational Errors*, Springer Optimization and Its Applications 108, DOI 10.1007/978-3-319-30921-7_8

be the subdifferential of f at x and let

$$\partial_\epsilon f(x) = \{l \in X : f(y) - f(x) \geq \langle l, y - x \rangle - \epsilon \text{ for all } y \in X\} \tag{8.2}$$

be the ϵ-subdifferential of f at x.

Let C be a closed nonempty subset of the space X.

Assume that

$$\lim_{\|x\| \to \infty} f(x) = \infty. \tag{8.3}$$

It means that for each $M_0 > 0$ there exists $M_1 > 0$ such that if a point $x \in X$ satisfies the inequality $\|x\| \geq M_1$, then $f(x) > M_0$.

Define

$$\inf(f; C) = \inf\{f(z) : z \in C\}. \tag{8.4}$$

Since the function f is Lipschitz on all bounded subsets of the space X it follows from (8.4) that $\inf(f; C)$ is finite.

Set

$$C_{\min} = \{x \in C : f(x) = \inf(f; C)\}. \tag{8.5}$$

It is well known that if the set C is convex, then the set $C_{\min}$ is nonempty. Clearly, the set $C_{\min} \neq \emptyset$ if the space X is finite-dimensional.

In this chapter we assume that

$$C_{\min} \neq \emptyset. \tag{8.6}$$

It is clear that $C_{\min}$ is a closed subset of X.

We suppose that the following assumption holds.

(A1)   For every positive number ϵ there exists $\delta > 0$ such that if a point $x \in C$ satisfies the inequality $f(x) \leq \inf(f; C) + \delta$, then $d(x, C_{\min}) \leq \epsilon$.

(It is clear that (A1) holds if the space X is finite-dimensional.)

We also suppose that the following assumption holds.

(A2)   There exists a continuous mapping $P_C : X \to X$ such that $P_C(X) = C$, $P_C(x) = x$ for all $x \in C$ and

$$\|x - P_C(y)\| \leq \|x - y\| \text{ for all } x \in C \text{ and all } y \in X.$$

For every number $\epsilon \in (0, \infty)$ let

$$\phi(\epsilon) = \sup\{\delta \in (0, 1] : \text{ if } x \in C \text{ satisfies } f(x) \leq \inf(f; C) + \delta,$$

$$\text{then } d(x, C_{\min}) \leq \min\{1, \epsilon\}\}. \tag{8.7}$$

In view of (A1), $\phi(\epsilon)$ is well defined for every positive number ϵ.

In this chapter we will prove the following two results obtained in [122].

Theorem 8.1. *Let* $\{\alpha_i\}_{i=0}^{\infty} \subset (0, 1]$ *satisfy*

$$\lim_{i \to \infty} \alpha_i = 0, \ \sum_{i=1}^{\infty} \alpha_i = \infty$$

and let $M, \epsilon > 0$. *Then there exist a natural number* n_0 *and* $\delta > 0$ *such that the following assertion holds.*

Assume that an integer $n \geq n_0$,

$$\{x_k\}_{k=0}^{n} \subset X, \ \|x_0\| \leq M,$$

$$v_k \in \partial_{\delta} f(x_k) \setminus \{0\}, \ k = 0, 1, \ldots, n-1,$$

$$\{\eta_k\}_{k=0}^{n-1}, \ \{\xi_k\}_{k=0}^{n-1} \subset B(0, \delta),$$

and that for $k = 0, \ldots, n-1$,

$$x_{k+1} = P_C(x_k - \alpha_k \|v_k\|^{-1} v_k - \alpha_k \xi_k) - \alpha_k \eta_k.$$

Then the inequality $d(x_k, C_{\min}) \leq \epsilon$ *hods for all integers* k *satisfying* $n_0 \leq k \leq n$.

Theorem 8.2. *Let* $M, \epsilon > 0$. *Then there exists* $\beta_0 \in (0, 1)$ *such that for each* $\beta_1 \in (0, \beta_0)$ *there exist a natural number* n_0 *and* $\delta > 0$ *such that the following assertion holds.*

Assume that an integer $n \geq n_0$,

$$\{x_k\}_{k=0}^{n} \subset X, \ \|x_0\| \leq M,$$

$$v_k \in \partial_{\delta} f(x_k) \setminus \{0\}, \ k = 0, 1, \ldots, n-1,$$

$$\{\alpha_k\}_{k=0}^{n-1} \subset [\beta_1, \beta_0],$$

$$\{\eta_k\}_{k=0}^{n-1}, \ \{\xi_k\}_{k=0}^{n-1} \subset B(0, \delta)$$

and that for $k = 0, \ldots, n-1$,

$$x_{k+1} = P_C(x_k - \alpha_k \|v_k\|^{-1} v_k - \alpha_k \xi_k) - \eta_k.$$

Then the inequality $d(x_k, C_{\min}) \leq \epsilon$ *holds for all integers* k *satisfying* $n_0 \leq k \leq n$.

In this chapter we use the following definitions and notation.

Define

$$X_0 = \{x \in X : f(x) \leq \inf(f; C) + 1\}. \tag{8.8}$$

In view of (8.3), there exists a number $\bar{K} > 0$ such that

$$. X_0 \subset B(0, \bar{K}). \tag{8.9}$$

Since the function f is Lipschitz on all bounded subsets of the space X there exists a number $\bar{L} > 1$ such that

$$|f(z_1) - f(z_2)| \leq \bar{L}\|z_1 - z_2\| \text{ for all } z_1, z_2 \in B(0, \bar{K} + 4). \tag{8.10}$$

8.2   Auxiliary Results

We use the notation and definitions introduced in Sect. 8.1 and suppose that all the assumptions posed in Sect. 8.1 hold.

Proposition 8.3. *Let* $\epsilon \in (0, 1]$. *Then for each* $x \in X$ *satisfying*

$$d(x, C) < \min\{\bar{L}^{-1}2^{-1}\phi(\epsilon/2), \ \epsilon/2\},$$
$$f(x) \leq \inf(f; C) + \min\{2^{-1}\phi(\epsilon/2), \ \epsilon/2\}, \tag{8.11}$$

the inequality $d(x, C_{\min}) \leq \epsilon$ *holds.*

Proof. In view of the definition of ϕ, $\phi(\epsilon/2) \in (0, 1]$ and

$$\text{if } x \in C \text{ satisfies } f(x) < \inf(f; C) + \phi(\epsilon/2),$$
$$\text{then } d(x, C_{\min}) \leq \min\{1, \epsilon/2\}. \tag{8.12}$$

Assume that a point $x \in X$ satisfies (8.11). There exists a point $y \in C$ which satisfies

$$\|x - y\| < 2^{-1}\bar{L}^{-1}\phi(\epsilon/2) \text{ and } \|x - y\| < \epsilon/2. \tag{8.13}$$

Relations (8.11), (8.8), (8.9), and (8.13) imply that

$$x \in B(0, \bar{K}), \ y \in B(0, \bar{K} + 1). \tag{8.14}$$

By (8.13), (8.14), and the definition of $\bar{L}$ [see (8.10)],

$$|f(x) - f(y)| \leq \bar{L}\|x - y\| < \phi(\epsilon/2)2^{-1}. \tag{8.15}$$

It follows from the choice of the point y, (8.11), and (8.15) that

$$y \in C$$

and

$$f(y) < f(x) + \phi(\epsilon/2)2^{-1} \leq \inf(f; C) + \phi(\epsilon/2).$$

Combined with (8.12) this implies that

$$d(y, C_{\min}) \leq \epsilon/2.$$

Together with (8.13) this implies that

$$d(x, C_{\min}) \leq \|x - y\| + d(y, C_{\min}) \leq \epsilon.$$

This completes the proof of Proposition 8.3.

Lemma 8.4. *Assume that $\epsilon > 0$, $x \in X$, $y \in X$,*

$$f(x) > \inf(f; C) + \epsilon, \; f(y) \leq \inf(f; C) + \epsilon/4, \tag{8.16}$$

$$v \in \partial_{\epsilon/4} f(x). \tag{8.17}$$

Then $\langle v, y - x \rangle \leq -\epsilon/2$.

Proof. In view of (8.2) and (8.17),

$$f(u) - f(x) \geq \langle v, u - x \rangle - \epsilon/4 \text{ for all } u \in X. \tag{8.18}$$

By (8.16) and (8.18),

$$-(3/4)\epsilon \geq f(y) - f(x) \geq \langle v, y - x \rangle - \epsilon/4.$$

The inequality above implies that

$$\langle v, y - x \rangle \leq -\epsilon/2.$$

This completes the proof of Lemma 8.4.

Lemma 8.5. *Let*

$$\bar{x} \in C_{\min}, \tag{8.19}$$

$K_0 > 0$, $\epsilon \in (0, 1]$, $\alpha \in (0, 1]$, *let a positive number δ satisfy*

$$\delta(K_0 + \bar{K} + 1) \leq (8\bar{L})^{-1}\epsilon, \tag{8.20}$$

let a point $x \in X$ satisfy

$$\|x\| \leq K_0, \ f(x) > \inf(f; C) + \epsilon, \tag{8.21}$$

$$\eta, \ \xi \in B(0, \delta), \ v \in \partial_{\epsilon/4} f(x) \setminus \{0\} \tag{8.22}$$

and let

$$y = P_C(x - \alpha \|v\|^{-1} v - \alpha \xi) - \eta. \tag{8.23}$$

Then

$$\|y - \bar{x}\|^2 \leq \|x - \bar{x}\|^2 - \alpha(4\bar{L})^{-1} \epsilon + 2\alpha^2 + \|\eta\|^2 + 2\|\eta\|(K_0 + \bar{K} + 2).$$

Proof. In view of (8.8)–(8.10) and (8.19), for every point $z \in B(\bar{x}, 4^{-1}\epsilon \bar{L}^{-1})$, we have

$$f(z) \leq f(\bar{x}) + \bar{L}\|z - \bar{x}\| \leq f(\bar{x}) + \epsilon/4 = \inf(f; C) + \epsilon/4. \tag{8.24}$$

Lemma 8.4, (8.21), (8.22), and (8.24) imply that for every point

$$z \in B(\bar{x}, 4^{-1}\epsilon \bar{L}^{-1}),$$

we have

$$\langle v, z - x \rangle \leq -\epsilon/2.$$

Combined with (8.22) the inequality above implies that

$$\langle \|v\|^{-1} v, z - x \rangle < 0 \text{ for all } z \in B(\bar{x}, (4\bar{L})^{-1}\epsilon). \tag{8.25}$$

Put

$$\tilde{z} = \bar{x} + 4^{-1}\bar{L}^{-1}\epsilon \|v\|^{-1} v. \tag{8.26}$$

It is easy to see that

$$\tilde{z} \in B(\bar{x}, 4^{-1}\bar{L}^{-1}\epsilon). \tag{8.27}$$

Relations (8.25), (8.26), and (8.27) imply that

$$0 > \langle \|v\|^{-1} v, \tilde{z} - x \rangle = \langle \|v\|^{-1} v, \bar{x} + 4^{-1}\bar{L}^{-1}\epsilon \|v\|^{-1} v - x \rangle. \tag{8.28}$$

By (8.28),

$$\langle \|v\|^{-1}v, \bar{x} - x \rangle < -4^{-1}\bar{L}^{-1}\epsilon. \tag{8.29}$$

Set

$$y_0 = x - \alpha\|v\|^{-1}v - \alpha\xi. \tag{8.30}$$

It follows from (8.30), (8.22), (8.21), (8.19), (8.8), (8.9), (8.29), and (8.20) that

$$
\begin{aligned}
\|y_0 - \bar{x}\|^2 &= \|x - \alpha\|v\|^{-1}v - \alpha\xi - \bar{x}\|^2 \\
&= \|x - \alpha\|v\|^{-1}v - \bar{x}\|^2 + \alpha^2\|\xi\|^2 \\
&\quad -2\alpha\langle\xi, x - \alpha\|v\|^{-1}v - \bar{x}\rangle \\
&\leq \|x - \alpha\|v\|^{-1}v - \bar{x}\|^2 \\
&\quad +\alpha^2\delta^2 + 2\alpha\delta(K_0 + \bar{K} + 1) \\
&\leq \|x - \bar{x}\|^2 - 2\langle x - \bar{x}, \alpha\|v\|^{-1}v\rangle \\
&\quad +\alpha^2 + \alpha^2\delta^2 + 2\alpha\delta(K_0 + \bar{K} + 1) \\
&< \|x - \bar{x}\|^2 - 2\alpha(4^{-1}\bar{L}^{-1}\epsilon) \\
&\quad +\alpha^2(1 + \delta^2) + 2\alpha\delta(K_0 + \bar{K} + 1) \\
&\leq \|x - \bar{x}\|^2 - \alpha(4\bar{L})^{-1}\epsilon + 2\alpha^2. \tag{8.31}
\end{aligned}
$$

In view of (8.8), (8.9), (8.19), (8.21), and (8.31),

$$\|y_0 - \bar{x}\|^2 \leq (K_0 + \bar{K})^2 + 2$$

and

$$\|y_0 - \bar{x}\| \leq K_0 + \bar{K} + 2. \tag{8.32}$$

By (8.23), (8.30), (8.19), (A2), (8.31), and (8.32),

$$
\begin{aligned}
\|y - \bar{x}\|^2 &= \|P_C(y_0) - \eta - \bar{x}\|^2 \\
&\leq \|P_C(y_0) - \bar{x}\|^2 + \|\eta\|^2 + 2\|\eta\|\|P_C(y_0) - \bar{x}\| \\
&\leq \|y_0 - \bar{x}\|^2 + \|\eta\|^2 + 2\|\eta\|\|y_0 - \bar{x}\| \\
&\leq \|x - \bar{x}\|^2 - \alpha(4\bar{L})^{-1}\epsilon \\
&\quad +2\alpha^2 + \|\eta\|^2 + 2\|\eta\|(K_0 + \bar{K} + 2).
\end{aligned}
$$

This completes the proof of Lemma 8.5.

Lemma 8.5 implies the following result.

Lemma 8.6. *Let $K_0 > 0$, $\epsilon \in (0, 1]$, $\alpha \in (0, 1]$, a positive number δ satisfy*

$$\delta(K_0 + \bar{K} + 1) \leq (8\bar{L})^{-1}\epsilon,$$

let $x \in X$ satisfy

$$\|x\| \leq K_0, \ f(x) > \inf(f; C) + \epsilon,$$

let

$$\eta, \ \xi \in B(0, \delta), \ v \in \partial_{\epsilon/4} f(x) \setminus \{0\}$$

and let

$$y = P_C(x - \alpha\|v\|^{-1}v - \alpha\xi) - \eta.$$

Then

$$d(y, C_{\min})^2 \leq d(x, C_{\min})^2 - \alpha(4\bar{L})^{-1}\epsilon$$
$$+ 2\alpha^2 + \|\eta\|^2 + 2\|\eta\|(K_0 + \bar{K} + 2).$$

8.3   Proof of Theorem 8.1

We may assume without loss of generality that $\epsilon < 1$. In view of Proposition 8.3, there exists a number

$$\bar{\epsilon} \in (0, \epsilon/8) \tag{8.33}$$

such that

$$\text{if } x \in X, \ d(x, C) \leq 2\bar{\epsilon} \text{ and } f(x) \leq \inf(f; C) + 2\bar{\epsilon},$$
$$\text{then } d(x, C_{\min}) \leq \epsilon. \tag{8.34}$$

Fix

$$\bar{x} \in C_{\min}. \tag{8.35}$$

Fix

$$\epsilon_0 \in (0, 4^{-1}\bar{\epsilon}). \tag{8.36}$$

Since $\lim_{i \to \infty} \alpha_i = 0$ there is an integer $p_0 > 0$ such that

$$\bar{K} + 4 < p_0 \tag{8.37}$$

and that for all integers $p \geq p_0$, we have

$$\alpha_p < (32\bar{L})^{-1}\epsilon_0. \tag{8.38}$$

Since $\sum_{i=0}^{\infty} \alpha_i = \infty$ there exists a natural number $n_0 > p_0 + 4$ such that

$$\sum_{i=p_0}^{n_0-1} \alpha_i > (4p_0 + M + \|\bar{x}\|)^2 \epsilon_0^{-1} 16\bar{L}. \tag{8.39}$$

Fix

$$K_* > \bar{K} + 4 + M + 4n_0 + 4\|\bar{x}\| \tag{8.40}$$

and a positive number δ such that

$$6\delta(K_* + 1) < (16\bar{L})^{-1}\epsilon_0. \tag{8.41}$$

Assume that an integer $n \geq n_0$ and that

$$\{x_k\}_{k=0}^{n} \subset X, \ \|x_0\| \leq M, \tag{8.42}$$

$$\{\eta_k\}_{k=0}^{n-1}, \ \{\xi_k\}_{k=0}^{n-1} \subset B(0, \delta), \tag{8.43}$$

$$v_k \in \partial_{\delta}f(x_k) \setminus \{0\}, \ k = 0, 1, \ldots, n-1 \tag{8.44}$$

and that for all integers $k = 0, \ldots, n-1$, we have

$$x_{k+1} = P_C(x_k - \alpha_k\|v_k\|^{-1}v_k - \alpha_k\xi_k) - \alpha_k\eta_k. \tag{8.45}$$

In order to prove the theorem it is sufficient to show that

$$d(x_k, C_{\min}) \leq \epsilon \text{ for all integers } k \text{ satisfying } n_0 \leq k \leq n. \tag{8.46}$$

Assume that an integer

$$k \in [p_0, n-1], \tag{8.47}$$

$$\|x_k\| \leq K_*, \ f(x_k) > \inf(f; C) + \epsilon_0. \tag{8.48}$$

In view of (8.35), (8.41), (8.48), (8.43), (8.44), and (8.45), the conditions of Lemma 8.5 hold with $K_0 = K_*$, $\epsilon = \epsilon_0$, $\alpha = \alpha_k$, $x = x_k$, $\xi = \xi_k$, $v = v_k$, $y = x_{k+1}$. $\eta = \alpha_k \eta_k$ and combined with (8.43), (847), (8.38), (8.41), and (8.40) this lemma implies that

$$
\begin{aligned}
\|x_{k+1} - \bar{x}\|^2 &\leq \|x_k - \bar{x}\|^2 - \alpha_k (4\bar{L})^{-1} \epsilon_0 \\
&\quad + 2\alpha_k^2 + \alpha_k^2 \|\eta_k\|^2 + 2\|\eta_k\| \alpha_k (K_* + \bar{K} + 2) \\
&\leq \|x_k - \bar{x}\|^2 - \alpha_k (4\bar{L})^{-1} \epsilon_0 + 2\alpha_k^2 + \alpha_k^2 \delta^2 + 2\delta \alpha_k (K_* + \bar{K} + 2) \\
&\leq \|x_k - \bar{x}\|^2 - \alpha_k (8\bar{L})^{-1} \epsilon_0 + 2\delta \alpha_k (K_* + \bar{K} + 3) \\
&\leq \|x_k - \bar{x}\|^2 - \alpha_k (16\bar{L})^{-1} \epsilon_0.
\end{aligned}
$$

Thus we have shown that the following property holds:

(P1)   If an integer $k \in [p_0, n-1]$ and (8.48) is valid, then we have

$$
\|x_{k+1} - \bar{x}\|^2 \leq \|x_k - \bar{x}\|^2 - (16\bar{L})^{-1} \alpha_k \epsilon_0.
$$

We claim that there exists an integer $j \in \{p_0, \ldots, n_0\}$ such that

$$
f(x_j) \leq \inf(f; C) + \epsilon_0
$$

Assume the contrary. Then

$$
f(x_j) > \inf(f; C) + \epsilon_0, \quad i = p_0, \ldots, n_0. \tag{8.49}
$$

It follows from (8.45), (8.43), (8.41), (8.35), and (A2) that for all integers $i = 0, \ldots, n-1$, we have

$$
\begin{aligned}
\|x_{i+1} - \bar{x}\| &\leq 1 + \|P_C(x_i - \alpha_i \|v_i\|^{-1} v_i - \alpha_i \xi) - \bar{x}\| \tag{8.50} \\
&\leq 1 + \|x_i - \alpha_i \|v_i\|^{-1} v_i - \alpha_i \xi - \bar{x}\| \\
&\leq 1 + \|x_i - \bar{x}\| + 2 = \|x_i - \bar{x}\| + 3.
\end{aligned}
$$

By (8.40), (8.42), and (8.50), for all integers $i = 0, \ldots, n_0$,

$$
\|x_i\| \leq \|x_0 - \bar{x}\| + 3i + \|\bar{x}\| \leq M + 3i + 2\|\bar{x}\| \leq M + 3n_0 + 2\|\bar{x}\| < K_*. \tag{8.51}
$$

Let

$$
i \in \{p_0, \ldots, n_0 - 1\}. \tag{8.52}
$$

It follows from (8.52), (8.51), (8.49), and property (P1) that

$$
\|x_{i+1} - \bar{x}\|^2 \leq \|x_i - \bar{x}\|^2 - (16\bar{L})^{-1} \alpha_i \epsilon_0. \tag{8.53}
$$

Relations (8.42), (8.50), (8.52), and (8.53) imply that

$$-(M + 3p_0 + \|\bar{x}\|)^2 \le \|x_{n_0} - \bar{x}\|^2 - \|x_{p_0} - \bar{x}\|^2$$
$$= \sum_{i=p_0}^{n_0-1} [\|x_{i+1} - \bar{x}\|^2 - \|x_i - \bar{x}\|^2] \le -(16\bar{L})^{-1}\epsilon_0 \sum_{i=p_0}^{n_0-1} \alpha_i$$

and

$$\sum_{i=p_0}^{n_0-1} \alpha_i \le 16\bar{L}\epsilon_0^{-1}(M + 3p_0 + \|\bar{x}\|)^2.$$

This contradicts (8.39). The contradiction we have reached proves that there exists an integer

$$j \in \{p_0, \dots, n_0\}$$

such that

$$f(x_j) \le \inf(f; C) + \epsilon_0. \tag{8.54}$$

By (8.45), (A2), (8.43), (8.41), and (8.36), we have

$$d(x_j, C) \le \alpha_{j-1}\delta < \bar{\epsilon}. \tag{8.55}$$

In view of (8.54), (8.55), (8.36), and (8.34),

$$d(x_j, C_{\min}) \le \epsilon. \tag{8.56}$$

We claim that for all integers i satisfying $j \le i \le n$,

$$d(x_i, C_{\min}) \le \epsilon.$$

Assume the contrary. Then there exists an integer $k \in [j, n]$ for which

$$d(x_k, C_{\min}) > \epsilon. \tag{8.57}$$

By (8.56) and (8.57), we have

$$k > j \ge p_0. \tag{8.58}$$

By (8.56) we may assume without loss of generality that

$$d(x_i, C_{\min}) \le \epsilon \text{ for all integers } i \text{ satisfying } j \le i < k. \tag{8.59}$$

Thus

$$d(x_{k-1}, C_{\min}) \leq \epsilon. \tag{8.60}$$

There are two cases:

$$f(x_{k-1}) \leq \inf(f; C) + \epsilon_0; \tag{8.61}$$

$$f(x_{k-1}) > \inf(f; C) + \epsilon_0. \tag{8.62}$$

Assume that (8.61) is valid. It follows from (8.61), (8.36), (8.33), (8.8), and (8.9) that

$$x_{k-1} \in X_0 \subset B(0, \bar{K}). \tag{8.63}$$

By (8.45) and (8.43) there exists a point $z \in C$ such that

$$\|x_{k-1} - z\| \leq \delta. \tag{8.64}$$

By (8.45), (8.43), (8.64), and (A2),

$$\|x_k - z\| \leq \alpha_{k-1}\delta + \|z - P_C(x_{k-1} - \alpha_{k-1}\|v_{k-1}\|^{-1}v_{k-1} - \alpha_{k-1}\xi_{k-1})\|$$
$$\leq \delta + \|z - x_{k-1}\| + \alpha_{k-1} + \delta = 3\delta + \alpha_{k-1}. \tag{8.65}$$

Combined with (8.41), (8.58), and (8.38) the relation above implies that

$$d(x_k, C) \leq 3\delta + \alpha_{k-1} < \epsilon_0. \tag{8.66}$$

In view of (8.64) and (8.65),

$$\|x_k - x_{k-1}\| \leq 4\delta + \alpha_{k-1}. \tag{8.67}$$

It follows from (8.60), (8.67), (8.41), (8.38), and (8.58) that

$$d(x_k, C_{\min}) \leq 2\epsilon. \tag{8.68}$$

Relations (8.63), (8.68), (8.8), and (8.9) imply that

$$x_{k-1}, x_k \in B(0, \bar{K} + 2).$$

Together with (8.10) and (8.67) the inclusion above implies that

$$|f(x_{k-1}) - f(x_k)| \leq \bar{L}\|x_{k-1} - x_k\| \leq \bar{L}(4\delta + \alpha_{k-1}). \tag{8.69}$$

In view of (8.69), (8.51), (8.41), (8.38), and (8.58), we have

$$f(x_k) \leq f(x_{k-1}) + \bar{L}(4\delta + \alpha_{k-1})$$
$$\leq \inf(f; C) + \epsilon_0 + \bar{L}(4\delta + \alpha_{k-1}) \leq \inf(f; C) + 2\epsilon_0. \tag{8.70}$$

It follows from (8.70), (8.66), (8.36), and (8.34) that

$$d(x_k, C_{\min}) \leq \epsilon.$$

This inequality contradicts (8.57). The contradiction we have reached proves (8.62).
By (8.60), (8.8), and (8.9), we have

$$\|x_{k-1}\| \leq \bar{K} + 1. \tag{8.71}$$

It follows from (8.40), (8.41), (8.43), (8.44), (8.71), and (8.62) that Lemma 8.6
holds with

$$x = x_{k-1}, \; y = x_k, \; \xi = \xi_{k-1}, \; v = v_{k-1}, \; \alpha = \alpha_{k-1}, \; K_0 = \bar{K} + 1,$$
$$\epsilon = \epsilon_0, \; \eta = \alpha_{k-1}\eta_{k-1}.$$

Combined with (8.38), (8.58), (8.43), (8.41), and (8.60) this implies that

$$d(x_k, C_{\min})^2 \leq d(x_{k-1}, C_{\min})^2 - \alpha_{k-1}(4\bar{L})^{-1}\epsilon_0 + 2\alpha_{k-1}^2 + 2\alpha_{k-1}^2\|\eta_{k-1}\|^2$$
$$+ 2\alpha_{k-1}\|\eta_{k-1}\|(2\bar{K} + 3)$$
$$\leq d(x_{k-1}, C_{\min})^2 - (8\bar{L})^{-1}\alpha_{k-1}\epsilon_0 + 2\delta^2\alpha_{k-1}^2 + 2\alpha_{k-1}\delta(2\bar{K} + 3)$$
$$\leq d(x_{k-1}, C_{\min})^2 - (8\bar{L})^{-1}\alpha_{k-1}\epsilon_0 + 2\delta\alpha_{k-1}(2\bar{K} + 4)$$
$$\leq d(x_{k-1}, C_{\min})^2 - (16\bar{L})^{-1}\alpha_{k-1}\epsilon_0 \leq d(x_{k-1}, C_{\min})^2 \leq \epsilon^2.$$

This contradicts (8.57).

The contradiction we have reached proves that $d(x_i, C_{\min}) \leq \epsilon$ for all integers i
satisfying $j \leq i \leq n$. Since $j \leq n_0$ this completes the proof of Theorem 8.1.

8.4 Proof of Theorem 8.2

We may assume that without loss of generality

$$\epsilon < 1, \; M > \bar{K} + 4. \tag{8.72}$$

Proposition 8.3 implies that there exists

$$\bar{\epsilon} \in (0, \epsilon/8) \tag{8.73}$$

such that

$$\text{if } x \in X, \ d(x, C) \leq 2\bar{\epsilon} \text{ and } f(x) \leq \inf(f; C) + 2\bar{\epsilon},$$
$$\text{then } d(x, C_{\min}) \leq \epsilon/4. \tag{8.74}$$

Put

$$\beta_0 = (64\bar{L})^{-1}\bar{\epsilon}. \tag{8.75}$$

Let

$$\beta_1 \in (0, \beta_0). \tag{8.76}$$

There exists an integer $n_0 \geq 4$ such that

$$\beta_1 n_0 > 16^2 (3 + 2M)^2 \bar{\epsilon}^{-1} \bar{L}. \tag{8.77}$$

Fix

$$K_* > 2M + 4 + 4n_0 + 2\bar{K} + 2M \tag{8.78}$$

and a positive number δ such that

$$6\delta K_* < (64\bar{L})^{-1}\bar{\epsilon}\beta_1. \tag{8.79}$$

Fix a point

$$\bar{x} \in C_{\min}. \tag{8.80}$$

Assume that an integer $n \geq n_0$,

$$\{x_k\}_{k=0}^{n} \subset X, \ \{\eta_k\}_{k=0}^{n-1} \subset X, \ \{\xi_k\}_{k=0}^{n-1} \subset X, \ \{\alpha_k\}_{k=0}^{n-1} \subset [\beta_1, \beta_0], \tag{8.81}$$

$$\|x_0\| \leq M, \ \|\eta_k\| \leq \delta, \ \|\xi_k\| \leq \delta, \ k = 0, \ldots, n-1, \tag{8.82}$$

$$v_k \in \partial_\delta f(x_k) \setminus \{0\}, \ k = 0, 1, \ldots, n-1 \tag{8.83}$$

and that for all integers $k = 0, \ldots, n-1$,

$$x_{k+1} = P_C(x_k - \alpha_k \|v_k\|^{-1} v_k - \alpha_k \xi_k) - \eta_k. \tag{8.84}$$

We claim that $d(x_k, C_{\min}) \leq \epsilon$ for all integers k satisfying $n_0 \leq k \leq n$.

Assume that an integer

$$k \in [0, n-1],$$

$$\|x_k\| \le K_* \ f(x_k) > \inf(f; C) + \bar{\epsilon}/4. \tag{8.85}$$

It follows from (8.75), (8.78)–(8.81), (8.83), (8.85), (8.82), and (8.74) that Lemma 8.5 holds with $\epsilon = \bar{\epsilon}/4$, $K_0 = K_*$, $\alpha = \alpha_k$, $x = x_k$, $\eta = \eta_k$, $\xi = \xi_k$, $v = v_k$, $y = x_{k+1}$ and combining with (8.79) this implies that

$$\|x_{k+1} - \bar{x}\|^2 \le \|x_k - \bar{x}\|^2 - \alpha_k (16\bar{L})^{-1}\bar{\epsilon}$$
$$+ 2\alpha_k^2 + \delta^2 + 2\delta(K_* + \bar{K} + 2)$$
$$\le \|x_k - \bar{x}\|^2 - \alpha_k (16\bar{L})^{-1}\bar{\epsilon} + 2\alpha_k^2 + 2\delta(K_* + \bar{K} + 3).$$

Together with (8.81), (8.75), (8.78), and (8.79) this implies that

$$\|x_{k+1} - \bar{x}\|^2 \le \|x_k - \bar{x}\|^2 - \alpha_k (32\bar{L})^{-1}\bar{\epsilon} + 2\delta(\bar{K} + 3 + K_*)$$
$$\le \|x_k - \bar{x}\|^2 - (32\bar{L})^{-1}\bar{\epsilon}\beta_1 + 2\delta(\bar{K} + 3 + K_*)$$
$$\le \|x_k - \bar{x}\|^2 - \beta_1 (64\bar{L})^{-1}\bar{\epsilon}.$$

Thus we have shown that the following property holds:
(P2) if an integer $k \in [0, n-1]$ and (8.85) is valid, then we have

$$\|x_{k+1} - \bar{x}\|^2 \le \|x_k - \bar{x}\|^2 - (64\bar{L})^{-1}\beta_1\bar{\epsilon}.$$

We claim that there exists an integer $j \in \{1, \dots, n_0\}$ for which

$$f(x_j) \le \inf(f; C) + \bar{\epsilon}/4.$$

Assume the contrary. Then we have

$$f(x_j) > \inf(f; C) + \bar{\epsilon}/4, \ j = 1, \dots, n_0. \tag{8.86}$$

It follows from (8.84), (8.82), (8.79), (A2), (8.80), (8.81), and (8.75) that for all integers $i = 0, \dots, n-1$, we have

$$\|x_{i+1} - \bar{x}\| \le 1 + \|x_i - \alpha_i \|v_i\|^{-1} v_i - \alpha_i \xi_i - \bar{x}\|$$
$$\le \|x_i - \bar{x}\| + 3. \tag{8.87}$$

By (8.80)–(8.82), (8.72), (8.87), and (8.78) for $i = 0, \dots, n_0$,

$$\|x_i - \bar{x}\| \le \|x_0 - \bar{x}\| + 3i, \tag{8.88}$$

$$\|x_i\| \le 2\|\bar{x}\| + M + 3n_0 < K_*. \tag{8.89}$$

Let $k \in \{1, \ldots, n_0 - 1\}$. It follows from (8.89), (8.86), and property (P2) that

$$\|x_{k+1} - \bar{x}\|^2 \leq \|x_k - \bar{x}\|^2 - (64\bar{L})^{-1}\beta_1\bar{\epsilon}. \tag{8.90}$$

Relations (8.72), (8.80), (8.88), (8.82), and (8.90) imply that

$$-(M + \|\bar{x}\| + 3)^2 \leq \|x_{n_0} - \bar{x}\|^2 - \|x_1 - \bar{x}\|^2$$

$$= \sum_{i=1}^{n_0-1} [\|x_{i+1} - \bar{x}\|^2 - \|x_i - \bar{x}\|^2]$$

$$\leq -(n_0 - 1)(64\bar{L})^{-1}\bar{\epsilon}\beta_1 \leq -\beta_1 n_0 / 2(64\bar{L})^{-1}\bar{\epsilon},$$

$$(n_0/2)(64\bar{L})^{-1}\bar{\epsilon}\beta_1 \leq (M + \|\bar{x}\| + 3)^2 \leq (2M + 3)^2.$$

This contradicts (8.77). The contradiction we have reached proves that there exists an integer

$$j \in \{1, \ldots, n_0\}$$

for which

$$f(x_j) \leq \inf(f; C) + \bar{\epsilon}/4. \tag{8.91}$$

By (8.84), (A2), and (8.82), we have

$$d(x_j, C) \leq \delta. \tag{8.92}$$

Relations (8.91), (8.92), (8.79), and (8.74) imply that

$$d(x_j, C_{\min}) \leq \epsilon. \tag{8.93}$$

We claim that for all integers i satisfying $j \leq i \leq n$, we have

$$d(x_i, C_{\min}) \leq \epsilon.$$

Assume the contrary. Then there exists an integer $k \in [j, n]$ for which

$$d(x_k, C_{\min}) > \epsilon. \tag{8.94}$$

It is easy to see that

$$k > j. \tag{8.95}$$

We may assume without loss of generality that

$$d(x_i, C_{\min}) \leq \epsilon \text{ for all integers } i \text{ satisfying } j \leq i < k. \qquad (8.96)$$

Then

$$d(x_{k-1}, C_{\min}) \leq \epsilon. \qquad (8.97)$$

There are two cases:

$$f(x_{k-1}) \leq \inf(f; C) + \bar{\epsilon}/4; \qquad (8.98)$$

$$f(x_{k-1}) > \inf(f; C) + \bar{\epsilon}/4. \qquad (8.99)$$

Assume that (8.98) is valid. In view of (8.98), (8.73), (8.8), and (8.9),

$$x_{k-1} \in X_0 \subset B(0, \bar{K}). \qquad (8.100)$$

By (8.82), (8.84), and (A2), there exists a point $z \in C$ such that

$$\|x_{k-1} - z\| \leq \delta. \qquad (8.101)$$

It follows from (8.82), (8.84), (8.101), and (A2) that

$$\begin{aligned}
\|x_k &- z\| \\
&\leq \delta + \|z - P_C(x_{k-1} - \alpha_{k-1}\|v_{k-1}\|^{-1}v_{k-1} - \alpha_{k-1}\xi_{k-1})\| \\
&\leq \delta + \|z - x_{k-1}\| + \alpha_{k-1} + \delta < 3\delta + \alpha_{k-1}.
\end{aligned} \qquad (8.102)$$

Relations (8.101), (8.98), (8.79), and (8.74) imply that

$$d(x_{k-1}, C_{\min}) \leq \epsilon/4. \qquad (8.103)$$

By (8.101), (8.102), (8.79), (8.81), (8.75), and (8.73),

$$\|x_k - x_{k-1}\| \leq 4\delta + \alpha_{k-1} < \bar{\epsilon} < \epsilon/8. \qquad (8.104)$$

In view of (8.103) and (8.104),

$$d(x_k, C_{\min}) < \epsilon.$$

This inequality contradicts (8.94). The contradiction we have reached proves (8.99).
In view of (8.97), (8.8), and (8.9),

$$\|x_{k-1}\| \leq \bar{K} + 1. \qquad (8.105)$$

It follows from (8.78), (8.79), (8.105), (8.99), and (8.82)–(8.84) that Lemma 8.6 holds with

$$x = x_{k-1}, \; y = x_k, \; \xi = \xi_{k-1}, \; \eta = \eta_{k-1},$$

$$v = v_{k-1}, \; \alpha = \alpha_{k-1}, \; K_0 = \bar{K} + 1$$

$\epsilon = 4^{-1}\bar{\epsilon}$ and combining with (8.81), (8.75), (8.79), and (8.97) this implies that

$$
\begin{aligned}
d(x_k, &C_{\min})^2 \\
&\leq d(x_{k-1}, C_{\min})^2 - \alpha_{k-1}(16\bar{L})^{-1}\bar{\epsilon} + 2\alpha_{k-1}^2 + \delta^2 + 2\delta(\bar{K} + 4) \\
&\leq d(x_{k-1}, C_{\min})^2 - (16\bar{L})^{-1}\alpha_{k-1}\bar{\epsilon} + 2\alpha_{k-1}^2 + 2\delta(\bar{K} + 5) \\
&\leq d(x_{k-1}, C_{\min})^2 - (32\bar{L})^{-1}\alpha_{k-1}\bar{\epsilon} + 2\delta(\bar{K} + 5) \\
&\leq d(x_{k-1}, C_{\min})^2 - (32\bar{L})^{-1}\beta_1\bar{\epsilon} - 2\delta(2\bar{K} + 5) \\
&< d(x_{k-1}, C_{\min})^2 \leq \epsilon^2.
\end{aligned}
$$

This contradicts (8.94). The contradiction we have reached proves that

$$d(x_i, C_{\min}) \leq \epsilon$$

for all integers i satisfying $j \leq i \leq n$. In view of inequality $n_0 \geq j$, Theorem 8.2 is proved.

Chapter 9
Proximal Point Method in Hilbert Spaces

In this chapter we study the convergence of a proximal point method under the presence of computational errors. Most results known in the literature show the convergence of proximal point methods when computational errors are summable. In this chapter the convergence of the method is established for nonsummable computational errors. We show that the proximal point method generates a good approximate solution if the sequence of computational errors is bounded from above by some constant.

9.1 Preliminaries and the Main Results

We analyze the behavior of the proximal point method in a Hilbert space which is an important tool in the optimization theory. See, for example, [15, 16, 31, 34, 36, 53, 55, 69, 77, 81, 87, 103, 104, 106, 107, 111, 113] and the references mentioned therein.

Let X be a Hilbert space equipped with an inner product $\langle \cdot, \cdot \rangle$ which induces the norm $\| \cdot \|$.

For each function $g : X \to R^1 \cup \{\infty\}$ set

$$\inf(g) = \inf\{g(y) : y \in X\}.$$

Suppose that $f : X \to R^1 \cup \{\infty\}$ is a convex lower semicontinuous function and a is a positive constant such that

$$\mathrm{dom}(f) := \{x \in X : f(x) < \infty\} \neq \emptyset,$$

$$f(x) \geq -a \text{ for all } x \in X \tag{9.1}$$

© Springer International Publishing Switzerland 2016
A.J. Zaslavski, *Numerical Optimization with Computational Errors*, Springer Optimization and Its Applications 108, DOI 10.1007/978-3-319-30921-7_9

and that

$$\lim_{\|x\| \to \infty} f(x) = \infty. \tag{9.2}$$

In view of (9.1) and (9.2), the set

$$\mathrm{argmin}(f) := \{z \in X : f(z) = \inf(f)\} \neq \emptyset. \tag{9.3}$$

Let a point

$$x^* \in \mathrm{argmin}(f) \tag{9.4}$$

and let M be any positive number such that

$$M > \inf(f) + 4. \tag{9.5}$$

In view of (9.2), there exists a number $M_0 > 1$ such that

$$f(z) > M + 4 \text{ for all } z \in X \text{ satisfying } \|z\| \geq M_0 - 1. \tag{9.6}$$

Clearly,

$$\|x^*\| < M_0 - 1. \tag{9.7}$$

Assume that

$$0 < \Lambda_1 < \Lambda_2 \leq M_0^{-2}/2. \tag{9.8}$$

The following theorem is the main result of this chapter.

Theorem 9.1. *Let*

$$\lambda_k \in [\Lambda_1, \Lambda_2], \; k = 0, 1, \ldots, \tag{9.9}$$

$\Delta \in (0, 1]$, *a natural number L satisfy*

$$L > 2(4M_0^2 + 1)\Lambda_2 \Delta^{-1} \tag{9.10}$$

and let a positive number ϵ satisfy

$$\epsilon^{1/2}(L + 1)(2\Lambda_1^{-1} + 8M_0\Lambda_1^{-1/2}) \leq 1 \text{ and } \epsilon(L + 1) \leq \Delta/4. \tag{9.11}$$

Assume that a sequence $\{x_k\}_{k=0}^{\infty} \subset X$ satisfies

$$f(x_0) \leq M \tag{9.12}$$

and

$$f(x_{k+1}) + 2^{-1}\lambda_k \|x_{k+1} - x_k\|^2 \le \inf(f + 2^{-1}\lambda_k \| \cdot -x_k\|^2) + \epsilon \qquad (9.13)$$

for all integers $k \ge 0$. Then for all integers $k > L$,

$$f(x_k) \le \inf(f) + \Delta.$$

By Theorem 9.1, for a given $\Delta > 0$, we obtain $\xi \in X$ satisfying

$$f(\xi) \le \inf(f) + \Delta$$

doing $\lfloor c_1 \Delta^{-1} \rfloor$ iterations [see (9.10)] with the computational error $\epsilon = c_2 \Delta^2$ [see (9.11)], where the constant $c_1 > 0$ depends only on M_0, Λ_2 and the constant $c_2 > 0$ depends only on $M_0, L, \Lambda_1, \Lambda_2$.

Theorem 9.1 implies the following result.

Theorem 9.2. *Let*

$$\lambda_k \in [\Lambda_1, \Lambda_2], \ k = 0, 1, \dots,$$

a natural number L satisfy

$$L > (4M_0^2 + 1)2\Lambda_2 \qquad (9.14)$$

and let a positive number $\bar{\epsilon}$ satisfy

$$\bar{\epsilon}^{1/2}(L+1)(2\Lambda_1^{-1} + 8M_0\Lambda_1^{-1/2}) \le 1 \text{ and } \bar{\epsilon}(L+1) \le 1/4. \qquad (9.15)$$

Assume that

$$\{\epsilon_i\}_{i=0}^{\infty} \subset (0, \bar{\epsilon}), \ \lim_{i \to \infty} \epsilon_i = 0 \qquad (9.16)$$

and that $\gamma > 0$. Then there exists a natural number T_0 such that for each sequence $\{x_k\}_{k=0}^{\infty} \subset X$ satisfying

$$f(x_0) \le M \qquad (9.17)$$

and

$$f(x_{k+1}) + 2^{-1}\lambda_k \|x_{k+1} - x_k\|^2 \le \inf(f + 2^{-1}\lambda_k \| \cdot -x_k\|^2) + \epsilon_k \qquad (9.18)$$

for all integers $k \ge 0$, the inequality

$$f(x_k) \le \inf(f) + \gamma$$

holds for all integers $k > T_0$.

Since the function f is convex and lower semicontinuous and satisfies (9.2), Theorem 9.2 easily implies the following result.

Corollary 9.3. *Suppose that all the assumptions of Theorem 9.2 hold and that the sequence $\{x_k\}_{k=0}^{\infty} \subset X$ satisfies (9.17) and (9.18) for all integers $k \geq 0$. Then $\lim_{k \to \infty} f(x_k) = \inf(f)$ and the sequence $\{x_k\}_{k=0}^{\infty}$ is bounded. Moreover, it possesses a weakly convergent subsequence and the limit of any weakly convergent subsequence of $\{x_k\}_{k=0}^{\infty}$ is a minimizer of f.*

Problem (P) is called well posed if the function f possesses a unique minimizer which is a limit in the norm topology of any minimizing sequence of f (see [60, 121] and the references mentioned therein).

Corollary 9.3 easily implies the following result.

Corollary 9.4. *Suppose that problem (P) is well posed, all the assumptions of Theorem 9.2 hold and that the sequence $\{x_k\}_{k=0}^{\infty} \subset X$ satisfies (9.17) and (9.18) for all integers $k \geq 0$. Then $\{x_k\}_{k=0}^{\infty}$ converges in the norm topology to a unique minimizer of f.*

Note that in [60] it was shown that most problems of type (P) (in the sense of Baire category) are well posed.

The results of the chapter were obtained in [120]. The chapter is organized as follows. Section 9.2 contains auxiliary results. Theorem 9.1 is proved in Sect. 9.3 and Theorem 9.2 is proved in Sect. 9.4.

9.2   Auxiliary Results

We use the notation and definitions introduced in Sect. 9.1 and suppose that all the assumptions made in the introduction holds.

Lemma 9.5. *Assume that*

$$\lambda_k \in [\Lambda_1, \Lambda_2], \ k = 0, 1, \dots \tag{9.19}$$

and that a sequence $\{x_k\}_{k=0}^{\infty}$ satisfies

$$f(x_0) \leq M, \tag{9.20}$$

$$f(x_{k+1}) + 2^{-1}\lambda_k\|x_{k+1} - x_k\|^2 \leq \inf(f + 2^{-1}\lambda_k\| \cdot -x_k\|^2) + 1 \tag{9.21}$$

for all integers $k \geq 0$. Then $\|x_k\| \leq M_0$ for all integers $k \geq 0$.

Proof. Relations (9.20) and (9.6) imply that

$$\|x_0\| \leq M_0.$$

Assume that an integer $k \geq 0$ and that

$$\|x_k\| \leq M_0. \tag{9.22}$$

It follows from (9.21), (9.19), (9.7), (9.8), (9.3), (9.4), (9.5), and (9.22) that

$$
\begin{aligned}
f(x_{k+1}) &\leq f(x^*) + 2^{-1}\lambda_k\|x^* - x_k\|^2 + 1 \\
&\leq f(x^*) + 2^{-1}\Lambda_2(2M_0)^2 + 1 \\
&\leq f(x^*) + 2 = \inf(f) + 2 < M.
\end{aligned}
$$

Together with (9.6) the inequality above implies that $\|x_{k+1}\| \leq M_0$. Thus we showed by induction that (9.22) holds for all integers $k \geq 0$. This completes the proof of Lemma 9.5.

Lemma 9.6. *Assume that*

$$\lambda_k \in [\Lambda_1, \Lambda_2], \ k = 0, 1, \ldots, \tag{9.23}$$

$\epsilon_k \in (0, 1], \ k = 0, 1, \ldots,$ *a sequence* $\{x_k\}_{k=0}^{\infty} \subset X$ *satisfies*

$$f(x_0) \leq M \tag{9.24}$$

and that for all integers $k \geq 0$,

$$f(x_{k+1}) + 2^{-1}\lambda_k\|x_{k+1} - x_k\|^2 \leq \inf(f + 2^{-1}\lambda_k\| \cdot -x_k\|^2) + \epsilon_k. \tag{9.25}$$

Then the following assertions hold.

1. For every integer $k \geq 0$,

$$
\begin{aligned}
&(2/\lambda_k)(f(x_{k+1}) - f(x^*)) + \|x_{k+1} - x_k\|^2 \\
&\leq 2\epsilon_k\Lambda_1^{-1} + \|x_k - x^*\|^2 - \|x_{k+1} - x^*\|^2 + 8M_0(\epsilon_k\Lambda_1^{-1})^{1/2}.
\end{aligned}
$$

2. For every pair of natural numbers $m > n$,

$$
\begin{aligned}
&\sum_{i=n}^{m} 2\Lambda_2^{-1}(f(x_i) - f(x^*)) + \sum_{i=n}^{m} \|x_{i-1} - x_i\|^2 \\
&\leq 4M_0^2 + \sum_{i=n-1}^{m-1} [2\Lambda_1^{-1}\epsilon_i + 8M_0(\epsilon_i\Lambda_1^{-1})^{1/2}].
\end{aligned}
$$

Proof. It follows from (9.24), (9.25), and Lemma 9.5 that

$$\|x_k\| \leq M_0 \text{ for all integers } k \geq 0. \tag{9.26}$$

In view of (9.25), for every integer $k \geq 0$,

$$f(x_{k+1}) \leq f(x_k) + \epsilon_k. \tag{9.27}$$

We will prove Assertion 1. Let $k \geq 0$ be an integer. There exists a point $y_{k+1} \in X$ such that

$$f(y_{k+1}) + 2^{-1}\lambda_k\|y_{k+1} - x_k\|^2 \leq f(x) + 2^{-1}\lambda_k\|x - x_k\|^2 \text{ for all } x \in X. \tag{9.28}$$

We estimate $\|x_{k+1} - y_{k+1}\|$. Set

$$z = 2^{-1}(x_{k+1} + y_{k+1}). \tag{9.29}$$

It is easy to see that

$$
\begin{aligned}
&2^{-1}\|y_{k+1} - x_k\|^2 + 2^{-1}\|x_{k+1} - x_k\|^2 - \|2^{-1}(x_{k+1} + y_{k+1}) - x_k\|^2 \\
&= 2^{-1}\|y_{k+1}\|^2 + 2^{-1}\|x_k\|^2 - \langle y_{k+1}, x_k \rangle + 2^{-1}\|x_{k+1}\|^2 + 2^{-1}\|x_k\|^2 \\
&\quad -\langle x_{k+1}, x_k \rangle - \|x_k\|^2 + \langle x_k, x_{k+1} + y_{k+1} \rangle - \|2^{-1}(y_{k+1} + x_{k+1})\|^2 \\
&= 2^{-1}\|y_{k+1}\|^2 + 2^{-1}\|x_{k+1}\|^2 - \|2^{-1}(y_{k+1} + x_{k+1})\|^2 \\
&= \|2^{-1}(y_{k+1} - x_{k+1})\|^2.
\end{aligned} \tag{9.30}
$$

In view of (9.29), convexity of the function f, (9.30), (9.28), and (9.25),

$$
\begin{aligned}
&f(z) + 2^{-1}\lambda_k\|z - x_k\|^2 \\
&\leq 2^{-1}f(x_{k+1}) + 2^{-1}f(y_{k+1}) \\
&\quad +2^{-1}\lambda_k(2^{-1}\|y_{k+1} - x_k\|^2 + 2^{-1}\|x_{k+1} - x_k\|^2 - \|2^{-1}(y_{k+1} - x_{k+1})\|^2) \\
&\leq \inf\{f(x) + 2^{-1}\lambda_k\|x - x_k\| : x \in X\} \\
&\quad +2^{-1}\epsilon_k - 2^{-1}\lambda_k\|2^{-1}(y_{k+1} - x_{k+1})\|^2.
\end{aligned}
$$

Combining with (9.19) the inequality above implies that

$$\|2^{-1}(y_{k+1} - x_{k+1})\|^2 \leq \epsilon_k\Lambda_1^{-1}$$

and that

$$\|y_{k+1} - x_{k+1}\| \leq 2(\epsilon_k\Lambda_1^{-1})^{1/2}. \tag{9.31}$$

Now we estimate $f(x^*) - f(x_{k+1})$. In view of (9.28),

$$0 \in \partial f(y_{k+1}) + \lambda_k(y_{k+1} - x_k)$$

and for every point $u \in X$,

$$f(u) - f(y_{k+1}) \geq \lambda_k \langle x_k - y_{k+1}, u - y_{k+1} \rangle. \tag{9.32}$$

By (9.32), we have

$$f(x^*) - f(y_{k+1}) \geq \lambda_k \langle x_k - y_{k+1}, x^* - y_{k+1} \rangle. \tag{9.33}$$

Relation (9.25) implies that

$$f(x_{k+1}) + 2^{-1}\lambda_k \|x_{k+1} - x_k\|^2 \leq f(y_{k+1}) + 2^{-1}\lambda_k \|y_{k+1} - x_k\|^2 + \epsilon_k$$

and

$$f(y_{k+1}) - f(x_{k+1}) \geq 2^{-1}\lambda_k(\|x_{k+1} - x_k\|^2 - \|y_{k+1} - x_k\|^2) - \epsilon_k.$$

Together with (9.33) the relation above implies that

$$
\begin{aligned}
f(x^*) &- f(x_{k+1}) \\
&= f(x^*) - f(y_{k+1}) + f(y_{k+1}) - f(x_{k+1}) \\
&\geq \lambda_k \langle x_k - y_{k+1}, x^* - y_{k+1} \rangle + f(y_{k+1}) - f(x_{k+1}) \\
&= 2^{-1}\lambda_k[\|y_{k+1} - x^*\|^2 - \|x_k - x^*\|^2 + \|x_k - y_{k+1}\|^2] + f(y_{k+1}) - f(x_{k+1}) \\
&\geq 2^{-1}\lambda_k[\|y_{k+1} - x^*\|^2 - \|x_k - x^*\|^2 + \|x_k - x_{k+1}\|^2] - \epsilon_k. \tag{9.34}
\end{aligned}
$$

It follows from (9.28), (9.23), (9.26), (9.7), (9.8), and (9.5) that for all integers $q \geq 1$,

$$f(y_q) \leq f(x^*) + 2^{-1}\Lambda_2 \|x^* - x_q\|^2 \leq f(x^*) + 1 < M.$$

Combined with (9.6) the inequality above implies that

$$\|y_q\| \leq M_0, \quad q = 1, 2, \ldots \tag{9.35}$$

Now we use (9.34) and (9.35) and obtain an estimation of $f(x^*) - f(x_{k+1})$ without terms which contain y_{k+1}.

In view of (9.26) and (9.31),

$$
\begin{aligned}
\|x_k - y_{k+1}\|^2 &= \|(x_k - x_{k+1}) - (y_{k+1} - x_{k+1})\|^2 \\
&= \|x_k - x_{k+1}\|^2 + \|y_{k+1} - x_{k+1}\|^2 - 2\langle x_k - x_{k+1}, y_{k+1} - x_{k+1} \rangle \\
&\geq \|x_k - x_{k+1}\|^2 - 2\|x_k - x_{k+1}\|\|y_{k+1} - x_{k+1}\| \\
&\geq \|x_k - x_{k+1}\|^2 - 8M_0(\epsilon_k \Lambda_1^{-1})^{1/2}. \tag{9.36}
\end{aligned}
$$

By (9.26), (9.7), and (9.31), we have

$$\|y_{k+1} - x^*\|^2 = \|(x_{k+1} - x^*) + (y_{k+1} - x_{k+1})\|^2$$
$$= \|x_{k+1} - x^*\|^2 + \|y_{k+1} - x_{k+1}\|^2 + 2\langle x_{k+1} - x^*, y_{k+1} - x_{k+1}\rangle$$
$$\geq \|x_{k+1} - x^*\|^2 - 8M_0(\epsilon_k \Lambda_1^{-1})^{1/2}. \tag{9.37}$$

Relations (9.34) and (9.37) imply that

$$f(x^*) - f(x_{k+1})$$
$$\geq -\epsilon_k + 2^{-1}\lambda_k[\|x_{k+1} - x^*\|^2 - \|x^* - x_k\|^2 + \|x_k - x_{k+1}\|^2 - 8M_0(\epsilon_k \Lambda_1^{-1})^{1/2}],$$
$$f(x_{k+1}) - f(x^*) + 2^{-1}\lambda_k\|x_k - x_{k+1}\|^2$$
$$\leq \epsilon_k + 2^{-1}\lambda_k[\|x_k - x^*\|^2 - \|x_{k+1} - x^*\|^2] + 2^{-1}\lambda_k 8M_0(\epsilon_k \Lambda_1^{-1})^{1/2}.$$

and by (9.23),

$$(2/\lambda_k)(f(x_{k+1}) - f(x^*)) + \|x_k - x_{k+1}\|^2$$
$$\leq 2\epsilon_k \Lambda_1^{-1} + \|x_k - x^*\|^2 - \|x_{k+1} - x^*\|^2 + 8M_0(\epsilon_k \Lambda_1^{-1})^{1/2}.$$

Thus Assertion 1 is proved.

Let us prove Assertion 2. It follows from Assertion 1, (9.23), (9.7), and (9.26) that for all pairs of natural numbers $m > n$,

$$\sum_{i=n}^{m}(2/\Lambda_2)(f(x_i) - f(x^*)) + \sum_{i=n}^{m}\|x_{i-1} - x_i\|^2$$
$$\leq \|x_{n-1} - x^*\|^2 + \sum_{i=n-1}^{m-1}[2\epsilon_i \Lambda_1^{-1} + 8M_0(\epsilon_i \Lambda_1^{-1})^{1/2}]$$
$$\leq 4M_0^2 + \sum_{i=n-1}^{m-1}[2\Lambda_1^{-1}\epsilon_i + 8M_0(\epsilon_i \Lambda_1^{-1})^{1/2}].$$

Assertion 2 is proved. This completes the proof of Lemma 9.6.

9.3 Proof of Theorem 9.1

It follows from (9.9), (9.10), (9.11), (9.12), (9.13), and Lemma 9.6 applied for a natural number n and $m = n + L$ that

$$\sum_{i=n}^{n+L}(2/\Lambda_2)(f(x_i)-f(x^*))$$

$$\leq 4M_0^2 + \sum_{i=n-1}^{n-1+L}[2\Lambda_1^{-1}\epsilon + 8M_0(\epsilon\Lambda_1^{-1})^{1/2}]$$

$$\leq 4M_0^2 + (L+1)\epsilon^{1/2}[2\Lambda_1^{-1} + 8M_0\Lambda_1^{-1/2}] \leq 4M_0^2 + 1. \tag{9.38}$$

Let $n \geq 1$ be an integer. In view of (9.38),

$$(L+1)2\Lambda_2^{-1}\min\{f(x_i)-f(x^*): i = n,\ldots,n+L\} \leq 4M_0^2 + 1$$

and by (9.10),

$$\min\{f(x_i)-f(x^*): i = n,\ldots,n+L\}$$

$$\leq (4M_0^2 + 1)(L+1)^{-1}2^{-1}\Lambda_2 < \Delta/4. \tag{9.39}$$

Since (9.39) holds for any natural number n there exists a strictly increasing sequence of natural numbers $\{S_i\}_{i=1}^{\infty}$ such that

$$S_1 \in \{1,\ldots,1+L\}, \ S_{i+1} - S_i \in [1,1+L], \ i = 1,2,\ldots, \tag{9.40}$$

$$f(x_{S_i})-f(x^*) \leq \Delta/4, \ i = 1,2,\ldots. \tag{9.41}$$

Let an integer $j \geq L+1$. In view of (9.40), there is an integer $i \geq 1$ such that

$$S_i \leq j < S_{i+1}$$

and

$$j - S_i \leq L+1. \tag{9.42}$$

By (9.13) for every integer $k \geq 0$, we have

$$f(x_{k+1}) \leq f(x_k) + \epsilon.$$

Combined with (9.42), (9.41) and (9.11) the inequality above implies that

$$f(x_j) \leq f(x_{S_i}) + \epsilon(L+1) \leq f(x^*) + \Delta/4 + \Delta/4 \leq f(x^*) + \Delta.$$

Theorem 9.1 is proved.

9.4   Proof of Theorem 9.2

By Theorem 9.1 the following property holds:

(P1) Let a sequence $\{x_k\}_{k=0}^{\infty} \subset X$ satisfy

$$f(x_0) \leq M,$$
$$f(x_{k+1}) + 2^{-1}\lambda_k \|x_{k+1} - x_k\|^2 \leq \inf(f + 2^{-1}\lambda_k \|\cdot - x_k\|^2) + \bar{\epsilon}, \ k = 0, 1, \ldots.$$

Then

$$f(x_k) \leq \inf(f) + 1 \text{ for all integers } k > L.$$

By Theorem 9.1 there exist $\delta \in (0, \bar{\epsilon})$ and an integer $L_0 \geq 1$ such that the following property holds:

(P2) For every sequence $\{y_i\}_{i=0}^{\infty} \subset X$ which satisfies

$$f(y_0) \leq \inf(f) + 1,$$
$$f(y_{k+1}) + 2^{-1}\lambda_k \|y_{k+1} - y_k\|^2 \leq \inf(f + 2^{-1}\lambda_k \|\cdot - y_k\|^2) + \delta$$

for all integers $k \geq 0$ we have

$$f(y_k) \leq \inf(f) + \gamma \text{ for all integers } k \geq L_0. \tag{9.43}$$

(Here γ is as in the statement of the theorem.)

In view of (9.16), there exists an integer $L_1 \geq 1$ such that

$$\epsilon_k < \delta \text{ for all natural numbers } k \geq L_1. \tag{9.44}$$

Fix a natural number

$$T_0 > L_0 + L_1 + L. \tag{9.45}$$

Assume that a sequence $\{x_i\}_{i=0}^{\infty} \subset X$ satisfies (9.17) and (9.18). It follows from property (P1), (9.17), (9.18), and (9.16) that

$$f(x_k) \leq \inf(f) + 1 \text{ for all integers } k > L. \tag{9.46}$$

For every nonnegative integer k set

$$y_k = x_{k+L+L_1}. \tag{9.47}$$

It follows from (9.47) and (9.46) that

$$f(y_0) \leq \inf(f) + 1. \tag{9.48}$$

By (9.18), (9.47), and (9.44), for all nonnegative integers k,

$$f(y_{k+1}) + 2^{-1}\lambda_k \|y_{k+1} - y_k\|^2 \leq \inf(f + 2^{-1}\lambda_k \| \cdot -y_k\|^2) + \delta. \tag{9.49}$$

In view of (9.48), (9.49), and property (P2),

$$f(y_k) \leq \inf(f) + \gamma \text{ for all integers } k \geq L_0. \tag{9.50}$$

Combined with (9.47) and (9.45) the inequality above implies that

$$f(x_k) \leq \inf(f) + \gamma \text{ for all integers } k > T_0.$$

This completes the proof of Theorem 9.2.

Chapter 10
Proximal Point Methods in Metric Spaces

In this chapter we study the local convergence of a proximal point method in a metric space under the presence of computational errors. We show that the proximal point method generates a good approximate solution if the sequence of computational errors is bounded from above by some constant. The principle assumption is a local error bound condition which relates the growth of an objective function to the distance to the set of minimizers, introduced by Hager and Zhang [55].

10.1 Preliminaries and the Main Results

Let X be a metric space equipped with a metric d.

For each $x \in X$ and each $r > 0$ set

$$B(x, r) = \{y \in X : d(x, y) \le r\}.$$

For each $x \in X$ and each nonempty set $A \subset X$ set

$$D(x, A) = \inf\{d(x, y) : y \in A\}.$$

For each $g : X \to R^1 \cup \{\infty\}$ put

$$\inf(g) = \inf\{g(x) : x \in X\}.$$

Let $f : X \to R^1 \cup \{\infty\}$ be a lower semicontinuous bounded from below function which is not identically infinity.

In this chapter we continue our study of proximal point methods which was began in the previous chapter. Literature connected with the analysis and development of proximal point methods and based on tools and methods of convex and variational

© Springer International Publishing Switzerland 2016
A.J. Zaslavski, *Numerical Optimization with Computational Errors*, Springer
Optimization and Its Applications 108, DOI 10.1007/978-3-319-30921-7_10

analysis includes [15, 16, 31, 34, 35, 42, 55, 56, 66, 67, 69, 77, 81, 87, 103, 104, 111, 113]. In the proximal point method iterates x_k, $k \geq 0$ are generated by the following rule:

$$x_{k+1} \in \operatorname{argmin}\{f(x) + 2^{-1}\mu_k d(x, x_k)^2\},$$

where $\{\mu_k\}_{k=0}^{\infty}$ is a sequence of positive numbers and $x_0 \in X$ is an initial point.

Most results, known in the literature, establish the convergence of proximal point methods when X is a Hilbert space and the function f is convex. Convergence of proximal point methods for a nonconvex objective function f was established in [55, 56]. In [56] convergence results were obtained for finite dimensional optimization problems. In a Hilbert space setting, convergence results for a nonconvex objective function were established in [55]. The principle assumption of [55, 56] is a local error bound condition [see (10.4)] which relates the growth of an objective function to the distance to the set of minimizers. Convergence results in Banach spaces were obtained in [7, 40, 61, 62, 100]. Variable-metric methods were used in order to obtain convergence results in [4, 26, 78].

Let $\bar{x} \in X$ satisfy

$$f(\bar{x}) = \inf(f). \tag{10.1}$$

Set

$$\Omega = \{x \in X : f(x) = \inf(f)\} \tag{10.2}$$

and let a set $\Omega_0 \subset X$ satisfy

$$\bar{x} \in \Omega_0 \subset \Omega \cap B(\bar{x}, \rho_*) \tag{10.3}$$

with a positive constant ρ_*.

Suppose that $\alpha > 0$, $\rho_0 > \rho_*$ and that

$$f(x) - f(\bar{x}) \geq \alpha D(x, \Omega_0)^2 \text{ for all } x \in B(\bar{x}, \rho_0). \tag{10.4}$$

Let

$$\rho \in (0, \rho_0), \quad \beta_0 \in (0, 2\alpha/3), \quad \beta_1 \in (0, \beta_0), \tag{10.5}$$

$$\gamma = 2\beta_0(2\alpha - \beta_0)^{-1}. \tag{10.6}$$

By (10.5) and (10.6),

$$\gamma < 1. \tag{10.7}$$

For each number Δ which satisfies

$$\Delta \in (0,1), \quad \Delta < \rho(1 + (1-\gamma)^{-1})^{-1}, \quad \Delta < (\rho_0 - \rho_*)/3, \tag{10.8}$$

choose a natural number $k_0(\Delta)$ which satisfies

$$\gamma^{k_0(\Delta)}\rho < \Delta/8 \tag{10.9}$$

and a positive number $\epsilon(\Delta)$ such that

$$2(k_0(\Delta))^2(2\epsilon(\Delta)\beta_1^{-1})^{1/2} + 4(k_0(\Delta))^2\epsilon(\Delta)\Delta^{-1}(2\alpha - \beta_0)^{-1} < \rho_0 - \rho, \tag{10.10}$$

$$(2\epsilon(\Delta)\beta_1^{-1})^{1/2} < 16^{-1}(k_0(\Delta))^{-1}\Delta(1-\gamma), \tag{10.11}$$

$$\epsilon(\Delta) < (1-\gamma)32^{-1}(k_0(\Delta))^{-1}\Delta^2(2\alpha - \beta_0). \tag{10.12}$$

The following theorem obtained in [124] is the main result of this chapter.

Theorem 10.1. *Let a number Δ satisfy (10.8), $k_0 = k_0(\Delta)$ and $\epsilon = \epsilon(\Delta)$. Assume that*

$$\{\mu_k\}_{k=0}^{\infty} \subset [\beta_1, \beta_0], \tag{10.13}$$

a sequence $\{x_k\}_{k=0}^{\infty} \subset X$ satisfies

$$d(x_0, \bar{x})(1 + (1-\gamma)^{-1}) < \rho \tag{10.14}$$

and that for all integers $k \geq 0$,

$$f(x_{k+1}) + 2^{-1}\mu_k d(x_{k+1}, x_k)^2$$
$$\leq f(z) + 2^{-1}\mu_k d(z, x_k)^2 + \epsilon \text{ for all } z \in X. \tag{10.15}$$

Then

$$x_j \in B(\bar{x}, \rho_0) \text{ for all integers } j \geq 0,$$

$$D(x_j, \Omega_0) \leq \Delta \text{ for all integers } j \geq k_0.$$

It is easy to see that $k_0(\cdot)$ in (10.9) and $\epsilon(\cdot)$ in (10.10)–(10.12) depend on Δ monotonically. Namely, $k_0(\cdot)$ ($\epsilon(\cdot)$, respectively) is an decreasing (increasing, respectively) function of Δ. Clearly, $k_0(\cdot)$ and $\epsilon(\cdot)$ also depend on the parameters γ and ρ. The influence of the choice of these parameters on the convergence is not simple. For example, according to (10.14) it is desirable to choose ρ which is sufficiently close to ρ_0. But in view of (10.10), if ρ is close to ρ_0, the value $\epsilon(\cdot)$ becomes very small. We can choose a small parameter β_0 [see (10.5)] and in this case in view of (10.5) and (10.6) γ and β_1 are small, by (10.9) the number of iterations $k_0(\Delta)$ is not large, but in view of (10.11) $\epsilon(\Delta)$ is small.

The following theorem is the second main result of this chapter.

Theorem 10.2. *There exists $\bar{\epsilon} > 0$ such that for each sequence $\{\epsilon_i\}_{i=0}^{\infty} \subset (0, \bar{\epsilon}]$ satisfying*

$$\lim_{i \to \infty} \epsilon_i = 0 \qquad\qquad (10.16)$$

the following assertion holds.

Let $\Delta > 0$. Then there exists a natural number k_1 such that for each sequence $\{\mu_k\}_{k=0}^{\infty} \subset [\beta_1, \beta_0]$ and for each sequence $\{x_k\}_{k=0}^{\infty} \subset X$ which satisfies

$$d(x_0, \bar{x})(1 + (1 - \gamma)^{-1})^{-1} < \rho$$

and for all integers $k \geq 0$

$$f(x_{k+1}) + 2^{-1} \mu_k d(x_{k+1}, x_k)^2$$
$$\leq f(z) + 2^{-1} \mu_k d(z, x_k)^2 + \epsilon_k \text{ for all } z \in X$$

the following relations hold:

$$x_j \in B(\bar{x}, \rho_0) \text{ for all integers } j \geq 0,$$

$$D(x_j, \Omega_0) \leq \Delta \text{ for all integers } j \geq k_1.$$

Theorem 10.2 establishes the convergence of the proximal algorithm in the presence of computational errors $\{\epsilon_i\}_{i=0}^{\infty}$ such that $\lim_{i \to \infty} \epsilon_i = 0$ without assuming their summability.

The local error bound condition (10.4) with parameters α, ρ_0, ρ_*, introduced in [55, 56], hold for many functions and can be verified in principle. In the three examples below we show that the parameters α, ρ_0, ρ_* can be calculated by investigating the function f in some cases. On the hand these parameters can be obtained as a result of numerical experiments.

Example 10.3. Let $(X, \langle \cdot, \cdot \rangle)$ be a Hilbert space with an inner product $\langle \cdot, \cdot \rangle$ which induces a complete norm $\| \cdot \|$. Assume that $f \in C^2$ is a real-valued convex function on X such that

$$\lim_{\|x\| \to \infty} f(x) = \infty$$

and

$$\inf\{\langle f''(x)h, h \rangle \|h\|^{-2} : x \in X \text{ and } h \in X \setminus \{0\}\} > 0,$$

where $f''(x)$ is the second order Frechet derivative of f at a point x. It is not difficult to see that the function f possesses a unique minimizer $\bar{x}$ and (10.4) holds with $\rho_* = 1$, $\Omega = \Omega_0 = \{\bar{x}\}$, $\rho_0 > 1$ and

$$\alpha = 2^{-1} \inf\{\langle f''(x)h, h \rangle \|h\|^{-2} : x \in X \text{ and } h \in X \setminus \{0\}\} > 0.$$

Example 10.4. Let $(X, \langle \cdot, \cdot \rangle)$ be a Hilbert space with an inner product $\langle \cdot, \cdot \rangle$ which induces a complete norm $\| \cdot \|$. Assume that $f : X \to R^1 \cup \{\infty\}$ is a lower semicontinuous function, $0 < a < b$, the restriction of f to the set $\{z \in X : \|z\| < b\}$ is a convex function which possesses a continuous second order Frechet derivative $f''(\cdot)$,

$$\inf\{\langle f''(x)h, h \rangle \|h\|^{-2} : x \in X, \|x\| < b \text{ and } h \in X \setminus \{0\}\} > 0$$

and that

$$f(z) > \inf\{f(x) : x \in X\} \text{ for all } z \in X \setminus B(0, a).$$

It is easy to see that there exists a unique minimizer $\bar{x}$ of f, $\|\bar{x}\| \leq a$ and that (10.4) holds with $\Omega = \Omega_0 = \{\bar{x}\}$, $\rho_0 = (b - a)/2$, any positive constant $\rho_* < \rho_0$ and

$$\alpha = 2^{-1} \inf\{< f''(x)h, h > \|h\|^{-2} : x \in X, \|x\| < b \text{ and } h \in X \setminus \{0\}\}.$$

Example 10.5. Assume that (X, d) is a metric space, $f : X \to R^1 \cup \{\infty\}$ is a lower semicontinuous function which is not identically infinity, $\bar{x} \in X$ satisfies (10.1), equations (10.2) and (10.3) hold with a positive constant ρ_* and (10.4) holds with constants $\alpha > 0$ and $\rho_0 > \rho_*$.

Assume that $T : X \to X$ satisfies

$$c_1 d(x, y) \leq d(T(x), T(y)) \leq c_2 d(x, y) \text{ for all } x, y \in X, \tag{10.17}$$

where c_1, c_2 are positive constants such that

$$c_2 c_1^{-1} \rho_* < \rho_0. \tag{10.18}$$

Set

$$g(x) = f(T(x)), \ x \in X. \tag{10.19}$$

It is easy to see that

$$g(T^{-1}(\bar{x})) = \inf(g) = \inf(f), \ T^{-1}(\Omega) = \{z \in X : g(z) = \inf(g)\}. \tag{10.20}$$

Let

$$x \in B(T^{-1}(\bar{x}), c_2^{-1}\rho_0). \tag{10.21}$$

By (10.17) and (10.21)

$$d(T(x), \bar{x}) \leq c_2 d(x, T^{-1}(\bar{x})) \leq \rho_0. \tag{10.22}$$

It follows from (10.4), (10.17), (10.19) and (10.22) that

$$
\begin{aligned}
g(x) &- g(T^{-1}(\bar{x})) \\
&= f(T(x)) - f(\bar{x}) \geq \alpha D(T(x), \Omega_0)^2 \\
&= \alpha \inf\{d(T(x), z) : z \in \Omega_0\}^2 \\
&\geq \alpha \inf\{c_1 d(x, T^{-1}(z)) : z \in \Omega_0\}^2 \\
&\geq \alpha c_1^2 \inf\{d(x, v) : v \in T^{-1}(\Omega_0)\}^2.
\end{aligned}
$$

Thus

$$g(x) - g(T^{-1}(\bar{x})) \geq \alpha c_1^2 D(x, T^{-1}(\Omega_0))^2$$

$$\text{for all } x \in B(T^{-1}(\bar{x}), c_2^{-1}\rho_0).$$

By (10.3), (10.17), and (10.18) for each $z \in T^{-1}(\Omega_0)$,

$$d(z, T^{-1}(\bar{x})) \leq c_1^{-1} d(T(z), \bar{x}) \leq c_1^{-1}\rho_*,$$

$$T^{-1}(\Omega_0) \subset T^{-1}(\Omega) \cap B(T^{-1}(\bar{x}), c_1^{-1}\rho_*).$$

Note that by (10.18), $c_1^{-1}\rho_* < c_2^{-1}\rho_0$. Thus the local error bound condition also holds for the function g.

Example 10.6. Consider the following constrained minimization problem

$$\int_0^{2\pi} |x(t) - \sin(t)|dt \to \min,$$

$x : [0, 2\pi] \to R^1$ is an absolutely continuous (a. c.) function such that

$$x(0) = x(2\pi) = 0, \ |x'(t)| \leq 1, \ t \in [0, 2\pi] \text{ almost everywhere (a.e.).} \tag{10.23}$$

Clearly, this problem possesses a unique solution $\bar{x}(t) = \sin(t)$, $t \in [0, 2\pi]$. Let us show that this constrained problem is a particular case of the problem considered in this session.

Denote by X the set of all a.c. functions $x : [0, 2\pi] \to R^1$ such that (10.23) holds. For all $x_1, x_2 \in X$ set

$$d(x_1, x_2) = \max\{|x_1(t) - x_2(t)| : t \in [0, 2\pi]\}.$$

Clearly, (X, d) is a metric space. For $x \in X$ put

$$f(x) = \int_0^{2\pi} |x(t) - \sin(t)| dt. \qquad (10.24)$$

Clearly, the functional $f : X \to R^1$ is continuous.

Let

$$x \in X \setminus \{\bar{x}\}, \ d(x, \bar{x}) \le 1. \qquad (10.25)$$

By (10.23),

$$|x(t)| \le 2\pi, \ t \in [0, 2\pi]. \qquad (10.26)$$

Clearly, there is $t_0 \in [0, 2\pi]$ such that

$$|x(t_0) - \sin(t_0)| = d(x, \bar{x}). \qquad (10.27)$$

By (10.23) and (10.27) for each $t \in [0, 2\pi]$ satisfying $|t_0 - t| \le d(x, \bar{x})/4$,

$$|x(t) - x(t_0)| \le |t - t_0| \le d(x, \bar{x})/4,$$

$$|\bar{x}(t) - \bar{x}(t_0)| \le |t - t_0| \le d(x, \bar{x})/4$$

and

$$|x(t) - \bar{x}(t)| \ge |x(t_0) - \bar{x}(t_0)| - |x(t) - x(t_0)| - |\bar{x}(t) - \bar{x}(t_0)|$$
$$\ge d(x, \bar{x}) - d(x, \bar{x})/2.$$

Together with (10.24) and (10.25) this implies that

$$f(x) = \int_0^{2\pi} |x(t) - \bar{x}(t)| dt \ge (d(x, \bar{x})/2) d(x, \bar{x})/4,$$

$$f(x) - f(\bar{x}) \ge 8^{-1} d(x, \bar{x})^2$$

and (10.4) holds with $\rho_0 = 1$, $\alpha = 8^{-1}$, $\rho_* = 1/2$, $\Omega = \Omega_0 = \{\bar{x}\}$.

The results of the chapter were obtained in [124]. The chapter is organized as follows. Section 10.2 contains auxiliary results. Section 10.3 contains the main lemma. Theorem 10.1 is proved in Sect. 10.4. Section 10.5 contains an auxiliary result for Theorem 10.2 which is proved in Sect. 10.6. In Sect. 10.7 we obtain extensions of the main results of the chapter for well-posed minimization problems. In Sect. 10.8 we construct an example of a function f for which under the conditions of Theorem 10.2 the sequence $\{x_k\}_{k=0}^{\infty}$ does not converge to a point.

10.2 Auxiliary Results

Lemma 10.7. *Let $\epsilon > 0$, $z_0, z_1 \in X$ and $\mu \in [\beta_1, \beta_0]$ satisfy*

$$f(z_1) + 2^{-1}\mu d(z_1, z_0)^2 \leq f(z) + 2^{-1}\mu d(z, z_0)^2 + \epsilon \text{ for all } z \in X \qquad (10.28)$$

and let

$$x \in \Omega. \qquad (10.29)$$

Then

$$d(z_1, z_0) \leq d(x, z_0) + (2\epsilon\mu^{-1})^{1/2}.$$

Proof. By (10.28), (10.29) and (10.2),

$$f(z_1) + 2^{-1}\mu d(z_1, z_0)^2$$
$$\leq f(x) + 2^{-1}\mu d(x, z_0)^2 + \epsilon$$
$$\leq f(z_1) + 2^{-1}\mu d(x, z_0)^2 + \epsilon.$$

This implies that

$$d(z_1, z_0)^2 \leq d(x, z_0)^2 + 2\epsilon\mu^{-1}$$

and

$$d(z_0, z_1) \leq d(x, z_0) + (2\epsilon\mu^{-1})^{1/2}.$$

Lemma 10.7 is proved.

Lemma 10.7 implies the following result.

Lemma 10.8. *Let $\epsilon > 0$, $z_0, z_1 \in X$ and $\mu \in [\beta_1, \beta_0]$ satisfy*

$$f(z_1) + 2^{-1}\mu d(z_1, z_0)^2 \leq f(z) + 2^{-1}\mu d(z, z_0)^2 + \epsilon \text{ for all } z \in X.$$

Then

$$d(z_1, z_0) \leq D(z_0, \Omega_0) + (2\epsilon\mu^{-1})^{1/2}.$$

Lemma 10.9. *Let*

$$\mu \in [\beta_1, \beta_0] \qquad (10.30)$$

and let $\epsilon > 0$, $z_0, z_1 \in X$ satisfy

$$z_1 \in B(\bar{x}, \rho_0), \quad D(z_1, \Omega_0) > 0, \tag{10.31}$$

$$f(z_1) + 2^{-1}\mu d(z_1, z_0)^2 \leq f(z) + 2^{-1}\mu d(z, z_0)^2 + \epsilon \text{ for all } z \in X. \tag{10.32}$$

Then

$$D(z_1, \Omega_0) \leq 2\epsilon(D(z_1, \Omega_0))^{-1}(2\alpha - \beta_0)^{-1} + \gamma D(z_0, \Omega_0) + (2\epsilon\mu^{-1})^{1/2}.$$

Proof. Let

$$x \in \Omega_0. \tag{10.33}$$

By (10.32),

$$f(z_1) + 2^{-1}\mu d(z_1, z_0)^2 \leq f(x) + 2^{-1}\mu d(x, z_0)^2 + \epsilon.$$

Together with (10.33), (10.3), and (10.2) this implies that

$$f(z_1) - f(\bar{x}) \leq 2^{-1}\mu(d(x, z_0)^2 - d(z_1, z_0)^2) + \epsilon$$
$$\leq \epsilon + 2^{-1}\mu d(x, z_1)(2d(z_1, z_0) + d(x, z_1)).$$

Since the inequality above holds for any $x \in \Omega_0$ we obtain that

$$f(z_1) - f(\bar{x})$$
$$\leq \epsilon + 2^{-1}\mu D(z_1, \Omega_0)(2d(z_1, z_0) + D(z_1, \Omega_0)). \tag{10.34}$$

By (10.31) and (10.4),

$$f(z_1) - f(\bar{x}) \geq \alpha D(z_1, \Omega_0)^2.$$

Together with (10.34) this implies that

$$\alpha D(z_1, \Omega_0)^2 \leq \epsilon + 2^{-1}\mu D(z_1, \Omega_0)(2d(z_1, z_0) + D(z_1, \Omega_0)).$$

The inequality above implies that

$$(\alpha - 2^{-1}\mu)(D(z_1, \Omega_0))^2 \leq \epsilon + \mu D(z_1, \Omega_0)d(z_1, z_0).$$

Combined with (10.30) and (10.31) this implies that

$$(\alpha - 2^{-1}\beta_0)D(z_1, \Omega_0) \leq \beta_0 d(z_1, z_0) + \epsilon(D(z_1, \Omega_0))^{-1}.$$

By the inequality above and (10.5)

$$D(z_1, \Omega_0) \le \beta_0(\alpha - 2^{-1}\beta_0)^{-1}d(z_1, z_0)$$

$$+ \epsilon(D(z_1, \Omega_0))^{-1}(\alpha - 2^{-1}\beta_0)^{-1}. \tag{10.35}$$

By Lemma 10.8, (10.30), and (10.32),

$$d(z_1, z_0) \le D(z_0, \Omega_0) + (2\epsilon\mu^{-1})^{1/2}. \tag{10.36}$$

In view of (10.35), (10.36), (10.6), and (10.7),

$$D(z_1, \Omega_0) \le 2\epsilon(D(z_1, \Omega_0))^{-1}(2\alpha - \beta_0)^{-1}$$

$$+\gamma D(z_0, \Omega_0) + \gamma(2\epsilon\mu^{-1})^{1/2}$$

$$\le 2\epsilon(D(z_1, \Omega_0))^{-1}(2\alpha - \beta_0)^{-1}$$

$$+\gamma D(z_0, \Omega_0) + (2\epsilon\mu^{-1})^{1/2}.$$

Lemma 10.9 is proved.

10.3   The Main Lemma

Lemma 10.10. *Let a number Δ satisfy (10.8), $k_0 = k_0(\Delta)$, $\epsilon = \epsilon(\Delta)$, a sequence $\{\mu_k\}_{k=0}^{\infty}$ satisfy (10.13) and let a sequence $\{y_k\}_{k=0}^{\infty} \subset X$ satisfy*

$$d(y_0, \bar{x})(1 + (1 - \gamma)^{-1}) < \rho \tag{10.37}$$

and for all integers $k \ge 0$,

$$f(y_{k+1}) + 2^{-1}\mu_k d(y_{k+1}, y_k)^2$$

$$\le f(z) + 2^{-1}\mu_k d(z, y_k)^2 + \epsilon \text{ for all } z \in X. \tag{10.38}$$

Then there exists an integer $k \in [0, k_0]$ such that

$$d(y_j, \bar{x}) \le \rho_0, \ j = 0, \ldots, k, \tag{10.39}$$

$$D(y_k, \Omega_0) \le \Delta/2. \tag{10.40}$$

Proof. For all integers $j \ge 0$ set

$$\Lambda_j = 4j\epsilon\Delta^{-1}(2\alpha - \beta_0)^{-1} + j(2\epsilon\beta_1^{-1})^{1/2}. \tag{10.41}$$

Assume that an integer k satisfies

$$0 \leq k < k_0 \tag{10.42}$$

and that for all integers $j = 0, \ldots, k$,

$$y_j \in B(\bar{x}, \rho_0), \quad D(y_j, \Omega_0) \leq \gamma^j D(y_0, \Omega_0) + \Lambda_j. \tag{10.43}$$

(Clearly for $k = 0$ this assumption holds.)

If there is an integer $k_1 \in [0, k+1]$ such that $D(y_{k_1}, \Omega_0) \leq \Delta/2$, then the assertion of the lemma holds.

Therefore we may assume without loss of generality that

$$D(y_j, \Omega_0) > \Delta/2 \text{ for al integers } j \in [0, k+1]. \tag{10.44}$$

For all integers $j = 0, \ldots, k$, it follows from (10.13), (10.38), and Lemma 10.8 applied with $z_0 = y_j, z_1 = y_{j+1}, \mu = \mu_j$ that

$$d(y_j, y_{j+1}) \leq D(y_j, \Omega_0) + (2\epsilon\mu_j^{-1})^{1/2}. \tag{10.45}$$

By (10.45) and (10.43),

$$d(y_{k+1}, y_0) \leq \sum_{j=0}^{k} d(y_{j+1}, y_j) \leq \sum_{j=0}^{k} [D(y_j, \Omega_0) + (2\epsilon\mu_j^{-1})^{1/2}]$$

$$\leq \sum_{j=0}^{k} [\gamma^j D(y_0, \Omega_0) + \Lambda_j + (2\epsilon\mu_j^{-1})^{1/2}]$$

$$\leq (1 - \gamma)^{-1} D(y_0, \Omega_0) + \sum_{j=0}^{k} \Lambda_j + (k+1)(2\epsilon\beta_1^{-1})^{1/2}. \tag{10.46}$$

By (10.46), (10.37), (10.3), (10.41), (10.42), and (10.10),

$$d(y_{k+1}, \bar{x}) \leq d(y_{k+1}, y_0) + d(y_0, \bar{x})$$

$$\leq d(y_0, \bar{x})(1 + (1 - \gamma)^{-1}) + \sum_{j=0}^{k} \Lambda_j + (k+1)(2\epsilon\beta_1^{-1})^{1/2}$$

$$\leq \rho + (k+1)\Lambda_k + (k+1)(2\epsilon\beta_1^{-1})^{1/2}$$

$$\leq \rho + 4k_0^2\epsilon\Delta^{-1}(2\alpha - \beta_0)^{-1}$$

$$+ k_0^2(2\epsilon\beta_1^{-1})^{1/2} + k_0(2\epsilon\beta_1^{-1})^{1/2} < \rho_0. \tag{10.47}$$

By (10.13), (10.47), (10.44), (10.38), (10.43), (10.7), (10.41), and Lemma 10.9 applied with $z_0 = y_k, z_1 = y_{k+1}, \mu = \mu_k$,

$$
\begin{aligned}
D(y_{k+1}, \Omega_0) &\leq 2\epsilon (D(y_{k+1}, \Omega_0))^{-1}(2\alpha - \beta_0)^{-1} \\
&\quad + \gamma D(y_k, \Omega_0) + (2\epsilon \mu_k^{-1})^{1/2} \\
&\leq 4\epsilon \Delta^{-1}(2\alpha - \beta_0)^{-1} + \gamma D(y_k, \Omega_0) + (2\epsilon \beta_1^{-1})^{1/2} \\
&\leq 4\epsilon \Delta^{-1}(2\alpha - \beta_0)^{-1} \\
&\quad + \gamma(\gamma^k D(y_0, \Omega_0) + \Lambda_k) + (2\epsilon \beta_1^{-1})^{1/2} \\
&\leq \gamma^{k+1} D(y_0, \Omega_0) + \Lambda_{k+1}.
\end{aligned}
\tag{10.48}
$$

In view of (10.47) and (10.48) we conclude that (10.43) holds for all $j = 0, \ldots, k+1$. Thus by induction we have shown that at least one of the following cases holds:

(i) there is an integer $k \in [0, k_0]$ such that (10.39) and (10.40) hold.
(ii) (10.43) holds for all $j = 0, \ldots, k_0$.

In the case (i) the assertion of the lemma holds.

Assume that the case (ii) holds. By (10.43) with $j = k_0$, (10.3), (10.37), (10.41), (10.9), (10.11), and (10.12),

$$
\begin{aligned}
D(y_{k_0}, \Omega_0) &\leq \gamma^{k_0} D(y_0, \Omega_0) + \Lambda_{k_0} \\
&\leq \gamma^{k_0} \rho + 4k_0 \epsilon \Delta^{-1}(2\alpha - \beta_0)^{-1} + k_0(2\epsilon \beta_1^{-1})^{1/2} < \Delta/2
\end{aligned}
$$

and the assertion of the lemma holds.

This completes the proof of Lemma 10.10.

10.4 Proof of Theorem 10.1

By Lemma 10.10 there is an integer $k \in [0, k_0]$ such that

$$
d(x_j, \bar{x}) \leq \rho_0, \ j = 0, \ldots, k,
\tag{10.49}
$$

$$
D(x_k, \Omega_0) \leq \Delta/2.
\tag{10.50}
$$

We show that for all integers $j \geq k$

$$
D(x_j, \Omega_0) \leq \Delta.
\tag{10.51}
$$

This will complete the proof of the theorem.

Assume that an integer $j \geq k$ satisfies (10.51). In view of (10.3) and (10.8), in order to complete the proof it is sufficient to show that

$$D(x_{j+1}, \Omega_0) \le \Delta.$$

We may assume without loss of generality that

$$D(x_{j+1}, \Omega_0) > \Delta/2. \tag{10.52}$$

By Lemma 10.8 applied with $z_0 = x_j$, $z_1 = x_{j+1}$, $\mu = \mu_j$, (10.13), (10.15), and (10.51),

$$d(x_j, x_{j+1}) \le D(x_j, \Omega_0) + (2\epsilon\mu_j^{-1})^{1/2} \le \Delta + (2\epsilon\beta_1^{-1})^{1/2}.$$

Together with (10.51) and (10.11) this implies that

$$D(x_{j+1}, \Omega_0) \le d(x_{j+1}, x_j) + D(x_j, \Omega_0) \le 2\Delta + (2\epsilon\beta_1^{-1})^{1/2} < 3\Delta.$$

Together with (10.3) and (10.8) this implies that

$$d(x_{j+1}, \bar{x}) < \rho_0. \tag{10.53}$$

By (10.13), (10.53), (10.31), (10.15), (10.52), (10.51), (10.11), (10.12), and Lemma 10.9 applied to $z_0 = x_j$, $z_1 = x_{j+1}$, $\mu = \mu_j$,

$$D(x_{j+1}, \Omega_0) \le 2\epsilon(D(x_{j+1}, \Omega_0))^{-1}(2\alpha - \beta_0)^{-1} + \gamma D(x_j, \Omega_0) + (2\epsilon\mu_j^{-1})^{1/2}$$

$$\le 4\epsilon\Delta^{-1}(2\alpha - \beta_0)^{-1} + \gamma\Delta + (2\epsilon\beta_1^{-1})^{1/2} < \Delta.$$

This completes the proof of Theorem 10.1.

10.5   An Auxiliary Result for Theorem 10.2

Let

$$\Delta \in (0, 1), \tag{10.54}$$

a natural number k_0 satisfy

$$\gamma^{k_0}\rho_0 < \Delta/8 \tag{10.55}$$

and let a positive number ϵ satisfy

$$4k_0\epsilon\Delta^{-1}(2\alpha - \beta_0)^{-1} < (1 - \gamma)\Delta/8, \tag{10.56}$$

$$k_0(2\epsilon\beta_1^{-1})^{1/2} < (1 - \gamma)\Delta/8. \tag{10.57}$$

Proposition 10.11. *Assume that*

$$\{\mu_k\}_{k=0}^{\infty} \subset [\beta_1, \beta_0], \tag{10.58}$$

a sequence $\{x_k\}_{k=0}^{\infty} \subset X$ *satisfies*

$$x_k \in B(\bar{x}, \rho_0) \text{ for all integers } k \geq 0 \tag{10.59}$$

and that for all integers $k \geq 0$

$$f(x_{k+1}) + 2^{-1}\mu_k d(x_{k+1}, x_k)$$
$$\leq f(z) + 2^{-1}\mu_k d(z, x_k) + \epsilon \text{ for all } z \in X. \tag{10.60}$$

Then

$$D(x_j, \Omega_0) \leq \Delta \text{ for all integers } j \geq k_0. \tag{10.61}$$

Proof. For all integers $j = 0, 1, \ldots$ put

$$\Lambda_j = j(2\epsilon\Delta^{-1}(2\alpha - \beta_0)^{-1} + (2\epsilon\beta_1^{-1})^{1/2}). \tag{10.62}$$

Assume that an integer $j \geq 0$ and that

$$D(x_j, \Omega_0) \leq \Delta. \tag{10.63}$$

We show that $D(x_{j+1}, \Omega_0) \leq \Delta$. We may assume without loss of generality that

$$D(x_{j+1}, \Omega_0) > \Delta/2. \tag{10.64}$$

By (10.58), (10.60), (10.64), (10.59), and Lemma 10.9 applied with $z_0 = x_j$, $z_1 = x_{j+1}$, $\mu = \mu_j$, (10.11), (10.12), (10.63), (10.56), and (10.57),

$$D(x_{j+1}, \Omega_0) \leq 2\epsilon(D(x_{j+1}, \Omega_0))^{-1}(2\alpha - \beta_0)^{-1} + \gamma D(x_j, \Omega_0) + (2\epsilon\mu_j^{-1})^{1/2}$$
$$\leq 4\epsilon\Delta^{-1}(2\alpha - \beta_0)^{-1} + \gamma\Delta + (2\epsilon\beta_1^{-1})^{1/2} < \Delta.$$

Thus we have shown that if an integer $j \geq 0$ satisfies (10.63), then

$$D(x_{j+1}, \Omega_0) \leq \Delta.$$

Therefore in order to prove the proposition it is sufficient to show that (10.63) holds with some integer $j \in [0, k_0]$.

Assume the contrary. Thus

$$D(x_j, \Omega_0) > \Delta \text{ for all integers } j \in [0, k_0]. \tag{10.65}$$

Assume that an integer k satisfies

$$0 \le k < k_0, \tag{10.66}$$

$$D(x_j, \Omega_0) \le \gamma^j D(x_0, \Omega_0) + \Lambda_j. \tag{10.67}$$

(Clearly, for $k = 0$ this assumption holds.)

By (10.58), (10.59), (10.65), (10.60), (10.67), (10.7), (10.62), (10.66), and Lemma 10.9 applied with $z_0 = x_k$, $z_1 = x_{k+1}$, $\mu = \mu_k$,

$$
\begin{aligned}
D(x_{k+1}, \Omega_0) &\le 2\epsilon (D(x_{k+1}, \Omega_0))^{-1}(2\alpha - \beta_0)^{-1} + \gamma D(x_k, \Omega_0) + (2\epsilon \mu_k^{-1})^{1/2} \\
&\le 2\epsilon \Delta^{-1}(2\alpha - \beta_0)^{-1} + \gamma^{k+1} D(x_0, \Omega_0) + \Lambda_k + (2\epsilon \beta_1^{-1})^{1/2} \\
&= \gamma^{k+1} D(x_0, \Omega_0) + \Lambda_{k+1}.
\end{aligned}
$$

Thus (10.67) holds for $j = k+1$. By induction we have shown that (10.67) holds for all $j = 0, \ldots, k_0$. Together with (10.59), (10.55), (10.57), and (10.62) this implies that

$$
\begin{aligned}
D(x_{k_0}, \Omega_0) &\le \gamma^{k_0} D(x_0, \Omega_0) + \Lambda_{k_0} \\
&\le \gamma^{k_0} \rho_0 + k_0 (2\epsilon \Delta^{-1}(2\alpha - \beta_0)^{-1} + (2\epsilon \beta_1^{-1})^{1/2}) < \Delta.
\end{aligned}
$$

This contradicts (10.65). The contradiction we have reached proves Proposition 10.11.

10.6   Proof of Theorem 10.2

By Theorem 10.1 there exists $\bar{\epsilon} > 0$ such that the following property holds:

(P1) For each sequence

$$\{\mu_k\}_{k=0}^{\infty} \subset [\beta_1, \beta_0] \tag{10.68}$$

and each sequence $\{x_k\}_{k=0}^{\infty} \subset X$ which satisfies

$$d(x_0, \bar{x})(1 + (1 - \gamma)^{-1}) < \rho \tag{10.69}$$

and for all integers $k = 0, 1, \ldots,$

$$f(x_{k+1}) + 2^{-1}\mu_k d(x_{k+1}, x_k)^2$$

$$\le f(z) + 2^{-1}\mu_k d(z, x_k)^2 + \bar{\epsilon} \text{ for all } z \in X$$

we have

$$x_j \in B(\bar{x}, \rho_0) \text{ for all integers } j \geq 0. \tag{10.70}$$

Assume that

$$\{\epsilon_i\}_{i=0}^{\infty} \subset (0, \bar{\epsilon}] \text{ and } \lim_{i \to \infty} \epsilon_i = 0 \tag{10.71}$$

and $\Delta > 0$. By Proposition 10.11 there are $\hat{\epsilon} > 0$ and a natural number q_1 such that the following property holds:

(P2)   Assume that (10.68) holds, a sequence $\{x_k\}_{k=0}^{\infty} \subset X$ satisfies (10.70) and that for all integers $k \geq 0$,

$$f(x_{k+1}) + 2^{-1}\mu_k d(x_{k+1}, x_k)^2$$
$$\leq f(z) + 2^{-1}\mu_k d(z, x_k)^2 + \hat{\epsilon} \text{ for all } z \in X.$$

Then $D(x_j, \Omega_0) \leq \Delta$ for all integers $j \geq q_1$.

By (10.71) there exists a natural number q_2 such that

$$\epsilon_j < \hat{\epsilon} \text{ for all integers } j \geq q_2. \tag{10.72}$$

Set

$$k_1 = q_1 + q_2. \tag{10.73}$$

Assume that (10.68) holds, a sequence $\{x_k\}_{k=0}^{\infty} \subset X$ satisfies (10.69) and that for all integers $k \geq 0$

$$f(x_{k+1}) + 2^{-1}\mu_k d(x_k, x_{k+1})^2$$
$$\leq f(z) + 2^{-1}\mu_k d(z, x_k)^2 + \epsilon_k \text{ for all } z \in X. \tag{10.74}$$

Equations (10.68), (10.69), (10.74), (10.71), and (P1) imply (10.70).

By (10.68), (10.70), (10.72), and (10.74) we can apply (P2) to the sequence $\{x_{q_2+j}\}_{j=0}^{\infty}$ and obtain that $D(x_j, \Omega_0) \leq \Delta$ for all integers $j \geq q_1 + q_2 = k_1$. Theorem 10.2 is proved.

10.7   Well-Posed Minimization Problems

We use the notation and definitions from Sect. 10.1. Suppose that

$$\Omega = \{\bar{x}\}$$

and that for each sequence $\{z_i\}_{i=0}^{\infty} \subset X$ satisfying $\lim_{i \to \infty} f(z_i) = \inf(f)$ we have

$$\lim_{i \to \infty} \rho(z_i, \bar{x}) = 0.$$

In other words the problem $f(x) \to \min$, $x \in X$ is well posed in the sense of [121].
  Fix $M > 1 + \rho_0$.

Proposition 10.12. *There exist $\Delta_*, \mu_* > 0$ such that for each $\mu \in (0, \mu_*]$, each*
$z_0 \in B(\bar{x}, M)$ *and each $z_1 \in X$ satisfying*

$$f(z_1) + 2^{-1} \mu d(z_1, z_0)^2$$
$$\leq f(z) + 2^{-1} \mu d(z, z_0)^2 + \Delta_* \text{ for all } z \in X \qquad (10.75)$$

the inequality

$$d(z_1, \bar{x}) \leq 2^{-1} \rho (1 + (1 - \gamma)^{-1})^{-1} \qquad (10.76)$$

holds.

Proof. Since the problem $f(z) \to \min$, $z \in X$ is well posed there is $\delta > 0$ such that

$$\text{if } z \in X \text{ satisfies } f(z) \leq \inf(f) + \delta,$$
$$\text{then } d(z, \bar{x}) \leq 2^{-1} \rho (1 + (1 - \gamma)^{-1})^{-1}. \qquad (10.77)$$

Choose positive numbers

$$\mu_* < (M^2 + 1)^{-1} \delta, \ \Delta_* \in (0, \delta/2). \qquad (10.78)$$

Let

$$\mu \in (0, \mu_*], \ z_0 \in B(\bar{x}, M) \qquad (10.79)$$

and let $z_1 \in X$ satisfy (10.75). By (10.79), (10.78), and (10.75),

$$f(z_1) \leq f(z_1) + 2^{-1} \mu d(z_1, z_0)^2$$
$$\leq f(\bar{x}) + 2^{-1} \mu_* d(\bar{x}, z_0)^2 + \Delta_*$$
$$\leq \inf(f) + 2^{-1} \mu_* M^2 + \Delta_* \leq \inf(f) + \delta.$$

Together with (10.77) this implies that (10.76). Proposition 10.12 is proved.

  Let $\Delta_*, \mu_* > 0$ be as guaranteed by Proposition 10.12. We suppose that

$$\beta_0 \leq \mu_*. \qquad (10.80)$$

We may assume without loss of generality that

$$\epsilon(\Delta) \in (0, \Delta_*] \text{ for all } \Delta \in (0, 1). \tag{10.81}$$

Theorem 10.13. *Let a number Δ satisfy*

$$\Delta \in (0, 1), \ \Delta < 2^{-1}\rho(1 + (1 - \gamma)^{-1})^{-1}, \ \Delta < \rho_0/3,$$

$k_0 = k_0(\Delta)$ *and let a positive number $\epsilon = \epsilon(\Delta)$. Assume that*

$$\{\mu_k\}_{k=0}^{\infty} \subset [\beta_1, \beta_0], \tag{10.82}$$

a sequence $\{x_k\}_{k=0}^{\infty} \subset X$ satisfies

$$d(x_0, \bar{x}) \leq M \tag{10.83}$$

and that for all integers $k \geq 0$,

$$f(x_{k+1}) + 2^{-1}\mu_k d(x_{k+1}, x_k)^2$$
$$\leq f(z) + 2^{-1}\mu_k d(z, x_k)^2 + \epsilon \text{ for all } z \in X. \tag{10.84}$$

Then

$$x_j \in B(\bar{x}, \rho_0) \text{ for all integers } j \geq 1,$$

$$x_j \in B(\bar{x}, \Delta) \text{ for all integers } j \geq k_0 + 1.$$

Proof. By the choice of Δ_* and μ_*, Proposition 10.12, (10.84), (10.82), (10.80), (10.83), and (10.81)

$$d(x_1, \bar{x}) \leq 2^{-1}\rho(1 + (1 - \gamma)^{-1})^{-1}.$$

Since ρ_* can be arbitrarily small positive number the assertion of the theorem now follows from Theorem 10.1.

Theorem 10.14. *Let $\bar{\epsilon} > 0$ be as guaranteed by Theorem 10.2,*

$$\hat{\epsilon} = \min\{\bar{\epsilon}, \Delta_*\}, \tag{10.85}$$

$$\{\epsilon_i\}_{i=0}^{\infty} \subset (0, \hat{\epsilon}], \ \lim_{i \to \infty} \epsilon_i = 0, \tag{10.86}$$

$\Delta > 0$ and let a natural number k_1 be as guaranteed by Theorem 10.2 with the sequence $\{\epsilon_{i+1}\}_{i=0}^{\infty}$. Assume that

$$\{\mu_k\}_{k=0}^{\infty} \subset [\beta_1, \beta_0], \tag{10.87}$$

a sequence $\{x_k\}_{k=0}^{\infty} \subset X$ satisfies

$$d(x_0, \bar{x}) \leq M \tag{10.88}$$

and that for all integers $k \geq 0$,

$$f(x_{k+1}) + 2^{-1}\mu_k d(x_{k+1}, x_k)^2$$
$$\leq f(z) + 2^{-1}\mu_k d(z, x_k)^2 + \epsilon_k \text{ for all } z \in X. \tag{10.89}$$

Then

$$x_j \in B(\bar{x}, \rho_0) \text{ for all integers } j \geq 1,$$

$$x_j \in B(\bar{x}, \Delta) \text{ for all integers } j \geq k_1 + 1.$$

Proof. By the choice of Δ_* and μ_*, Proposition 10.12, (10.88), (10.86), (10.85), (10.87), and (10.80),

$$d(x_1, \bar{x}) \leq 2^{-1}\rho(1 + (1 - \gamma)^{-1})^{-1}.$$

The assertion of the theorem now follows from Theorem 10.2.

10.8 An Example

Let $X = R^n$ be equipped with the Euclidean norm $\|\cdot\|$ which induces the metric $d(x, y) = \|x - y\|$, $x, y \in R^n$.
  Set

$$\Omega = B(0, 1),$$

$$f(x) = D(x, B(0, 1))^2, \ x \in R^n.$$

Clearly, all the assumptions made in Sect. 10.1 hold with $\bar{x} = 0$, $\Omega = \Omega_0 = B(0, 1)$, $\rho_* = 1$, $\alpha = 1$ and any positive constant $\rho_0 > 1$. Thus Theorems 10.1 and 10.2 hold for the function f.
  We prove the following result.

Proposition 10.15. *Assume that $\mu > 0$, a sequence $\{\epsilon_i\}_{i=0}^{\infty} \subset (0, 1]$ satisfies*

$$\sum_{i=0}^{\infty} \epsilon_i^{1/2} = \infty \tag{10.90}$$

and that $x_0 \in B(0, 1)$. Then there exists a sequence $\{x_k\}_{k=0}^{\infty} \subset B(0, 1)$ such that for all integers $k \geq 0$,

$$f(x_{k+1}) + 2^{-1}\mu\|x_{k+1} - x_k\|^2$$
$$\leq f(z) + 2^{-1}\mu\|z - x_k\|^2 + \epsilon_k \text{ for all } z \in R^n \tag{10.91}$$

and that for all $z \in B(0, 1)$,

$$\liminf_{k \to \infty} \|x_k - z\| = 0.$$

Proposition 10.15 easily follows from the following auxiliary result.

Lemma 10.16. *Assume that $\mu > 0$, a sequence $\{\epsilon_i\}_{i=0}^{\infty} \subset (0, 1]$ satisfies (10.90) and that $y_0, y_1 \in B(0, 1)$. Then there exist a natural number q and a sequence $\{x_k\}_{k=0}^q \subset B(0, 1)$ such that $x_0 = y_0$, $x_q = y_1$ ant that for all integers $k \in [0, q - 1]$, Eq. (10.91) holds.*

Proof. Set

$$F = \{ty_1 + (1 - t)y_0 : t \in [0, 1]\}.$$

Set $t_0 = 0$ and for all integers $i \geq 0$

$$t_{i+1} = \min\{t_i + 2^{-1}(2\epsilon_i\mu^{-1})^{1/2}, 1\}. \tag{10.92}$$

By (10.90) and (10.92) there exists a natural number q such that

$$t_q = 1, \ t_i < 1 \text{ for all nonnegative integers } i < q. \tag{10.93}$$

For any integer $k \in [0, q]$ set

$$x_k = t_k y_1 + (1 - t_k)y_0. \tag{10.94}$$

Clearly,

$$\{x_k\}_{k=0}^q \subset F \subset B(0, 1). \tag{10.95}$$

Let an integer k satisfy $0 \leq k < q$. By (10.92), (10.94) and (10.95)

$$f(x_{k+1}) + 2^{-1}\mu\|x_{k+1} - x_k\|^2$$
$$= 2^{-1}\mu\|t_{k+1}y_1 + (1 - t_{k+1})y_0 - (t_k y_1 + (1 - t_k)y_0)\|^2$$
$$= 2^{-1}\mu\|(t_{k+1} - t_k)(y_1 - y_0)\|^2 \leq 2\mu(t_{k+1} - t_k)^2 \leq \epsilon_k.$$

This implies (10.91) and completes the proof of Lemma 10.16.

Chapter 11
Maximal Monotone Operators and the Proximal Point Algorithm

In a finite-dimensional Euclidean space, we study the convergence of a proximal point method to a solution of the inclusion induced by a maximal monotone operator, under the presence of computational errors. The convergence of the method is established for nonsummable computational errors. We show that the proximal point method generates a good approximate solution, if the sequence of computational errors is bounded from above by a constant.

11.1 Preliminaries and the Main Results

Let R^n be the n-dimensional Euclidean space equipped with an inner product $\langle \cdot, \cdot \rangle$ which induces the norm $\| \cdot \|$.

A multifunction $T : R^n \to 2^{R^n}$ is called a monotone operator if

$$\langle z - z', w - w' \rangle \geq 0 \quad \forall z, z', w, w' \in R^n$$

$$\text{such that } w \in T(z) \text{ and } w' \in T(z'). \tag{11.1}$$

It is called maximal monotone if, in addition, the graph

$$\{(z, w) \in R^n \times R^n : w \in T(z)\}$$

is not strictly contained in the graph of any other monotone operator $T' : R^n \to R^n$. A fundamental problem consists in determining an element z such that $0 \in T(z)$.

A proximal point algorithm is an important tool for solving this problem. This algorithm has been studied extensively because of its role in convex analysis and optimization. See, for example, [15–17, 31, 34, 36, 53, 55, 69, 81–83, 87, 103, 104, 106, 107, 111, 113] and the references mentioned therein.

© Springer International Publishing Switzerland 2016

A.J. Zaslavski, *Numerical Optimization with Computational Errors*, Springer Optimization and Its Applications 108, DOI 10.1007/978-3-319-30921-7_11

Let $T : R^n \to R^n$ be a maximal monotone operator. The algorithm for solving the inclusion $0 \in T(z)$ is based on the fact established by Minty [82], who showed that, for each $z \in R^n$ and each $c > 0$, there is a unique $u \in R^n$ such that

$$z \in (I + cT)(u),$$

where $I : R^n \to R^n$ is the identity operator ($Ix = x$ for all $x \in R^n$).

The operator

$$P_c := (I + cT)^{-1} \tag{11.2}$$

is therefore single-valued from all of R^n onto R^n (where c is any positive number). It is also nonexpansive:

$$\|P_c(z) - P_c(z')\| \le \|z - z'\| \text{ for all } z, z' \in R^n \tag{11.3}$$

and $P_c(z) = z$ if and only if $0 \in T(z)$. Following the terminology of Moreau [87] P_c is called the proximal mapping associated with cT.

The proximal point algorithm generates, for any given sequence $\{c_k\}_{k=0}^\infty$ of positive real numbers and any starting point $z^0 \in R^n$, a sequence $\{z^k\}_{k=0}^\infty \subset R^n$, where

$$z^{k+1} := P_{c_k}(z^k), \ k = 0, 1, \dots \tag{11.4}$$

We study the convergence of a proximal point method to the set of solutions of the inclusion $0 \in T(x)$ under the presence of computational errors. We show that the proximal point method generates a good approximate solution, if the sequence of computational errors is bounded from above by some constant.

More precisely, we show (Theorem 11.2) that, for given positive numbers M, ϵ, there exist a natural number n_0 and $\delta > 0$ such that, if the computational errors do not exceed δ for any iteration and if $\|x_0\| \le M$, then the algorithm generates a sequence $\{x_k\}_{k=0}^{n_0}$ such that $\|x_{n_0} - \bar{x}\| \le \epsilon$, where $\bar{x} \in R^n$ satisfies $0 \in T(\bar{x})$.

The results of this chapter were obtained in [125].

Let $T : R^n \to R^n$ be a maximal monotone operator. It is not difficult to see that its graph

$$\text{graph} T := \{(x, w) \in R^n \times R^n : w \in T(x)\}$$

is closed.

Assume that

$$F := \{z \in R^n : 0 \in T(z)\} \ne \emptyset.$$

For each $x \in R^n$ and each nonempty set $A \subset R^n$ put

$$d(x, A) = \inf\{\|x - y\| : y \in A\}.$$

Fix

$$\bar{\lambda} > 0. \tag{11.5}$$

For each $x \in R^n$ and each $r > 0$ set

$$B(x, r) = \{y \in R^n : \|x - y\| \le r\}.$$

We prove the following result, which establishes the convergence of the proximal point algorithm without computational errors.

Theorem 11.1. *Let $M, \epsilon > 0$. Then there exists a natural number n_0 such that, for each sequence $\{\lambda_k\}_{k=0}^{\infty} \subset [\bar{\lambda}, \infty)$ and each sequence $\{x_k\}_{k=0}^{\infty} \subset R^n$ such that*

$$\|x_0\| \le M,$$

$$x_{k+1} = P_{\lambda_k}(x_k) \text{ for all integers } k \ge 0, \tag{11.6}$$

the inequality $d(x_k, F) \le \epsilon$ holds for all integers $k \ge n_0$.

Since n_0 depends only on M, ϵ we can say that Theorem 11.1 establishes the uniform convergence of the proximal point algorithm without computational errors on bounded sets.

Theorem 11.1 is proved in Sect. 11.3. The following theorem is one of the main results of this chapter.

Theorem 11.2. *Let $M, \epsilon > 0$. Then there exist a natural number n_0 and a positive number δ such that, for each sequence $\{\lambda_k\}_{k=0}^{n_0-1} \subset [\bar{\lambda}, \infty)$ and each sequence $\{x_k\}_{k=0}^{n_0} \subset R^n$ such that*

$$\|x_0\| \le M,$$

$$\|x_{k+1} - P_{\lambda_k}(x_k)\| \le \delta, \ k = 0, 1, \dots, n_0 - 1,$$

the following inequality holds:

$$d(x_{n_0}, F) \le \epsilon.$$

Theorem 11.2 easily follows from the following result, which is proved in Sect. 11.4.

Theorem 11.3. *Let $M, \epsilon_0 > 0$, let a natural number n_0 be as guaranteed by Theorem 11.1 with $\epsilon = \epsilon_0/2$, and let $\delta = \epsilon_0(2n_0)^{-1}$.*

Then, for each sequence $\{\lambda_k\}_{k=0}^{n_0-1} \subset [\bar{\lambda}, \infty)$ and each sequence $\{x_k\}_{k=0}^{n_0} \subset R^n$ such that

$$\|x_0\| \leq M,$$

$$\|x_{k+1} - P_{\lambda_k}(x_k)\| \leq \delta, \ k = 0, 1, \ldots .n_0 - 1,$$

the following inequality holds:

$$d(x_{n_0}, F) \leq \epsilon.$$

Theorem 11.2 easily implies the following result.

Theorem 11.4. *Let $M, \epsilon > 0$ and let a natural number n_0 and $\delta > 0$ be as guaranteed by Theorem 11.2. Assume that $\{\lambda_k\}_{k=0}^{\infty} \subset [\bar{\lambda}, \infty)$ and that a sequence $\{x_k\}_{k=0}^{\infty} \subset B(0, M)$ satisfies*

$$\|x_{k+1} - P_{\lambda_k}(x_k)\| \leq \delta, \ k = 0, 1, \ldots$$

Then $d(x_k, F) \leq \epsilon$ for all integers $k \geq n_0$.

Next result is proved in Sect. 11.4. It establishes a convergence of the proximal point algorithm with computational errors, which converge to zero under an assumption that all the iterates are bounded by the same prescribed bound. This convergence is uniform since n_ϵ depends only on M, ϵ, and $\{\delta_k\}_{k=0}^{\infty}$.

Theorem 11.5. *Let $M > 0$, $\{\delta_k\}_{k=0}^{\infty}$ be a sequence of positive numbers such that $\lim_{k \to \infty} \delta_k = 0$ and let $\epsilon > 0$. Then there exists a natural number n_ϵ such that, for each sequence $\{\lambda_k\}_{k=0}^{\infty} \subset [\bar{\lambda}, \infty)$ and each sequence $\{x_k\}_{k=0}^{\infty} \subset B(0, M)$ satisfying*

$$\|x_{k+1} - P_{\lambda_k}(x_k)\| \leq \delta_k \text{ for all integers } k \geq 0,$$

the inequality

$$d(x_k, F) \leq \epsilon$$

holds for all integers $k \geq n_\epsilon$.

In the last two theorems, which are proved in Sect. 11.4, we consider the case when the set F is bounded. In Theorem 11.6 it is assumed that computational errors do not exceed a certain positive constant, while in Theorem 11.7 computational errors tend to zero.

Theorem 11.6. *Suppose that the set F be bounded. Let $M, \epsilon > 0$. Then there exists $\delta_* > 0$ and a natural number n_0 such that, for each $\{\lambda_k\}_{k=0}^{\infty} \subset [\bar{\lambda}, \infty)$ and each sequence $\{x_k\}_{k=0}^{\infty} \subset R^n$ satisfying*

$$\|x_0\| \leq M,$$

$$\|x_{k+1} - P_{\lambda_k}(x_k)\| \leq \delta_*, \ k = 0, 1, \ldots$$

the inequality $d(x_k, F) \leq \epsilon$ holds for all integers $k \geq n_0$.

Theorem 11.7. *Suppose that the set F be bounded and let M > 0. Then there exists* δ > 0 *such that the following assertion holds.*
   Assume that $\{\delta_k\}_{k=0}^{\infty} \subset (0, \delta]$ *satisfies*

$$\lim_{k \to \infty} \delta_k = 0$$

and that $\epsilon > 0$. *Then there exists a natural number* n_ϵ *such that, for each sequence* $\{\lambda_k\}_{k=0}^{\infty} \subset [\bar{\lambda}, \infty)$ *and each sequence* $\{x_k\}_{k=0}^{\infty} \subset R^n$ *satisfying*

$$\|x_0\| \le M,$$

$$\|x_{k+1} - P_{\lambda_k}(x_k)\| \le \delta_k, \ k = 0, 1, \ldots,$$

the inequality $d(x_k, F) \le \epsilon$ *holds for all integers* $k \ge n_\epsilon$.

Note that in Theorem 11.6 δ_* depends on ϵ, while in Theorem 11.7 δ does not depend on ϵ.

11.2 Auxiliary Results

It is easy to see that the following lemma holds.

Lemma 11.8. *Let* $z, x_0, x_1 \in R^n$. *Then*

$$2^{-1}\|z - x_0\|^2 - 2^{-1}\|z - x_1\|^2 - 2^{-1}\|x_0 - x_1\|^2 = \langle x_0 - x_1, x_1 - z \rangle.$$

Lemma 11.9. *Let* $\{\lambda_k\}_{k=0}^{\infty} \subset (0, \infty)$, $\{x_k\}_{k=0}^{\infty} \subset R^n$ *satisfy for all integers* $k \ge 0$,

$$x_{k+1} = P_{\lambda_k}(x_k) = (I + \lambda_k T)^{-1}(x_k) \tag{11.7}$$

and let $z \in R^n$ *satisfies*

$$0 \in T(z). \tag{11.8}$$

Then, for all integers $k \ge 0$,

$$\|z - x_k\|^2 - \|z - x_{k+1}\|^2 - \|x_k - x_{k+1}\|^2 \ge 0.$$

Proof. Let $k \ge 0$ be an integer. By Lemma 11.8,

$$2^{-1}\|z - x_k\|^2 - 2^{-1}\|z - x_{k+1}\|^2 - 2^{-1}\|x_k - x_{k+1}\|^2 = \langle x_k - x_{k+1}, x_{k+1} - z \rangle. \tag{11.9}$$

By (11.7),

$$x_k - x_{k+1} \in \lambda_k T(x_{k+1}).$$

Together with (11.9) and (11.8) this completes the proof of Lemma 11.8.

Using (11.3) we can easily deduce the following lemma.

Lemma 11.10. *Assume that $z \in R^n$ satisfies (11.8), $M > 0$,*

$$\{\lambda_k\}_{k=0}^\infty \subset (0, \infty), \ \{x_k\}_{k=0}^\infty \subset R^n,$$

$\|x_0 - z\| \le M$ and that (11.7) holds for all integers $k \ge 0$. Then $\|x_k - z\| \le M$ for all integers $k \ge 0$.

Lemma 11.11. *Let $M, \epsilon > 0$. Then there exists $\delta > 0$ such that, for each $x \in B(0, M)$, each $\lambda \ge \bar{\lambda}$ and each $z \in B(0, \delta)$ satisfying $z \in \lambda T(x)$ the inequality $d(x, F) \le \epsilon$ holds.*

Proof. Assume the contrary. Then, for each natural number k there exist

$$x_k \in B(0, M), \ z_k \in B(0, k^{-1}), \ \lambda_k \ge \bar{\lambda} \tag{11.10}$$

such that

$$d(x_k, F) > \epsilon, \ z_k \in \lambda_k T(x_k). \tag{11.11}$$

By (11.11) and (11.10), for all integers $k \ge 1$,

$$\lambda_k^{-1} z_k \in T(x_k) \tag{11.12}$$

and

$$\|\lambda_k^{-1} z_k\| \le \bar{\lambda}^{-1} \|z_k\| \le \bar{\lambda}^{-1} k^{-1} \to 0 \text{ as } k \to \infty. \tag{11.13}$$

By (11.10), extracting a subsequence and re-indexing, we can assume that these exists

$$x := \lim_{k \to \infty} x_k. \tag{11.14}$$

Since graph T is closed, then (11.12), (11.13) and (11.14) imply that $0 \in T(x)$ and that $x \in F$. Together with (11.14) this implies that $d(x_k, F) \le \epsilon/2$ for all sufficiently large natural numbers k. This contradicts (11.11) and proves Lemma 11.11.

Lemma 11.12. *Assume that the integers p, q, with $0 \le p < q$, are such that*

$$\{\epsilon_k\}_{k=p}^{q-1} \subset (0, \infty), \ \{\lambda_k\}_{k=p}^{q-1} \subset (0, \infty), \tag{11.15}$$

$$\{x_k\}_{k=p}^q \subset R^n, \ \{y_k\}_{k=p}^q \subset R^n, \ y_p = x_p,$$

and that for all integers $k \in \{p, \ldots, q-1\}$,

$$y_{k+1} = P_{\lambda_k}(y_k), \quad \|x_{k+1} - P_{\lambda_k}(x_k)\| \le \epsilon_k. \tag{11.16}$$

Then, for any integer $k \in \{p+1\ldots, q\}$,

$$\|y_k - x_k\| \le \sum_{i=p}^{k-1} \epsilon_i. \tag{11.17}$$

Proof. We prove the lemma by induction. In view of (11.16) and (11.15), equation (11.17) holds for $k = p+1$.

Assume that an integer j satisfies $p+1 \le j \le q$, (11.17) holds for all $k = p+1, \ldots, j$ and that $j < q$.

By (11.16), (11.3) and (11.17) with $k = j$

$$\|y_{j+1} - x_{j+1}\| \le \|P_{\lambda_j} y_j - x_{j+1}\| \le \|P_{\lambda_j} y_j - P_{\lambda_j} x_j\| + \|P_{\lambda_j} x_j - x_{j+1}\|$$

$$\le \|y_j - x_j\| + \epsilon_j \le \sum_{i=p}^{j-1} \epsilon_i + \epsilon_j = \sum_{i=p}^{j} \epsilon_i$$

and (11.17) holds for all $k = p+1, \ldots, j+1$. Therefore we showed by induction that (11.17) holds for all $k = p+1, \ldots, q$. This completes the proof of Lemma 11.12. $\blacksquare$

11.3 Proof of Theorem 11.1

Fix

$$z \in F. \tag{11.18}$$

By Lemma 11.11, there exists $\delta \in (0, 1)$ such that the following property holds:

(P1) For each $x \in B(0, M + 2\|z\|)$, each $\lambda \ge \bar{\lambda}$ and each $z \in B(0, \delta)$ satisfying $z \in \lambda T(x)$ we have $d(x, F) \le \epsilon/2$.

Choose a natural number n_0 such that

$$(\|z\| + M)^2 n_0^{-1} < \delta^2. \tag{11.19}$$

Assume that

$$\{\lambda_k\}_{k=0}^{\infty} \subset [\bar{\lambda}, \infty), \quad \{x_k\}_{k=0}^{\infty} \subset R^n, \quad \|x_0\| \le M, \tag{11.20}$$

$$x_{k+1} = P_{\lambda_k}(x_k) \text{ for all integers } k \ge 0.$$

By Lemma 11.9, (11.20) and (11.18) for all integers $k \geq 0$,

$$\|x_k\| \leq \|z\| + \|x_k - z\| \leq \|z\| + \|x_0 - z\| \leq 2\|z\| + \|x_0\| \leq 2\|z\| + M. \quad (11.21)$$

By Lemma 11.9, (11.18) and (11.20) for any integer $k \geq 0$,

$$\|x_k - x_{k+1}\|^2 \leq \|z - x_k\|^2 - \|z - x_{k+1}\|^2$$

and this implies that

$$\sum_{k=0}^{n_0-1} \|x_k - x_{k+1}\|^2 \leq \|z - x_0\|^2 - \|z - x_{n_0}\|^2 \leq \|z - x_0\|^2 \leq (\|z\| + M)^2.$$

Together with (11.19) this implies that

$$\min\{\|x_k - x_{k+1}\|^2 : k = 0, \dots, n_0 - 1\} \leq (\|z\| + M)^2 n_0^{-1} < \delta^2.$$

Therefore there is an integer j such that

$$0 \leq j \leq n_0 - 1, \ \|x_j - x_{j+1}\| \leq \delta. \quad (11.22)$$

In view of (11.20)

$$x_j - x_{j+1} \in \lambda_j T(x_{j+1}). \quad (11.23)$$

It follows from (11.22), (11.23), (11.21), (11.20), and (P1) that

$$d(x_{j+1}, F) \leq \epsilon/2. \quad (11.24)$$

By (11.24), (11.20), and Lemma 11.9,

$$d(x_i, F) \leq \epsilon/2 \text{ for all integers } i \geq j + 1$$

$$\text{and for all integers } i \geq n_0.$$

This completes the proof of Theorem 11.1.

11.4   Proofs of Theorems 11.3, 11.5, 11.6, and 11.7

Proof of Theorem 11.3. Assume that

$$\{\lambda_k\}_{k=0}^{n_0-1} \subset [\bar{\lambda}, \infty), \ \{x_k\}_{k=0}^{n_0} \subset R^n, \quad (11.25)$$

$$\|x_0\| \leq M, \ \|x_{k+1} - P_{\lambda_k}(x_k)\| \leq \delta = \epsilon_0(2n_0)^{-1}, \ k = 0, \ldots, n_0 - 1. \tag{11.26}$$

Put

$$y_0 = x_0, \ y_{k+1} = P_{\lambda_k}(y_k), \ k = 0, \ldots, n_0 - 1. \tag{11.27}$$

By the choice of n_0, (11.27), (11.26) and (11.25),

$$d(y_{n_0}, F) \leq \epsilon_0/2. \tag{11.28}$$

By Lemma 11.12 and (11.25)–(11.27),

$$\|y_{n_0} - x_{n_0}\| \leq n_0\delta \leq \epsilon_0/2.$$

Combined with (11.28) this implies that

$$d(x_{n_0}, F) \leq \epsilon_0.$$

Theorem 11.3 is proved.

Proof of Theorem 11.5. Let $\delta > 0$ and a natural number n_0 be as guaranteed by Theorem 11.2. There is a natural number p such that

$$\delta_k \leq \delta \text{ for all integers } k \geq p. \tag{11.29}$$

Put

$$n_\epsilon = n_0 + p. \tag{11.30}$$

Assume that

$$\{\lambda_k\}_{k=0}^{\infty} \subset [\bar{\lambda}, \infty), \ \{x_k\}_{k=0}^{\infty} \subset B(0, M), \tag{11.31}$$

$$\|x_{k+1} - P_{\lambda_k}(x_k)\| \leq \delta_k \text{ for all integers } k \geq 0.$$

In view of (11.29) and (11.31),

$$\|x_{k+1} - P_{\lambda_k}(x_k)\| \leq \delta \text{ for al integers } k \geq p. \tag{11.32}$$

By (11.32), (11.31), the choice of δ, n_0, Theorem 11.4 and (11.30),

$$d(x_k, F) \leq \epsilon \text{ for all integers } k \geq p + n_0 = n_\epsilon.$$

Theorem 11.5 is proved.

Proof of Theorem 11.6. We may assume without loss of generality that

$$M > 1 + \sup\{\|z\| : z \in F\}, \ \epsilon < 1. \tag{11.33}$$

By Theorem 11.2 there exist $\delta > 0$ and a natural number n_0 such that the following property holds:
(P2)
For each sequence $\{\lambda_k\}_{k=0}^{n_0-1} \subset [\bar{\lambda}, \infty)$ and each sequence $\{x_k\}_{k=0}^{n_0} \subset R^n$ which satisfies

$$\|x_0\| \le M,$$

$$\|x_{k+1} - P_{\lambda_k}(x_k)\| \le \delta, \ k = 0, \ldots, n_0 - 1$$

we have

$$d(x_{n_0}, F) \le \epsilon/4.$$

Put

$$\delta_* = \min\{\delta, (\epsilon/4)n_0^{-1}\}. \tag{11.34}$$

Assume that

$$\{\lambda_k\}_{k=0}^{\infty} \subset [\bar{\lambda}, \infty), \ \{x_k\}_{k=0}^{\infty} \subset R^n \tag{11.35}$$

and

$$\|x_0\| \le M,$$

$$\|x_{k+1} - P_{\lambda_k}(x_k)\| \le \delta_*, \ k = 0, 1, \ldots \tag{11.36}$$

By (11.35), (11.36), (11.34) and (P2),

$$d(x_{n_0}, F) \le \epsilon/4. \tag{11.37}$$

In view of (11.37) and (11.33),

$$\|x_{n_0}\| \le M. \tag{11.38}$$

We show by induction that for any natural number j,

$$d(x_{jn_0}, F) \le \epsilon/4, \ \|x_{jn_0}\| \le M. \tag{11.39}$$

Equations (11.37) and (11.38) imply that (11.39) is valid for $j = 1$.

Assume that j is a natural number and (11.39) holds. By (11.39), (11.35), (11.36), (11.34), and (P2),

$$d(x_{(j+1)n_0}, F) \leq \epsilon/4.$$

Together with (11.33) this implies that $\|x_{(j+1)n_0}\| \leq M$. Thus (11.39) holds for all natural numbers j.

Let j be a natural number. Put

$$y_{jn_0} = x_{jn_0}, \; y_{k+1} = P_{\lambda_k}(y_k), \; k = jn_0, \ldots, 2jn_0 - 1. \tag{11.40}$$

By Lemma 11.12, (11.35), (11.36), (11.40), and (11.34) for all $k = jn_0 + 1, \ldots$, $2(j+1)n_0$,

$$\|y_k - x_k\| \leq n_0 \delta_* \leq \epsilon/4. \tag{11.41}$$

Since the set F is closed and bounded there is $z \in F$ such that

$$d(x_{jn_0}, F) = \|x_{jn_0} - z\|. \tag{11.42}$$

It follows from (11.39) and (11.42) that

$$\|x_{jn_0} - z\| \leq \epsilon/4. \tag{11.43}$$

By (11.35), (11.40), the inclusion $z \in F$, Lemma 11.9 and (11.43),

$$\|y_k - z\| \leq \|y_{jn_0} - z\| = \|x_{jn_0} - z\| \leq \epsilon/4$$

$$\text{for all integers } k = jn_0 + 1, \ldots, 2jn_0. \tag{11.44}$$

By (11.41), (11.44) and the inclusion $z \in F$ for all integers $k = jn_0 + 1, \ldots$, $2(j+1)n_0$,

$$d(x_k, F) \leq \|x_k - z\| \leq \|x_k - y_k\| + \|y_k - z\| \leq \epsilon/4 + \epsilon/4.$$

Since j is any natural number we conclude that

$$d(x_k, F) \leq \epsilon/2 \text{ for all integers } j \geq n_0.$$

Theorem 11.6 is proved.

Proof of Theorem 11.7. We may assume without loss of generality that

$$M > 2 + \sup\{\|z\| : z \in F\}. \tag{11.45}$$

By Theorem 11.6 there are $\delta > 0$ and a natural number n_0 such that the following property holds:

(P3) for each $\{\lambda_k\}_{k=0}^{\infty} \subset [\bar{\lambda}, \infty)$ and each $\{x_k\}_{k=0}^{\infty} \subset R^n$ satisfying

$$\|x_0\| \leq M,$$

$$\|x_{k+1} - P_{\lambda_k}(x_k)\| \leq \delta, \ k = 0, 1, \dots$$

the inequality $d(x_k F) \leq 1$ holds for all integers $k \geq n_0$.

Assume that

$$\{\delta_k\}_{k=0}^{\infty} \subset (0, \delta], \ \lim_{k \to \infty} \delta_k = 0, \ \epsilon > 0. \tag{11.46}$$

We may assume without loss of generality that

$$\epsilon < 1. \tag{11.47}$$

By Theorem 11.6 there are $\delta_* \in (0, \delta)$ and a natural number n_* such that the following property holds:

(P4) for each $\{\lambda_k\}_{k=0}^{\infty} \subset [\bar{\lambda}, \infty)$ and each $\{x_k\}_{k=0}^{\infty} \subset R^n$ satisfying $\|x_0\| \leq M$,

$$\|x_{k+1} - P_{\lambda_k}(x_k)\| \leq \delta_*, \ k = 0, 1, \dots$$

we have $d(x_k, F) \leq \epsilon$ for all integers $k \geq n_*$.

By (11.46) there is a natural number p such that

$$\delta_k < \delta_* \text{ for all integers } k \geq p. \tag{11.48}$$

Put

$$n_\epsilon = n_0 + p + n_*. \tag{11.49}$$

Assume that

$$\{\lambda_k\}_{k=0}^{\infty} \subset [\bar{\lambda}, \infty), \ \{x_k\}_{k=0}^{\infty} \subset R^n, \tag{11.50}$$

$$\|x_0\| \leq M,$$

$$\|x_{k+1} - P_{\lambda_k}(x_k)\| \leq \delta_k, \ k = 0, 1, \dots \tag{11.51}$$

By (11.50), (11.51), (11.46) and (P3),

$$d(x_k, F) \leq 1 \text{ for all integers } k \geq n_0. \tag{11.52}$$

It follows from (11.52) and (11.45) that

$$\|x_k\| \le M \text{ for all integers } k \ge n_0. \tag{11.53}$$

By (11.48) and (11.51) for all integers $k \ge n_0 + p$,

$$\|x_{k+1} - P_{\lambda_k}(x_k)\| \le \delta_*. \tag{11.54}$$

It follows from (11.50), (11.53), (11.54), (11.49), and property (P4) applied to the sequence $\{x_k\}_{k=n_0+p}^{\infty}$ that

$$d(x_k, F) \le \epsilon \text{ for all integers } k \ge n_0 + p + n_* = n_\epsilon.$$

This completes the proof of Theorem 11.7.

Chapter 12
The Extragradient Method for Solving Variational Inequalities

In a Hilbert space, we study the convergence of the subgradient method to a solution of a variational inequality, under the presence of computational errors. The convergence of the subgradient method for solving variational inequalities is established for nonsummable computational errors. We show that the subgradient method generates a good approximate solution, if the sequence of computational errors is bounded from above by a constant.

12.1 Preliminaries and the Main Results

The studies of gradient-type methods and variational inequalities are important topics in optimization theory. See, for example, [3, 12, 30, 31, 37, 44, 52, 54, 68, 71–74] and the references mentioned therein. In this chapter we study convergence of the subgradient method, introduced in [75] and known in the literature as the extragradient method, to a solution of a variational inequality in a Hilbert space, under the presence of computational errors.

Let $(X, \langle \cdot, \cdot \rangle)$ be a Hilbert space with an inner product $\langle \cdot, \cdot \rangle$ which induces a complete norm $\| \cdot \|$. For each $x \in X$ and each $r > 0$ set

$$B(x, r) = \{ y \in X : \|x - y\| \leq r \}.$$

Let C be a nonempty closed convex subset of X. By Lemma 2.2, for each $x \in X$ there is a unique point $P_C(x) \in C$ satisfying

$$\|x - P_C(x)\| = \inf\{\|x - y\| : y \in C\}.$$

Moreover,

$$\|P_C(x) - P_C(y)\| \leq \|x - y\| \text{ for all } x, y \in X$$

© Springer International Publishing Switzerland 2016

A.J. Zaslavski, *Numerical Optimization with Computational Errors*, Springer Optimization and Its Applications 108, DOI 10.1007/978-3-319-30921-7_12

and

$$\langle z - P_C(x), x - P_C(x) \rangle \leq 0$$

for each $x \in X$ and each $z \in C$.

Consider a mapping $f : X \to X$. We say that the mapping f is monotone on C if

$$\langle f(x) - f(y), x - y \rangle \geq 0 \text{ for all } x, y \in C.$$

We say that f is pseudo-monotone on C if for each $x, y \in C$ the inequality

$$\langle f(y), x - y \rangle \geq 0 \text{ implies that } \langle f(x), x - y \rangle \geq 0.$$

Clearly, if f is monotone on C, then f is pseudo-monotone on C. Denote by S the set of all $x \in C$ such that

$$\langle f(x), y - x \rangle \geq 0 \text{ for all } y \in C. \tag{12.1}$$

We suppose that

$$S \neq \emptyset. \tag{12.2}$$

For each $\epsilon > 0$ denote by S_ϵ the set of all $x \in C$ such that

$$\langle f(x), y - x \rangle \geq -\epsilon \|y - x\| - \epsilon \text{ for all } y \in C. \tag{12.3}$$

In the sequel, we present examples which provide simple and clear estimations for the sets S_ϵ in some important cases. These examples show that elements of S_ϵ can be considered as ϵ-approximate solutions of the variational inequality.

In this chapter, in order to solve the variational inequality (to find $x \in S$), we use the algorithm known in the literature as the extragradient method [75]. In each iteration of this algorithm, in order to get the next iterate x_{k+1}, two orthogonal projections onto C are calculated, according to the following iterative step. Given the current iterate x_k calculate $y_k = P_C(x_k - \tau_k f(x_k))$ and then

$$x_{k+1} = P_C(x_k - \tau_k f(y_k)),$$

where τ_k is some positive number. It is known that this algorithm generates sequences which converge to an element of S. In this chapter, we study the behavior of the sequences generated by the algorithm taking into account computational errors which are always present in practice. Namely, in practice the algorithm generates sequences $\{x_k\}_{k=0}^{\infty}$ and $\{y_k\}_{k=0}^{\infty}$ such that for each integer $k \geq 0$,

$$\|y_k - P_C(x_k - \tau_k f(x_k))\| \leq \delta$$

and

$$\|x_{k+1} - P_C(x_k - \tau_k f(y_k))\| \le \delta,$$

with a constant $\delta > 0$ which depends only on our computer system. Surely, in this situation one cannot expect that the sequence $\{x_k\}_{k=0}^{\infty}$ converges to the set S. The goal is to understand what subset of C attracts all sequences $\{x_k\}_{k=0}^{\infty}$ generated by the algorithm. The main result of this chapter (Theorem 12.2) shows that this subset of C is the set S_ϵ with some $\epsilon > 0$ depending on δ [see (12.9) and (12.10)]. The examples considered in this section show that one cannot expect to find an attracting set smaller than S_ϵ, whose elements can be considered as approximate solutions of the variational inequality.

The results of this chapter were obtained [127].

We suppose that the mapping f is Lipschitz on all bounded subsets of X and that

$$\langle f(y), y - x \rangle \ge 0 \text{ for all } y \in C \text{ and all } x \in S. \tag{12.4}$$

Remark 12.1. Note that (12.4) holds if f is pseudo-monotone on C.

Usually algorithms, studied in the literature, generate sequences which converge weakly to an element of S. In this chapter, for a given $\epsilon > 0$, we are interested to find a point $x \in X$ such that $\inf\{\|x - y\| : y \in S_\epsilon\} \le \epsilon$. This point x is considered as an ϵ-approximate solution of the variational inequality. We will prove the following result, which shows that an ϵ-approximate solution can be obtained after k iterations of the subgradient method and under the presence of computational errors bounded from above by a constant δ, where δ and k are constants depending on ϵ.

Theorem 12.2. *Let* $\epsilon \in (0, 1)$, $M_0 > 0$, $M_1 > 0$, $L > 0$ *be such that*

$$B(0, M_0) \cap S \ne \emptyset, \tag{12.5}$$

$$f(B(0, 3M_0 + 1)) \subset B(0, M_1), \tag{12.6}$$

$$\|f(z_1) - f(z_2)\| \le L\|z_1 - z_2\|$$

for all $z_1, z_2 \in B(0, 3M_0 + M_1 + 1),$ \tag{12.7}

$$0 < \tilde{\tau} < \tau_* \le 1, \ \tau_* L < 1, \tag{12.8}$$

$\epsilon_0 > 0$ *satisfy*

$$3\epsilon_0(M_1 + \tilde{\tau}^{-1}(1 + M_0 + M_1) + (L + \tilde{\tau}^{-1})) < \epsilon, \tag{12.9}$$

let $\delta \in (0, 1)$ *satisfy*

$$4\delta(1 + 2M_0) < \epsilon_0^2(1 - \tau_*^2 L^2)/2 \tag{12.10}$$

and let an integer

$$k > 8M_0^2\epsilon_0^{-2}(1 - \tau_*^2 L^2)^{-1}. \tag{12.11}$$

Assume that

$$\{x_i\}_{i=0}^\infty \subset X, \ \{y_i\}_{i=0}^\infty \subset X, \ \{\tau_i\}_{i=0}^\infty \subset [\tilde{\tau}, \tau_*], \tag{12.12}$$

$$\|x_0\| \le M_0, \tag{12.13}$$

and that for each integer $i \ge 0$,

$$\|y_i - P_C(x_i - \tau_i f(x_i))\| \le \delta, \tag{12.14}$$

$$\|x_{i+1} - P_C(x_i - \tau_i f(y_i))\| \le \delta. \tag{12.15}$$

Then there is an integer $j \in [0, k]$ such that

$$x_i \in B(0, 3M_0), \ i = 0, \dots, j, \tag{12.16}$$

$$\|x_j - y_j\| \le 2\epsilon_0,$$

$$\|x_i - y_i\| > 2\epsilon_0 \text{ for all integers } i \text{ satisfying } 0 \le i < j. \tag{12.17}$$

Moreover, if an integer $j \in [0, k]$ satisfies (12.17), then

$$\langle f(x_j), \xi - x_j \rangle \ge -\epsilon - \epsilon\|\xi - x_j\| \text{ for all } \xi \in C \tag{12.18}$$

and there is $y \in S_\epsilon$ such that $\|x_j - y\| \le \epsilon$.

Note that Theorem 12.2 provides the estimations for the constants δ and k, which follow from (12.9)–(12.11). Namely, $\delta = c_1\epsilon^2$ and $k = c_2\epsilon^{-2}$, where c_1 and c_2 are positive constants depending on M_0.

Let us consider the following particular example.

Example 12.3. Assume that $f(x) = x$ for all $x \in X$ and $C = X$. Then $S = \{0\}$. Let $\epsilon \in (0, 1/2)$ and $M_0 = 10^2$. Clearly, in this case

$$S_\epsilon \subset \{y \in X : \|y\| \le 2\epsilon\}$$

and the assertion of Theorem 12.2 holds with $M_1 = 400$ [see (12.6)], $L = 1$ [see (12.7)], $\tilde{\tau} = 2^{-1}$, $\tau_* = 3/4$ [see (12.8)],

$$\epsilon_0 = 5^{-1} \cdot 10^{-3}\epsilon$$

[see (12.9)],

$$\delta = 2^{-1} \cdot 10^{-11} \epsilon^2$$

[see (12.10)] and with k, which is the smallest integer larger than $10^{12} \cdot 16\epsilon^{-2}$ [see (12.11)].

The following example demonstrates that the set S_ϵ can be easily calculated if the mapping f is strongly monotone.

Example 12.4. Let $\bar{r} \in (0, 1)$. Set

$$C_{\bar{r}} = \{x \in X : \|x - P_C(x)\| \leq \bar{r}\}.$$

We say that f is strongly monotone with a constant $\alpha > 0$ on $C_{\bar{r}}$ if

$$\langle f(x) - f(y), x - y \rangle \geq \alpha \|x - y\|^2 \text{ for all } x, y \in C_{\bar{r}}.$$

Fix $u_* \in S$. We suppose that there is $\alpha \in (0, 1)$ such that

$$\langle f(x), x - u_* \rangle \geq \alpha \|x - u_*\|^2 \text{ for all } x \in C_{\bar{r}}. \tag{12.19}$$

Remark 12.5. Note that inequality (12.19) holds, if f is strongly monotone with a constant α on $C_{\bar{r}}$.

Let $\epsilon > 0$ and $x \in S_\epsilon$. Then for all $y \in C$,

$$\langle f(x), y - x \rangle \geq -\epsilon \|y - x\| - \epsilon$$

and in particular

$$-\epsilon \|u_* - x\| - \epsilon \leq \langle f(x), u_* - x \rangle \leq -\alpha \|x - u_*\|^2.$$

This implies that

$$\alpha \|x - u_*\|^2 \leq 2 \max\{\epsilon, \ \epsilon \|x - u_*\|\},$$

$$\|x - u_*\| \leq \max\{2\epsilon\alpha^{-1}, \ (2\epsilon\alpha^{-1})^{1/2}\}$$

and if $\epsilon \leq 2^{-1}\alpha$, then

$$\|x - u_*\| \leq (2\epsilon\alpha^{-1})^{1/2}$$

and

$$S_\epsilon \subset \{x \in X : \|x - u_*\| \leq (2\epsilon\alpha^{-1})^{1/2}\}.$$

According to Theorem 12.2, under its assumptions, there is an integer $j \in [0, k]$ such that

$$\|x_j - u_*\| \leq \epsilon + (2\epsilon\alpha^{-1})^{1/2}.$$

Note that the constant α can be obtained by analyzing an explicit form of the mapping f.

In next example we show what is the set S_ϵ when $C = X$.

Example 12.6. Assume that $C = X$. It is easy to see that

$$S = \{x \in X : f(x) = 0\}.$$

Let $\epsilon > 0$ and $x \in S_\epsilon$. Then for all $z \in B(0, 1)$,

$$\langle f(x), z \rangle \geq -\epsilon\|z\| - \epsilon \geq -2\epsilon$$

and $\|f(x)\| \leq 2\epsilon$. Thus

$$S_\epsilon \subset \{x \in X : \|f(x)\| \leq 2\epsilon\}.$$

In the following example, we demonstrate that, if computational errors made by our computer system are $\delta > 0$, then in principle any element of the set S_ϵ, where ϵ is a positive constant depending on δ, can be a limit point of the sequence $\{x_i\}_{i=0}^\infty$ generated by the subgradient method. This means that Theorem 12.2 cannot be improved.

Example 12.7. Assume that $f(x) = x$ for all $x \in X$ and $C = X$. Clearly, f is strongly monotone on X with a constant 1 and $S = \{0\}$. According to Example 12.6 for any $\epsilon \in (0, 2^{-1})$, $S_\epsilon \subset \{y \in X : \|y\| \leq 2\epsilon\}$.

Let $\tau \in (0, 1)$, $\delta \in (0, 1)$,

$$v \in B(0, \delta)$$

and let sequences $\{x_i\}_{i=0}^\infty$, $\{y_i\}_{i=0}^\infty \subset X$ satisfy for all integers $i \geq 0$,

$$y_i = x_i - \tau f(x_i) = (1 - \tau)x_i,$$

$$x_{i+1} = x_i - \tau f(y_i) + v = (1 - \tau + \tau^2)x_i + v. \tag{12.20}$$

By induction it follows from (12.20) that for all integers $n \geq 1$,

$$x_n = (1 - \tau + \tau^2)^n x_0 + \sum_{i=0}^{n-1} (1 - \tau + \tau^2)^i v \to (\tau - \tau^2)^{-1} v \text{ as } n \to \infty.$$

Thus any

$$\xi \in B(0, \delta(\tau - \tau^2)^{-1})$$

can be a limit of a sequence $\{x_n\}_{n=0}^{\infty}$ generated by the subgradient method under the present of computational errors δ.

In next example we obtain an estimation for the sets S_ϵ when $f(\cdot)$ is the Gâteaux derivative of a convex function.

Example 12.8. Assume that $F : X \to R^1$ is a convex Gâteaux differentiable function and $f(x) = F'(x)$ for all $x \in X$, where $F'(x)$ is the Gâteaux derivative of F at the point $x \in X$. We suppose that $F' = f$ be Lipschitz on all bounded subsets of X and that $x_* \in C$ satisfies

$$F(x_*) \le F(z) \text{ for all } z \in C.$$

Assume that a constant $\tilde{M} > \|x_*\|$ be given. (Note that $\tilde{M}$ can be known a priori or obtained by analyzing an explicit form of the function F.) Let $\epsilon \in (0, 1)$ and

$$x \in S_\epsilon \cap B(0, \tilde{M}).$$

Then for each $y \in C$,

$$F(y) - F(x) \ge \langle F'(x), y - x \rangle$$
$$= \langle f(x), y - x \rangle \ge -\epsilon \|y - x\| - \epsilon$$

and, in particular,

$$F(x_*) - F(x)$$
$$\ge -\epsilon \|x - x_*\| - \epsilon \ge -2\epsilon \tilde{M} - \epsilon.$$

Thus

$$S_\epsilon \cap B(0, \tilde{M}) \subset \{x \in C : F(x) \le F(x_*) + \epsilon(2\tilde{M} + 1)\}.$$

The chapter is organized as follows. Section 12.2 contains auxiliary results. Theorem 12.2 is proved in Sect. 12.3. Convergence results for the finite-dimensional space X are obtained in Sects. 12.4 and 12.5.

12.2 Auxiliary Results

We use the assumptions, definitions, and the notation introduced in Sect. 12.1.

Lemma 12.9. *Assume that*

$$\tau > 0, \ u_* \in S, \ M_0 > 0, \ M_1 > 0, \ L > 0, \tag{12.21}$$

$$f(B(u_*, M_0)) \subset B(0, M_1), \tag{12.22}$$

$$\|f(z_1) - f(z_2)\| \leq L\|z_1 - z_2\|$$

for all $z_1, z_2 \in B(u_*, M_0 + \tau M_1)$. $\tag{12.23}$

Let

$$u \in B(u_*, M_0), \ v = P_C(u - \tau f(u)), \tag{12.24}$$

$$T := \{w \in X : \ \langle u - \tau f(u) - v, w - v \rangle \leq 0\}, \tag{12.25}$$

D be a convex and closed subset of X such that

$$C \subset D \subset T \tag{12.26}$$

(by Lemma 2.2, $C \subset T$) and let

$$\tilde{u} = P_D(u - \tau f(v)). \tag{12.27}$$

Then

$$\|\tilde{u} - u_*\|^2 \leq \|u - u_*\|^2 - (1 - \tau^2 L^2)\|u - v\|^2. \tag{12.28}$$

Proof. By (12.22) and (12.24),

$$\|f(u)\| \leq M_1.$$

Together with (12.21), (12.24), and Lemma 2.2, this implies that

$$\|u_* - v\| \leq \|u_* - (u - \tau f(u))\| \leq M_0 + \tau M_1. \tag{12.29}$$

By (12.4), (12.21), and (12.24),

$$\langle f(v), \tilde{u} - u_* \rangle \geq \langle f(v), \tilde{u} - v \rangle. \tag{12.30}$$

In view of (12.25), (12.26), and (12.27),

$$\langle \tilde{u} - v, (u - \tau f(u)) - v \rangle \leq 0.$$

This implies that

$$\langle \tilde{u} - v, (u - \tau f(v)) - v \rangle \leq \tau \langle \tilde{u} - v, f(u) - f(v) \rangle. \tag{12.31}$$

Set

$$z = u - \tau f(v). \tag{12.32}$$

By (12.27) and (12.32),

$$||\tilde{u} - u_*||^2 = ||z - u_*||^2 + ||z - P_D(z)||^2$$
$$+ 2\langle P_D(z) - z, z - u_* \rangle. \tag{12.33}$$

By (12.21), (12.26) and Lemma 2.2,

$$2\|z - P_D(z)\|^2 + 2\langle P_D(z) - z, z - u_* \rangle$$
$$= 2\langle z - P_D(z), u_* - P_D(z) \rangle \leq 0.$$

Together with (12.27), (12.30), (12.32), and (12.33) this implies that

$$\|\tilde{u} - u_*\|^2 \leq \|z - u_*\|^2 - \|z - P_D(z)\|^2$$
$$= \|u - \tau f(v) - u_*\|^2 - \|u - \tau f(v) - \tilde{u}\|^2$$
$$= \|u - u_*\|^2 - \|u - \tilde{u}\|^2$$
$$+ 2\tau \langle u_* - \tilde{u}, f(v) \rangle$$
$$\leq \|u - u_*\|^2 - \|u - \tilde{u}\|^2 + 2\tau \langle v - \tilde{u}, f(v) \rangle.$$

Together with (12.31) this implies that

$$\|\tilde{u} - u_*\|^2 \leq \|u - u_*\|^2$$
$$+ 2\tau \langle v - \tilde{u}, f(v) \rangle - \langle u - v + v - \tilde{u}, u - v + v - \tilde{u} \rangle$$
$$= \|u - u_*\|^2 - \|u - v\|^2 - \|v - \tilde{u}\|^2$$
$$+ 2\langle \tilde{u} - v, u - v - \tau f(v) \rangle$$
$$\leq \|u - u_*\|^2 - \|u - v\|^2 - \|v - \tilde{u}\|^2$$
$$+ 2\tau \langle \tilde{u} - v, f(u) - f(v) \rangle. \tag{12.34}$$

By Cauchy-Schwarz inequality, (12.23), (12.24), (12.29), and (12.34),

$$\|\tilde{u} - u_*\|^2 \leq \|u - u_*\|^2 - \|v - u\|^2 - \|v - \tilde{u}\|^2 + 2\tau L\|\tilde{u} - v\|\|u - v\|$$
$$\leq \|u - u_*\|^2 - \|u - v\|^2 - \|v - \tilde{u}\|^2 + \tau^2 L^2\|u - v\|^2 + \|\tilde{u} - v\|^2$$
$$\leq \|u - u_*\|^2 - (1 - \tau^2 L^2)\|u - v\|^2.$$

Lemma 12.9 is proved.

Lemma 12.10. *Let*

$$u_* \in S,\ M_0 > 0,\ M_1 > 0,\ L > 0, \delta \in (0,1), \tag{12.35}$$

$$f(B(u_*, M_0)) \subset B(0, M_1), \tag{12.36}$$

$$\|f(z_1) - f(z_2)\| \le L\|z_1 - z_2\|$$

for all $z_1, z_2 \in B(u_*, M_0 + M_1 + 1),$ \tag{12.37}

$$\tau \in (0, 1],\ L\tau < 1. \tag{12.38}$$

Assume that

$$x \in B(u_*, M_0),\ y \in X,\ \|y - P_C(x - \tau f(x))\| \le \delta, \tag{12.39}$$

$$\tilde{x} \in X,\ \|\tilde{x} - P_C(x - \tau f(y))\| \le \delta. \tag{12.40}$$

Then

$$\|\tilde{x} - u_*\|^2 \le 4\delta(1 + M_0) + \|x - u_*\|^2 - (1 - \tau^2 L^2)\|x - P_C(x - \tau f(x))\|^2.$$

Proof. Set

$$v = P_C(x - \tau f(x)),\ z = P_C(x - \tau f(v)),\ \tilde{z} = P_C(x - \tau f(y)). \tag{12.41}$$

By Lemma 12.9, (12.41), (12.35), (12.36), (12.37), (12.38), and (12.39),

$$\|z - u_*\|^2 \le \|x - u_*\|^2 - (1 - \tau^2 L^2)\|x - v\|^2. \tag{12.42}$$

Clearly,

$$\begin{aligned}
\|\tilde{x} - u_*\|^2 &= \|\tilde{x} - z + z - u_*\|^2 \\
&= \|\tilde{x} - z\|^2 + 2\langle \tilde{x} - z, z - u_* \rangle + \|z - u_*\|^2 \\
&\le \|\tilde{x} - z\|^2 + 2\|\tilde{x} - z\|\|z - u_*\| + \|z - u_*\|^2.
\end{aligned} \tag{12.43}$$

By (12.39) and (12.41),

$$\|v - y\| \le \delta. \tag{12.44}$$

It follows from (12.41), (12.35), Lemma 2.2, (12.39), and (12.36) that

$$\|u_* - v\| \le \|u_* - x\| + \tau\|f(x)\| \le M_0 + \tau M_1. \tag{12.45}$$

By (12.45), (12.44), and (12.35),

$$\|u_* - y\| \le \|u_* - v\| + \|v - y\| \le M_0 + \tau M_1 + 1. \tag{12.46}$$

By (12.46), (12.41), Lemma 2.2, (12.45), (12.40), (12.37), (12.38), and (12.44),

$$\begin{aligned}
\|\tilde{x} - z\| &\le \|\tilde{x} - \tilde{z}\| + \|\tilde{z} - z\| \\
&\le \delta + \|\tilde{z} - z\| \le \delta + \tau \|f(y) - f(v)\| \\
&\le \delta + \tau L \delta = \delta(1 + \tau L).
\end{aligned} \tag{12.47}$$

By (12.43), (12.47), (12.42), (12.41), (12.38), and (12.39),

$$\begin{aligned}
\|\tilde{x} - u_*\|^2 &\le (\delta(1 + \tau L))^2 + 2\delta(1 + \tau L)\|z - u_*\| + \|z - u_*\|^2 \\
&\le (\delta(1 + \tau L))^2 + 2\delta(1 + \tau L)\|x - u_*\| + \|z - u_*\|^2 \\
&\le \delta^2(1 + \tau L)^2 + 2\delta(1 + \tau L)\|x - u_*\| \\
&\quad + \|x - u_*\|^2 - (1 - \tau^2 L^2)\|x - P_C(x - \tau f(x))\|^2 \\
&\le 4\delta^2 + 4\delta M_0 + \|x - u_*\|^2 - (1 - \tau^2 L^2)\|x - P_C(x - \tau f(x))\|^2.
\end{aligned}$$

This completes the proof of Lemma 12.10.

12.3   Proof of Theorem 12.2

By (12.5) there is

$$u_* \in S \cap B(0, M_0). \tag{12.48}$$

By (12.13) and (12.48),

$$\|x_0 - u_*\| \le 2M_0. \tag{12.49}$$

Assume that $i \ge 0$ is an integer and that

$$x_i \in B(u_*, 2M_0). \tag{12.50}$$

(Note that for $i = 0$ (12.50) holds.) By Lemma 12.10 applied with $x = x_i$, $y = y_i$, $\tilde{x} = x_{i+1}$ and $\tau = \tau_i$, (12.48), (12.50), (12.6), (12.7), (12.12), (12.14), and (12.15),

$$\|x_{i+1} - u_*\|^2 \le 4\delta(1 + 2M_0) + \|x_i - u_*\|^2$$

$$- (1 - \tau_i^2 L^2)\|x_i - P_C(x_i - \tau_i f(x_i))\|^2. \tag{12.51}$$

There are two cases:

$$\|x_i - y_i\| \leq 2\epsilon_0;$$ (12.52)

$$\|x_i - y_i\| > 2\epsilon_0.$$ (12.53)

Assume that (12.53) holds. Then by (12.53), (12.14), (12.10), and the inequality $\epsilon_0 < 1$,

$$\|x_i - P_C(x_i - \tau_i f(x_i))\|$$
$$\geq \|x_i - y_i\| - \| - y_i + P_C(x_i - \tau_i f(x_i))\|$$
$$> 2\epsilon_0 - \delta > \epsilon_0 > \epsilon_0^2.$$ (12.54)

Then in view of (12.54), (12.51), (12.12), and (12.10),

$$\|x_{i+1} - u_*\|^2$$
$$\leq 4\delta(1 + 2M_0) + \|x_i - u_*\|^2 - \epsilon_0^2(1 - \tau_*^2 L^2)$$
$$\leq \|x_i - u_*\|^2 - \epsilon_0^2(1 - \tau_*^2 L^2)2^{-1}.$$ (12.55)

Thus we have shown that the following property holds:

(P) if an integer $i \geq 0$ satisfies (12.50) and (12.53), then (12.55) holds.

Property (P), (12.52), (12.53), and (12.49) imply that at least one of the following cases holds:

(a) for all integers $i = 0, \ldots, k$ the relations (12.50), (12.53), and (12.55) are true;
(b) there is an integer $j \in \{0, \ldots, k\}$ such that for all integers $i = 0, \ldots, j$, (12.50) is valid, for all integers i satisfying $0 \leq i < j$ (12.53) holds and that

$$\|x_j - y_j\| \leq 2\epsilon_0.$$ (12.56)

Assume that case (a) holds. Then by (12.49) and (12.55),

$$4M_0^2 \geq \|u_* - x_0\|^2 - \|u_* - x_k\|^2$$
$$= \sum_{i=0}^{k-1}[\|u_* - x_i\|^2 - \|u_* - x_{i+1}\|^2] \geq 2^{-1}k\epsilon_0^2(1 - \tau_*^2 L^2)$$

and

$$k \leq 8M_0^2\epsilon_0^{-2}(1 - \tau_*^2 L^2)^{-1}.$$

This contradicts (12.11). The contradiction we have reached proves that case (a) does not hold. Then case (b) holds and there is an integer $j \in \{0, \ldots, k\}$ guaranteed by (b). Then (12.16) and (12.17) hold.

Assume that an integer $j \in [0, k]$ satisfies (12.17). (Clearly, in view of (b) such integer j is unique.) Then

$$\|x_j - u_*\| \leq 2M_0, \ \|x_j - y_j\| \leq 2\epsilon_0. \tag{12.57}$$

By (12.57), (12.10), and (12.14),

$$\|x_j - P_C(x_j - \tau_j f(x_j))\|$$

$$\leq \|x_j - y_j\| + \|y_j - P_C(x_j - \tau_j f(x_j))\| \leq 2\epsilon_0 + \delta \leq 3\epsilon_0. \tag{12.58}$$

By Lemma 2.2, for each $\xi \in C$,

$$0 \geq \langle x_j - \tau_j f(x_j) - P_C(x_j - \tau_j f(x_j)), \xi - P_C(x_j - \tau_j f(x_j)) \rangle. \tag{12.59}$$

By (12.58) and (12.59) for each $\xi \in C$,

$$0 \geq \langle x_j - P_C(x_j - \tau_j f(x_j)), \xi - P_C(x_j - \tau_j f(x_j)) \rangle$$
$$\quad -\tau_j \langle f(x_j), \xi - P_C(x_j - \tau_j f(x_j)) \rangle$$
$$\geq -\|x_j - P_C(x_j - \tau_j f(x_j))\|(\|\xi - x_j\| + \|x_j - P_C(x_j - \tau_j f(x_j))\|)$$
$$\quad -\tau_j \langle f(x_j), \xi - x_j \rangle - \tau_j \langle f(x_j), x_j - P_C(x_j - \tau_j f(x_j)) \rangle$$
$$\geq -3\epsilon_0 \|\xi - x_j\| - 9\epsilon_0^2 - \tau_j \langle f(x_j), \xi - x_j \rangle - 3\tau_j \|f(x_j)\| \epsilon_0. \tag{12.60}$$

By (12.60), (12.12), (12.6), (12.9), (12.57), (12.48) for each $\xi \in C$,

$$\tau_j \langle f(x_j), \xi - x_j \rangle \geq -3\epsilon_0(1 + \tau_* M_1) - 3\epsilon_0 \|\xi - x_j\|$$

and in view of (12.8) and (12.12)

$$\langle f(x_j), \xi - x_j \rangle \geq -3\epsilon_0 \tau_j^{-1}(1 + \tau_* M_1) - 3\epsilon_0 \tau_j^{-1} \|\xi - x_j\|$$
$$\geq -3\epsilon_0 \tilde{\tau}^{-1}(1 + M_1) - 3\epsilon_0 \tilde{\tau}^{-1} \|\xi - x_j\|. \tag{12.61}$$

By (12.9) and (12.61),

$$\langle f(x_j), \xi - x_j \rangle \geq -\epsilon - \epsilon \|\xi - x_j\| \text{ for all } \xi \in C. \tag{12.62}$$

Clearly, (12.62) is the claimed (12.18). Set

$$\bar{y} = P_C(x_j - \tau_j f(x_j)). \tag{12.63}$$

By (12.63) and (12.58),

$$\bar{y} \in C, \ \|x_j - \bar{y}\| \leq 3\epsilon_0 \leq \epsilon < 1. \tag{12.64}$$

By (12.57), (12.48), (12.64), and (12.7),

$$\|f(x_j) - f(\bar{y})\| \leq L\|x_j - \bar{y}\| \leq 3\epsilon_0 L. \tag{12.65}$$

By (12.64), (12.57), (12.48), (12.6), (12.9), (12.61), (12.65), and (12.4) for each $\xi \in C$,

$$\langle f(\bar{y}), \xi - \bar{y} \rangle$$
$$\geq \langle f(\bar{y}), \xi - x_j \rangle - \|f(\bar{y})\| \|x_j - \bar{y}\|$$
$$\geq \langle f(\bar{y}), \xi - x_j \rangle - 3M_1\epsilon_0$$
$$\geq \langle f(x_j), \xi - x_j \rangle - \|f(\bar{y}) - f(x_j)\| \|\xi - x_j\| - 3M_1\epsilon_0$$
$$\geq -3\epsilon_0 \tilde{\tau}^{-1}(1 + M_1) - 3\epsilon_0 \tilde{\tau}^{-1}\|\xi - x_j\| - 3\epsilon_0 L\|\xi - x_j\| - 3M_1\epsilon_0$$
$$\geq -3\epsilon_0(M_1 + \tilde{\tau}^{-1}(1 + M_1)) - 3\epsilon_0(L + \tilde{\tau}^{-1})(\|\xi - \bar{y}\| + \|\bar{y} - x_j\|)$$
$$\geq -3\epsilon_0(L + \tilde{\tau}^{-1})(\|\xi - \bar{y}\|) - 3\epsilon_0(M_1 + \tilde{\tau}^{-1}(1 + M_1) + (L + \tilde{\tau}^{-1}))$$
$$\geq -\epsilon\|\xi - \bar{y}\| - \epsilon$$

for all $\xi \in C$. Thus $\bar{y} \in S_\epsilon$. This completes the proof of Theorem 12.2.

12.4   The Finite-Dimensional Case

We use the assumptions, definitions, and notation introduced in Sect. 12.1 and we prove the following result.

Theorem 12.11. *Let* $X = R^n$, $\epsilon \in (0, 1)$, $M_0 > 0$ *be such that*

$$B(0, M_0) \cap S \neq \emptyset,$$

$M_1 > 0$ *be such that*

$$f(B(0, 3M_0 + 1)) \subset B(0, M_1),$$

$L > 0$ *be such that*

$$\|f(z_1) - f(z_2)\| \leq L\|z_1 - z_2\|$$

for all $z_1, z_2 \in B(0, 3M_0 + M_1 + 1)$,

$$0 < \tau \leq 1, \ \tau L < 1.$$

Then there exist $\delta \in (0, \epsilon)$ and an integer $k \geq 1$ such that for each $\{x_i\}_{i=0}^{\infty} \subset R^n$ and each $\{y_i\}_{i=0}^{\infty} \subset R^n$ which satisfy $\|x_0\| \leq M_0$ and for each integer $i \geq 0$,

$$\|y_i - P_C(x_i - \tau f(x_i))\| \leq \delta,$$

$$\|x_{i+1} - P_C(x_i - \tau_i f(y_i))\| \leq \delta$$

there is an integer $j \in [0, k]$ such that

$$\|x_j\| \leq 3M_0 \text{ and } \inf\{\|x_j - z\| : z \in S\} \leq \epsilon.$$

Theorem 12.11 follows immediately from Theorem 12.2 and the following result.

Lemma 12.12. *Let $M_0 > 0$, $\gamma > 0$. Then there exists $\epsilon \in (0, \gamma)$ such that for each*

$$z \in S_\epsilon \cap B(0, M_0)$$

the following relation holds:

$$\inf\{\|z - u\| : u \in S\} \leq \gamma.$$

Proof. Assume the contrary. Then there exists a sequence $\{\epsilon_k\}_{k=1}^{\infty} \subset (0, \gamma)$ which converges to zero and a sequence

$$z^{(k)} \in B(0, M_0) \cap S_{\epsilon_k}, \ k = 1, 2, \ldots$$

such that for each integer $k \geq 1$,

$$\inf(\{\|z^{(k)} - u\| : u \in S\}) > \gamma. \tag{12.66}$$

We may assume without loss of generality that there is

$$z = \lim_{k \to \infty} z^{(k)}.$$

By definition, for each integer $k \geq 1$ and each $\xi \in C$,

$$\langle f(z^{(k)}), \xi - z^{(k)} \rangle \geq -\epsilon_k \|\xi - z^{(k)}\| - \epsilon_k.$$

This implies that for each $\xi \in C$,

$$\langle f(z), \xi - z \rangle = \lim_{k \to \infty} \langle f(z^{(k)}), \xi - z^{(k)} \rangle$$

$$\geq \lim_{k \to \infty} (-\epsilon_k \|\xi - z^{(k)}\| - \epsilon_k) = 0.$$

Thus $z \in S$ and for all natural numbers k large enough

$$\|z - z^{(k)}\| < \gamma/2.$$

This contradicts (12.66). The contradiction we have reached proves Lemma 12.12.

12.5 A Convergence Result

We use the assumptions, definitions, and notation introduced in Sect. 12.1. Let $X = R^n$. For each $x \in R^n$ and each $A \subset R^n$ set

$$d(x, A) = \inf\{\|x - y\| : y \in A\}.$$

Suppose that the set S is bounded and choose $\bar{M}_0 > 2, \bar{M}_1, \bar{M}_2 > 0$ such that

$$S \subset B(0, \bar{M}_0 - 2), \tag{12.67}$$

$$f(B(0, 3\bar{M}_0 + 1)) \subset B(0, \bar{M}_1), \ f(B(0, 3\bar{M}_0 + 3\bar{M}_1 + 1)) \subset B(0, \bar{M}_2). \tag{12.68}$$

Assume that

$$M_0 > \bar{M}_0 + \bar{M}_1 + \bar{M}_2, \ M_1 > 0, \ L > 0, \tag{12.69}$$

$$f(B(0, 3M_0 + 1)) \subset B(0, M_1), \tag{12.70}$$

$$\|f(z_1) - f(z_2)\| \leq L\|z_1 - z_2\| \text{ for all } z_1, z_2 \in B(0, 3M_0 + M_1 + 1), \tag{12.71}$$

$$0 < \tau \leq 1 \text{ and } \tau L < 1. \tag{12.72}$$

By Theorem 12.11 there exist

$$\gamma \in (0, 4^{-1}) \tag{12.73}$$

and a natural number $\bar{k}$ such that the following property holds:

(P2) for each pair of sequences $\{x_i\}_{i=0}^{\infty} \subset R^n$ and $\{y_i\}_{i=0}^{\infty} \subset R^n$ with $\|x_0\| \leq M_0$ and for each integer $i \geq 0$ with

$$\|y_i - P_C(x_i - \tau f(x_i))\| \leq \gamma, \ \|x_{i+1} - P_C(x_i - \tau f(y_i))\| \leq \gamma \tag{12.74}$$

there is an integer $j \in [0, \bar{k}]$ such that $d(x_j, S) \leq 1/4$.

We prove the following result.

Theorem 12.13. *Let*

$$\{\delta_i\}_{i=0}^{\infty} \subset (0, \gamma], \ \lim_{i \to \infty} \delta_i = 0 \tag{12.75}$$

and let $\epsilon \in (0, 1)$. Then there exists a natural number k_0 such that for each pair of sequences $\{x_i\}_{i=0}^{\infty} \subset R^n$ and $\{y_i\}_{i=0}^{\infty} \subset R^n$ which satisfies

$$\|x_0\| \le M_0 \tag{12.76}$$

and for each integer $i \ge 0$ with

$$\|y_i - P_C(x_i - \tau f(x_i))\| \le \delta_i, \ \|x_{i+1} - P_C(x_i - \tau f(y_i))\| \le \delta_i \tag{12.77}$$

the inequality

$$d(x_i, S) \le \epsilon$$

holds for all integers $i \ge k_0$.

Proof. By Theorem 12.11, there exist a positive number

$$\bar{\delta} < \epsilon^2 (1 + M_0)^{-1} 64^{-1} \tag{12.78}$$

and an integer $k_1 \ge 1$ such that the following property holds:
 (P3) for each pair of sequences $\{u_i\}_{i=0}^{\infty} \subset R^n$ and $\{v_i\}_{i=0}^{\infty} \subset R^n$ such that

$$\|u_0\| \le M_0 \tag{12.79}$$

and that for each integer $i \ge 0$,

$$\|v_i - P_C(u_i - \tau f(u_i))\| \le \bar{\delta}, \ \|u_{i+1} - P_C(u_i - \tau f(v_i))\| \le \bar{\delta} \tag{12.80}$$

there is an integer $j \in [0, k_1]$ such that

$$d(u_j, S) \le \epsilon/8. \tag{12.81}$$

By (12.75) there is an integer $k_2 \ge 1$ such that

$$\delta_i < k_1^{-2} \bar{\delta} \text{ for all integers } i \ge k_2. \tag{12.82}$$

Set

$$k_0 = 2 + \bar{k} + k_1 + k_2. \tag{12.83}$$

Assume that sequences $\{x_i\}_{i=0}^{\infty} \subset R^n$ and $\{y_i\}_{i=0}^{\infty} \subset R^n$ satisfy (12.76) and that for each integer $i \geq 0$ equation (12.77) holds. Assume that an integer $j \geq 0$ satisfies

$$\|x_j\| \leq M_0. \tag{12.84}$$

We show that there exists an integer $i \in [1+j, 1+j+\bar{k}]$ such that $\|x_i\| \leq M_0$.
By (12.75), (12.77), (12.84), and (P2) there is an integer $p \in [j, j+\bar{k}]$ such that

$$d(x_p, S) \leq 1/4. \tag{12.85}$$

In view of (12.67), (12.69), and (12.85),

$$\|x_p\| \leq M_0.$$

If $p > j$ we put $i = p$ and obtain

$$i \in [j+1, j+\bar{k}], \ \|x_i\| \leq M_0. \tag{12.86}$$

Assume that $p = j$. Then in view of (12.67) and (12.85),

$$\|x_j\| \leq \bar{M}_0 - 2 + 1/4. \tag{12.87}$$

By Lemma 2.2, (12.73), (12.85), (12.87), (12.68), (12.72), (12.77), and (12.75),

$$\begin{aligned}
\|y_j\| &\leq \gamma + \|P_C(x_j - \tau f(x_j))\| \\
&\leq \gamma + \|P_C(x_j)\| + \tau \|f(x_j)\| \\
&\leq \|x_j\| + 1/2 + \tau \|f(x_j)\| \\
&\leq \bar{M}_0 - 2 + 3/4 + \bar{M}_1.
\end{aligned} \tag{12.88}$$

By (12.88) and (12.68),

$$\|f(y_j)\| \leq \bar{M}_2. \tag{12.89}$$

By (12.69), (12.77), (12.75), (12.73), (12.85), (12.72), (12.87), (12.89), and Lemma 2.2,

$$\begin{aligned}
\|x_{j+1}\| &\leq 1/4 + \|P_C(x_j - \tau f(y_j))\| \\
&\leq \|P_C(x_j)\| + \tau \|f(y_j)\| + 1/4 \\
&\leq \|x_j\| + 1/2 + \|f(y_j)\| \leq \bar{M}_0 + \bar{M}_2 < M_0.
\end{aligned} \tag{12.90}$$

By (12.90), (12.77), (12.75), and (P2) there exists an integer $i \in [j + 1, j + 1 + \bar{k}]$ such that $d(x_i, S) \leq 1/4$ and together with (12.67) and (12.69) the equation above implies that $\|x_i\| < M_0$. Thus we have shown that the following property holds:

- (P4) if an integer $j \geq 0$ satisfies $\|x_j\| \leq M_0$, then there is an integer $i \in [j + 1, j + 1 + \bar{k}]$ such that $\|x_i\| \leq M_0$.

Set

$$j_0 = \sup\{i : i \text{ is an integer}, i \leq k_2 \text{ and } \|x_i\| \leq M_0\}. \tag{12.91}$$

By (12.76) the number j_0 is well defined and satisfies

$$0 \leq j_0 \leq k_2. \tag{12.92}$$

In view of (P4) and (12.91),

$$j_0 + 1 + \bar{k} \geq k_2. \tag{12.93}$$

By (P4) and (12.91) there is an integer j_1 such that

$$j_1 \in [j_0 + 1, j_0 + 1 + \bar{k}] \text{ and } \|x_{j_1}\| \leq M_0. \tag{12.94}$$

By (12.91) and (12.94),

$$j_1 > k_2, \; j_1 - k_2 \leq j_1 - j_0 \leq 1 + \bar{k}. \tag{12.95}$$

Assume that an integer $j \geq j_1$ satisfies

$$\|x_j\| \leq M_0. \tag{12.96}$$

We show that there is an integer $i \in [j + 1, j + 1 + k_1]$ such that $d(x_i, S) \leq \epsilon/8$.

By (P3), (12.96), (12.77), (12.95), and (12.82) there is an integer $p \in [j, j + k_1]$ such that

$$d(x_p, S) \leq \epsilon/8. \tag{12.97}$$

If $p > j$, then we set $i = p$. Assume that $p = j$. Clearly, (12.87)–(12.90) hold and

$$\|x_{j+1}\| \leq M_0. \tag{12.98}$$

By (P3), (12.98), (12.77), (12.95), and (12.82) there is an integer $i \in [j+1, j+k_1+1]$ for which $d(x_i, S) \leq \epsilon/8$. Thus we have shown that the following property holds:

(P5) If an integer $j \geq j_1$ and $\|x_j\| \leq M_0$, then there is an integer $i \in [j+1, j+1+k_1]$ such that $d(x_i, S) \leq \epsilon/8$.

(P5), (12.94), (12.67), and (12.69) imply that there exists a sequence of natural numbers $\{j_p\}_{p=1}^{\infty}$ such that for each integer $p \geq 1$,

$$1 \leq j_{p+1} - j_p \leq 1 + k_1 \tag{12.99}$$

and for each integer $p \geq 2$,

$$d(x_{j_p}, S) \leq \epsilon/8. \tag{12.100}$$

We show that

$$d(x_i, S) \leq \epsilon \text{ for all integers } i \geq j_2.$$

Set

$$\Delta_0 = 4^{-1} k_1^{-1} \epsilon. \tag{12.101}$$

Let $p \geq 2$ be an integer. We show that for each integer l satisfying $0 \leq l < j_{p+1} - j_p$,

$$d(x_{j_p+l}, S) \leq (\epsilon/8) + l\Delta_0. \tag{12.102}$$

By (12.100) estimate (12.102) holds for $l = 0$. Assume that an integer l satisfies

$$0 \leq l < j_{p+1} - j_p \tag{12.103}$$

and that (12.102) holds. By (12.102), (12.103), (12.99), and (12.101),

$$d(x_{j_p+l}, S) \leq (\epsilon/8) + k_1 \Delta_0 < \epsilon/2. \tag{12.104}$$

By (12.102) there is $u_* \in S$ such that

$$d(x_{j_p+l}, S) = \|x_{j_p+l} - u_*\| \leq (\epsilon/8) + l\Delta_0. \tag{12.105}$$

It follows from (12.67), (12.69)–(12.72), (12.105), (12.99), (12.101)–(12.103), (12.82), (12.78), (12.77), (12.95), and Lemma 12.10 applied with $u_*, M_0, M_1, L,$

$$\delta = \delta_{j_p+l}, \ x = x_{j_p+l}, \ y = y_{j_p+l}, \ \tilde{x} = x_{j_p+l+1}$$

that

$$\|x_{j_p+l+1} - u_*\|$$
$$\leq \|x_{j_p+l} - u_*\| + (4\delta_{j_p+l}(1 + M_0))^{1/2} \leq \epsilon/8 + l\Delta_0$$
$$+ (4\bar{\delta} k_1^{-2}(1 + M_0))^{1/2} \leq \epsilon/8 + l\Delta_0 + k_1^{-1}\epsilon/4 = \epsilon/8 + (l+1)\Delta_0.$$

This implies that

$$d(x_{j_p+l+1}, S) \leq (\epsilon/8) + (l+1)\Delta_0.$$

Thus by induction we have shown that for all $l = 0, \ldots, j_{p+1} - j_p$ relation (12.102) holds and it follows from (12.102), (12.99), and (12.101) that for all integers $l = 0, \ldots, j_{p+1} - j_p - 1$,

$$d(x_{j_p+l}, S) \leq (\epsilon/8) + l\Delta_0 \leq \epsilon/8 + k_1\Delta_0 \leq \epsilon/2.$$

Since the inequality above holds for all integers $p \geq 2$, we conclude that

$$d(x_i, S) \leq \epsilon/2 \text{ for all integers } i \geq j_2. \tag{12.106}$$

By (12.99), (12.95), and (12.83),

$$j_2 \leq k_1 + j_1 + 1 \leq k_1 + 2 + \bar{k} + k_2 = k_0.$$

Together with (12.106) this implies that

$$d(x_i, S) \leq \epsilon/2$$

for all integers $i \geq k_0$. Theorem 12.13 is proved.

Chapter 13
A Common Solution of a Family of Variational Inequalities

In a Hilbert space, we study the convergence of the subgradient method to a common solution of a finite family of variational inequalities and of a finite family of fixed point problems under the presence of computational errors. The convergence of the subgradient method is established for nonsummable computational errors. We show that the subgradient method generates a good approximate solution, if the sequence of computational errors is bounded from above by a constant.

13.1 Preliminaries and the Main Result

Let $(X, \langle \cdot, \cdot \rangle)$ be a Hilbert space equipped with an inner product $\langle \cdot, \cdot \rangle$ which induces a complete norm $\| \cdot \|$. We denote by $\mathrm{Card}(A)$ the cardinality of the set A. For every point $x \in X$ and every nonempty set $A \subset X$ define

$$d(x, A) := \inf\{\|x - y\| : y \in A\}.$$

For every point $x \in X$ and every positive number r put

$$B(x, r) = \{y \in X : \|x - y\| \le r\}.$$

Let $\bar{c} \in (0, 1)$ and $0 < \tau_* < \tau^* \le 1$.

Let C be a nonempty closed convex subset of X. In view of Lemma 2.2, for every point $x \in X$ there is a unique point $P_C(x) \in C$ satisfying

$$\|x - P_C(x)\| = \inf\{\|x - y\| : y \in C\}.$$

Moreover,

$$\|P_C(x) - P_C(y)\| \le \|x - y\|$$

© Springer International Publishing Switzerland 2016
A.J. Zaslavski, *Numerical Optimization with Computational Errors*, Springer Optimization and Its Applications 108, DOI 10.1007/978-3-319-30921-7_13

for all $x, y \in X$ and for each $x \in X$ and each $z \in C$,

$$\langle z - P_C(x), x - P_C(x) \rangle \leq 0.$$

Let $\mathcal{L}_1$ be a finite set of pairs (f, C) where C is a nonempty closed convex subset of X, and $f : X \to X$ and $\mathcal{L}_2$ be a finite set of mappings $T : X \to X$. We suppose that the set $\mathcal{L}_1 \cup \mathcal{L}_2$ is nonempty. (Note that one of the sets $\mathcal{L}_1$ or $\mathcal{L}_2$ may be empty.)

We suppose that for each $(f, C) \in \mathcal{L}_1$ the mapping f is Lipschitz on all bounded subsets of X, the set

$$S(f, C) := \{x \in C : \langle f(x), y - x \rangle \geq 0 \text{ for all } y \in C\} \tag{13.1}$$

is nonempty and that

$$\langle f(y), y - x \rangle \geq 0 \text{ for all } y \in C \text{ and all } x \in S(f, C). \tag{13.2}$$

Evidently, every point $x \in S(f, C)$ is considered as a solution of the variational inequality associated with the pair $(f, C) \in \mathcal{L}_1$:

Find $x \in C$ such that $\langle f(x), y - x \rangle \geq 0$ for all $y \in C$.

For every pair $(f, C) \in \mathcal{L}_1$ and every positive number ϵ define

$$S_\epsilon(f, C) = \{x \in C : \langle f(x), y - x \rangle \geq -\epsilon \|y - x\| - \epsilon \text{ for all } y \in C\}. \tag{13.3}$$

Note that this set was introduced in Chap. 12 and that every point $x \in S_\epsilon(f, C)$ is an ϵ-approximate solution of the variational inequality associated with the pair $(f, C) \in \mathcal{L}_1$. The examples considered in Chap. 12, show that elements of $S_\epsilon(f, C)$ can be considered as ϵ-approximate solutions of the corresponding variational inequality.

We suppose that for every mapping $T \in \mathcal{L}_2$ the set

$$\text{Fix}(T) := \{z \in X : T(z) = z\} \neq \emptyset, \tag{13.4}$$

$$\|T(z_1) - T(z_2)\| \leq \|z_1 - z_2\| \text{ for all } z_1, z_2 \in X, \tag{13.5}$$

$$\|z - x\|^2 \geq \|z - T(x)\|^2 + \bar{c}\|x - T(x)\|^2$$

$$\text{for all } x \in X \text{ and all } z \in \text{Fix}(T). \tag{13.6}$$

For every mapping $T \in \mathcal{L}_2$ and every positive number ϵ define

$$\text{Fix}_\epsilon(T) := \{z \in X : \|T(z) - z\| \leq \epsilon\}. \tag{13.7}$$

Suppose that the set

$$S := [\cap_{(f,C) \in \mathcal{L}_1} S(f, C)] \cap [\cap_{T \in \mathcal{L}_2} \text{Fix}(T)] \neq \emptyset. \tag{13.8}$$

Let $\tau > 0$ and $(f, C) \in \mathcal{L}_1$. For every point $x \in X$ define

$$Q_{\tau,f,C}(x) = P_C(x - \tau f(x)), \tag{13.9}$$
$$P_{\tau,f,C}(x) = P_C(x - \tau f(Q_{\tau,f,C}(x))). \tag{13.10}$$

Let a natural number

$$l \geq \mathrm{Card}(\mathcal{L}_1 \cup \mathcal{L}_2).$$

Denote by $\mathcal{R}$ the set of all mappings

$$A : \{0, 1, 2, \ldots\} \to \mathcal{L}_2 \cup \{P_{\tau,f,C} : (f, C) \in \mathcal{L}_1, \tau \in [\tau_*, \tau^*]\}$$

such that the following properties hold:

(P1) for every nonnegative integer p and every mapping $T \in \mathcal{L}_2$ there exists an integer $i \in \{p, \ldots, p + l - 1\}$ such that $A(i) = T$;

(P2) for each integer $p \geq 0$ and every pair $(f, C) \in \mathcal{L}_1$ there exist an integer $i \in \{p, \ldots, p + l - 1\}$ and $\tau \in [\tau_*, \tau^*]$ such that $A(i) = P_{\tau,f,C}$.

We are interested to find solutions of the inclusion $x \in S$. In order to meet this goal we apply algorithms generated by $A \in \mathcal{R}$. More precisely, we associate with any $A \in \mathcal{R}$ the algorithm which generates, for any starting point $x_0 \in X$, a sequence $\{x_k\}_{k=0}^{\infty} \subset X$, where

$$x_{k+1} := [A(k)](x_k), k = 0, 1, \ldots.$$

According to the results known in the literature, this sequence should converge weakly to an element of S. In this chapter, we study the behavior of the sequences generated by $A \in \mathcal{R}$ taking into account computational errors which are always present in practice. Namely, in practice the algorithm associated with $A \in \mathcal{R}$ generates a sequence $\{x_k\}_{k=0}^{\infty}$ such that for each integer $k \geq 0$,

$$\text{if } A(k) = T \in \mathcal{L}_2, \text{ then } \|x_{k+1} - A(k)(x_k)\| \leq \delta$$

and if

$$A(k) = P_{\tau,f,C} \text{ with } \tau \in [\tau_*, \tau^*], (f, C) \in \mathcal{L}_1,$$

then there is $v_k \in X$ such that

$$\|v_k - Q_{\tau,f,C}(x_k)\| \leq \delta,$$
$$\|x_{k+1} - P_C(x_k - \tau f(v_k))\| \leq \delta$$

with a constant $\delta > 0$ which depends only on our computer system. Surely, in this situation one cannot expect that the sequence $\{x_k\}_{k=0}^{\infty}$ converges to the set S. Our goal is to understand what subset of X attracts all sequences $\{x_k\}_{k=0}^{\infty}$ generated by algorithms associated with $A \in \mathcal{R}$. In Chap. 12 we showed that in the case when $\mathcal{L}_2 = \emptyset$ and the set $\mathcal{L}_1$ is a singleton, this subset of X is the set of ϵ-approximate solutions of the corresponding variational inequality with some $\epsilon > 0$ depending on δ. In this chapter we generalize the main result of Chap. 12 and show that in the general case (see Theorem 13.1 stated below) this subset of X is the set

$$S_\epsilon := \{x \in X : x \in \text{Fix}_\epsilon(T) \text{ for each } T \in \mathcal{L}_2$$

$$\text{and } d(x, S_\epsilon(f, C)) \le \epsilon \text{ for all } (f, C) \in \mathcal{L}_1\}$$

with some $\epsilon > 0$ depending on δ [see (13.15) and (13.17)].

Our goal is also, for a given $\epsilon > 0$, to find a point $x \in S_\epsilon$. This point x is considered as an ϵ-approximate common solution of the problems associated with the family of operators $\mathcal{L}_1 \cup \mathcal{L}_2$. We will prove the following result (Theorem 13.1), which shows that an ϵ-approximate common solution can be obtained after $l(n_0 - 1)$ iterations of the algorithm associated with $A \in \mathcal{R}$ and under the presence of computational errors bounded from above by a constant δ, where δ and n_0 are constants depending on ϵ [see (13.15)–(13.17)]. This result was obtained in [128].

Theorem 13.1. *Let $\epsilon \in (0, 1]$, $M_0 > 0$ be such that*

$$B(0, M_0) \cap S \ne \emptyset \tag{13.11}$$

and let $M_1 > 0$, $L > 0$ be such that for each $(f, C) \in \mathcal{L}_1$,

$$f(B(0, 3M_0 + 2)) \subset B(0, M_1), \tag{13.12}$$

$$\|f(z_1) - f(z_2)\| \le L\|z_1 - z_2\|$$

$$\text{for all } z_1, z_2 \in B(0, 5M_0 + M_1 + 2) \tag{13.13}$$

and

$$\tau^* L < 1. \tag{13.14}$$

Let $\gamma_1 \in (0, 2^{-1})$ satisfy

$$4\gamma_1(5l\tau_*^{-1} + 2\tau^* \tau_*^{-1} M_1 + 4L + 2M_1) \le \epsilon, \tag{13.15}$$

an integer

$$n_0 > 16M_0^2 \bar{c}^{-1}(1 - (\tau^*)^2 L^2)^{-1} \gamma_1^{-2} \tag{13.16}$$

and a number $\delta \in (0, 1)$ satisfy

$$4\delta(2 + 2M_0)l < 16^{-1}\bar{c}\gamma_1^2(1 - (\tau^*)^2 L^2). \tag{13.17}$$

Assume that

$$A \in \mathcal{R}, \ \{x_k\}_{k=0}^{\infty} \subset X, \ \|x_0\| \leq M_0 \tag{13.18}$$

and that for each integer $k \geq 0$,

$$\text{if } A(k) = T \in \mathcal{L}_2, \text{ then } \|x_{k+1} - A(k)(x_k)\| \leq \delta, \tag{13.19}$$

and if

$$A(k) = P_{\tau f, C} \text{ with } \tau \in [\tau_*, \tau^*], \ (f, C) \in \mathcal{L}_1, \tag{13.20}$$

then there is $v_k \in X$ such that

$$\|v_k - Q_{\tau f, C}(x_k)\| \leq \delta, \ \|x_{k+1} - P_C(x_k - \tau f(v_k))\| \leq \delta. \tag{13.21}$$

Then there is an integer $p \in [0, n_0 - 1]$ such that

$$\|x_i\| \leq 3M_0 + 1, \ i = 0, \ldots, (p + 1)l \tag{13.22}$$

and for each integer $i \in \{pl, \ldots, (p + 1)l - 1\}$,

(P3)   If $A(i) = T \in \mathcal{L}_2$, then $\|x_{i+1} - x_i\| \leq \gamma_1$;
(P4)   If $A(i) = P_{\tau f, C}$ with $\tau \in [\tau_*, \tau^*]$, $(f, C) \in \mathcal{L}_1$, then $\|x_i - v_i\| \leq \gamma_1$.

Moreover, if an integer $p \in [0, n_0 - 1]$ be such that for each integer $i \in [pl, (p+1)l-1]$ properties (P3) and (P4) hold and $\|x_i\| \leq 3M_0 + 1$, then for each pair $i, j \in \{pl, \ldots, (p + 1)l\}$,

$$\|x_i - x_j\| \leq \epsilon$$

and for each $i \in \{pl, \ldots, (p + 1)l\}$,

$$x_i \in Fix_\epsilon(T) \text{ for all } T \in \mathcal{L}_2,$$

$$d(x_i, S_\epsilon(f, C)) \leq \epsilon \text{ for all } (f, C) \in \mathcal{L}_1.$$

Note that Theorem 13.1 provides the estimations for the constants δ and n_0, which follow from relations (13.15)–(13.17). Namely, $\delta = c_1 \epsilon^2$ and $n_0 = c_2 \epsilon^{-2}$, where c_1 and c_2 are positive constants depending only on M_0.

Let $\epsilon \in (0, 1]$, a positive number δ be defined by relations (13.15) and (13.17), and let an integer $n_0 \geq 1$ satisfy inequality (13.16). Assume that we apply an algorithm associated with a mapping $A \in \mathcal{R}$ under the presence of computational errors bounded from above by a positive constant δ and that our goal is to find an ϵ-approximate solution $x \in S_\epsilon$. It is not difficult to see that Theorem 13.1 also answers an important question how we can find an iteration number i such that $x_i \in S_\epsilon$. According to Theorem 13.1, we should find the smallest integer $q \in [0, n_0 - 1]$ such that for every integer $i \in [ql, (q+1)l - 1]$ properties (P3) and (P4) hold and that the relation $\|x_i\| \leq 3M_0 + 1$ is true. Then the inclusion $x_i \in S_\epsilon$ is valid for all integers $i \in [ql, (q+1)l]$.

Consider the following convex feasibility problem. Suppose that $C_1, \ldots, C_m$ are nonempty closed convex subsets of X, where m is a natural number, such that the set $C = \cap_{i=1}^{m} C_i$ is also nonempty. We are interested to find a solution of the feasibility problem $x \in C$.

For every point $x \in X$ and every integer $i = 1, \ldots, m$ there exists a unique element $P_i(x) \in C_i$ such that

$$\|x - P_i(x)\| = \inf\{\|x - y\| : y \in C_i\}.$$

The feasibility problem is a particular case of the problem discussed above with $\mathcal{L}_1 = \emptyset$ and $\mathcal{L}_2 = \{P_i : i = 1, \ldots, m\}$.

13.2   Auxiliary Results

The next result follows from Lemma 12.10.

Lemma 13.2. *Let* $(f, C) \in \mathcal{L}_1$,

$$u_* \in S, \ m_0 > 0, \ M_1 > 0, \ L > 0, \delta \in (0, 1),$$

$$f(B(u_*, m_0)) \subset B(0, M_1),$$

$$\|f(z_1) - f(z_2)\| \leq L\|z_1 - z_2\|$$

$$\text{for all } z_1, z_2 \in B(u_*, m_0 + M_1 + 1)$$

and let

$$\tau \in (0, 1], \ L\tau < 1.$$

Assume that

$$x \in B(u_*, m_0), \ y \in X, \ \|y - P_C(x - \tau f(x))\| \leq \delta,$$

$$\tilde{x} \in X, \ \|\tilde{x} - P_C(x - \tau f(y))\| \leq \delta.$$

Then

$$\|\tilde{x} - u_*\|^2 \leq 4\delta(1 + m_0) + \|x - u_*\|^2 - (1 - \tau^2 L^2)\|x - P_C(x - \tau f(x))\|^2.$$

Lemma 13.3. *Let* $(f, C) \in \mathcal{L}_1$,

$$M_0 > 0, \ M_1 > 0, \ L > 0, \delta \in (0, 1),$$

$$B(0, M_0) \cap S \neq \emptyset, \tag{13.23}$$

$$f(B(0, 3M_0 + 1)) \subset B(0, M_1), \tag{13.24}$$

$$\|f(z_1) - f(z_2)\| \leq L\|z_1 - z_2\|$$

$$\text{for all } z_1, z_2 \in B(0, 5M_0 + M_1 + 2) \tag{13.25}$$

and let

$$\tau \in (0, 1], \ L\tau < 1. \tag{13.26}$$

Let

$$x \in B(0, 3M_0 + 1), \ y \in X, \ \|y - P_C(x - \tau f(x))\| \leq \delta, \tag{13.27}$$

$$\tilde{x} \in X, \ \|\tilde{x} - P_C(x - \tau f(y))\| \leq \delta. \tag{13.28}$$

Then

$$\|\tilde{x} - x\| \leq 2\delta + (1 + \tau L)\|x - P_C(x - \tau f(x))\|.$$

Proof. In view of (13.28) and Lemma 2.2,

$$\|\tilde{x} - P_C(x - \tau f(x))\| \leq \delta + \|P_C(x - \tau f(y)) - P_C(x - \tau f(x))\|$$

$$\leq \delta + \tau\|f(y) - f(x)\|. \tag{13.29}$$

It follows from (13.23) that there exists a point

$$u_* \in B(0, M_0) \cap S. \tag{13.30}$$

Lemma 2.2, (13.27), and (13.30) imply that

$$\|y - u_*\| \leq \delta + \|P_C(x - \tau f(x)) - u_*\| \leq \delta + \|x - \tau f(x) - u_*\|$$

$$\leq \delta + \|x\| + \|u_*\| + \tau\|f(x)\|.$$

Combined with (13.24), (13.27), and (13.30) the relation above implies that

$$\|y\| \leq 1 + 3M_0 + 1 + 2M_0 + \tau M_1. \tag{13.31}$$

In view of (13.25), (13.26), (13.27), (13.29), and (13.31),

$$\|\tilde{x} - P_C(x - \tau f(x))\| \leq \delta + \tau L\|y - x\|. \tag{13.32}$$

It follows from (13.32) that

$$\|\tilde{x} - x\| \le \|\tilde{x} - P_C(x - \tau f(x))\| + \|x - P_C(x - \tau f(x))\|$$
$$\le \delta + \tau L\|y - x\| + \|x - P_C(x - \tau f(x))\|.$$

Combined with (13.26) and (13.27) the relation above implies that

$$\|\tilde{x} - x\| \le 2\delta + \|y - x\|(1 + \tau L),$$
$$\|\tilde{x} - x\| \le 2\delta + (1 + \tau L)\|x - P_C(x - \tau f(x))\|.$$

Lemma 13.3 is proved.

Lemma 13.4. *Suppose that all the assumptions of Theorem 13.1 hold. Then*

$$\delta < \gamma_1/4, \tag{13.33}$$

$$2\delta(2M_0 + 1) < 16^{-1}\bar{c}\gamma_1^2, \tag{13.34}$$

$$4\delta(2 + 2M_0) \le 16^{-1}\gamma_1^2(1 - (\tau^*)^2 L^2). \tag{13.35}$$

Proof. It is not difficult to see that inequality (13.33) follows from (13.14), (13.17) and the relations $l \ge 1$, $\bar{c} < 1$ and $\gamma_1 < 1/2$. Relation (13.34) follows from (13.17), (13.14) and the inequality $l \ge 1$. Relation (13.35) follows from (13.17) and the inequalities $l \ge 1$ and $\bar{c} < 1$.

13.3   Proof of Theorem 13.1

In view of (13.11), there exists

$$u_* \in B(0, M_0) \cap S. \tag{13.36}$$

It follows from (13.18) and (13.36) that

$$\|u_* - x_0\| \le 2M_0. \tag{13.37}$$

Assume that a nonnegative integer p satisfies

$$\|u_* - x_{pl}\| \le 2M_0. \tag{13.38}$$

(Clearly, in view of (13.37), inequality (13.38) holds with $p = 0$.)
Assume that an integer $i \in \{pl, \dots, (p+1)l - 1\}$ satisfies

$$\|x_i - u_*\| \le 2M_0 + \gamma_1(i - pl). \tag{13.39}$$

Then one of the following two cases hold:

$$A(i) = T \in \mathcal{L}_2; \tag{13.40}$$

$$A(i) = P_{\tau,f,C} \text{ with } \tau \in [\tau_*, \tau^*], \ (f, C) \in \mathcal{L}_1. \tag{13.41}$$

Assume that (13.40) is valid. Then it follows from (13.19), (13.36), (13.40), (13.5), (13.39), Lemma 13.4, and (13.33) that

$$\|x_{i+1} - u_*\| \le \|x_{i+1} - A(i)(x_i)\| + \|A(i)(x_i) - u_*\|$$
$$\le \delta + \|x_i - u_*\| \le 2M_0 + (i + 1 - pl)\gamma_1. \tag{13.42}$$

In view of (13.39), (13.42), the inclusions $i \in [pl, (p + 1)l - 1]$ and $\epsilon \in (0, 1)$ and (13.15),

$$\|x_i - u_*\| \le 2M_0 + l\gamma_1 \le 2M_0 + 1,$$
$$\|x_{i+1} - u_*\| \le 2M_0 + l\gamma_1 \le 2M_0 + 1.$$

Combined with (13.42) these inequalities above imply that

$$\|x_{i+1} - u_*\|^2 - \|x_i - u_*\|^2$$
$$\le \delta(\|x_{i+1} - u_*\| + \|x_i - u_*\|) \le 2\delta(2M_0 + 1). \tag{13.43}$$

Assume that (13.41) is valid. In view of (13.36), (13.39), (13.15), (13.12), (13.13), (13.17), (13.41),

$$\|x_i - u_*\| \le 2M_0 + 1$$

and all the assumptions of Lemma 13.2 hold with $x = x_i$, $y = v_i$, $\tilde{x} = x_{i+1}$, $m_0 = 2M_0 + 1$, and this implies that

$$\|x_{i+1} - u_*\|^2 \le 4\delta(2 + 2M_0) + \|x_i - u_*\|^2 \tag{13.44}$$

and

$$\|x_{i+1} - u_*\| \le \|x_i - u_*\| + 2(\delta(2 + 2M_0))^{1/2}$$
$$\le \|x_i - u_*\| + \gamma_1 \le 2M_0 + (i + 1 - pl)\gamma_1. \tag{13.45}$$

(Note that the first inequality of (13.45) follows from (13.44), the second inequality follows from (13.17) and the inequalities $\bar{c} < 1$ and $l \ge 1$, and the third inequality follows from (13.39).)

It follows from (13.42)–(13.45) that in both cases

$$\|x_{i+1} - u_*\| \leq 2M_0 + (i + 1 - pl)\gamma_1,$$
$$\|x_{i+1} - u_*\|^2 \leq \|x_i - u_*\|^2 + 4\delta(2 + 2M_0).$$

Thus by induction we proved that for all integers $i = pl, \ldots, (p+1)l$ the inequality

$$\|x_i - u_*\| \leq 2M_0 + (i - pl)\gamma_1$$

holds and that for all integers $i = pl, \ldots, (p+1)l - 1$ the inequality

$$\|x_{i+1} - u_*\|^2 \leq \|x_i - u_*\|^2 + 4\delta(2 + 2M_0)$$

is valid.

By (13.15), we have shown that the following property holds:

(P5)   If a nonnegative integer p satisfies the inequality

$$\|u_* - x_{pl}\| \leq 2M_0,$$

then we have

$$\|x_i - u_*\| < 2M_0 + 1 \text{ for all } i = pl, \ldots, (p+1)l \tag{13.46}$$

and

$$\|x_i - u_*\|^2 - \|x_{i+1} - u_*\|^2 \geq -4\delta(2 + 2M_0)$$
$$\text{for all } i = pl, \ldots, (p+1)l - 1. \tag{13.47}$$

Assume that an integer $\tilde{q} \in [0, n_0 - 1]$ and that for every integer $p \in [0, \tilde{q}]$ the following property holds:

(P6)   there exists $i \in \{pl, \ldots, (p+1)l - 1\}$ such that (P3) and (P4) do not hold.

Assume now that an integer $p \in [0, \tilde{q}]$ satisfies

$$\|u_* - x_{pl}\| \leq 2M_0. \tag{13.48}$$

In view of property (P5) and relation (13.48), inequalities (13.46) and (13.47) are valid.

Property (P6) implies that there exists an integer $j \in \{pl, \ldots, (p+1)l - 1\}$ such that properties (P3) and (P4) do not hold with $i = j$. Evidently, one of the following cases holds:

$$A(j) = T \in \mathcal{L}_2; \tag{13.49}$$

$$A(j) = P_{\tau, f, C} \text{ with } \tau \in [\tau_*, \tau^*], \ (f, C) \in \mathcal{L}_1. \tag{13.50}$$

Assume that relation (13.49) holds. Since property (P3) does not hold with $i = j$ we have

$$\|x_{j+1} - x_j\| > \gamma_1. \tag{13.51}$$

It follows from (13.49), (13.51), (13.19), and Lemma 13.4 that

$$\|x_j - T(x_j)\| \geq \|x_{j+1} - x_j\| - \|T(x_j) - x_{j+1}\|$$
$$\geq \gamma_1 - \delta \geq (3/4)\gamma_1. \tag{13.52}$$

Relations (13.49), (13.6), (13.36), (13.8), and (13.52) imply that

$$\|u_* - x_j\|^2 \geq \|u_* - T(x_j)\|^2 + \bar{c}\|x_j - T(x_j)\|^2$$
$$\geq \|u_* - T(x_j)\|^2 + \bar{c}(9/16)\gamma_1^2$$
$$= \|u_* - x_{j+1}\|^2 + \|x_{j+1} - T(x_j)\|^2$$
$$+2\langle u_* - x_{j+1}, x_{j+1} - T(x_j)\rangle + \bar{c}(9/16)\gamma_1^2$$
$$\geq \|u_* - x_{j+1}\|^2 - 2\|u_* - x_{j+1}\|\|x_{j+1} - T(x_j)\| + \bar{c}(9/16)\gamma_1^2. \tag{13.53}$$

In view of (13.53), (13.19), (13.46), and (13.34),

$$\|u_* - x_j\|^2 - \|u_* - x_{j+1}\|^2$$
$$\geq \bar{c}(9/16)\gamma_1^2 - 2\delta(2M_0 + 1) \geq (\bar{c}/2)\gamma_1^2.$$

Hence

$$\text{if (13.49) holds, then } \|u_* - x_j\|^2 - \|u_* - x_{j+1}\|^2 \geq (\bar{c}/2)\gamma_1^2. \tag{13.54}$$

Assume that relation (13.50) is valid. Then relations (13.50), (13.36), (13.46), (13.12), (13.13), (13.20), and (13.21) imply that all the assumptions of Lemma 13.2 hold with

$$x = x_j, \ y = v_j, \ \tilde{x} = x_{j+1}, \ m_0 = 2M_0 + 1$$

and this implies that

$$\|u_* - x_{j+1}\|^2$$
$$\leq 4\delta(2 + 2M_0) + \|u_* - x_j\|^2 - (1 - (\tau L)^2)\|x_j - P_C(x_j - \tau f(x_j))\|^2. \tag{13.55}$$

Since property (P4) does not hold with $i = j$ we conclude that

$$\|x_j - v_j\| > \gamma_1. \tag{13.56}$$

In view of (13.46), (13.21), and (13.56),

$$\|x_j - P_C(x_j - \tau f(x_j))\|$$
$$\geq \|x_j - v_j\| - \|P_C(x_j - \tau f(x_j)) - v_j\| > \gamma_1 - \delta > (3/4)\gamma_1. \tag{13.57}$$

It follows from (13.35), (13.55), and (13.57) that

$$\|x_j - u_*\|^2 - \|x_{j+1} - u_*\|^2$$
$$\geq (1 - (\tau^* L)^2)(9/16)\gamma_1^2 - 4\delta(2 + 2M_0)$$
$$\geq (1 - (\tau^* L)^2)\gamma_1^2/2. \tag{13.58}$$

By relations (13.54) and (13.58), in both cases we have

$$\|x_j - u_*\|^2 - \|x_{j+1} - u_*\|^2 \geq \bar{c}(1 - (\tau^* L)^2)\gamma_1^2/2. \tag{13.59}$$

It follows from (13.17), (13.47), and (13.59) that

$$\|u_* - x_{pl}\|^2 - \|u_* - x_{(p+1)l}\|^2$$
$$= \sum_{i=pl}^{(p+1)l-1} [\|u_* - x_i\|^2 - \|u_* - x_{i+1}\|^2]$$
$$\geq \bar{c}(1 - (\tau^* L)^2)\gamma_1^2/2 - 4\delta(2 + 2M_0)l$$
$$> \bar{c}(1 - (\tau^* L)^2)\gamma_1^2/4. \tag{13.60}$$

By (13.48) and (13.60),

$$\|u_* - x_{(p+1)l}\| \leq \|u_* - x_{pl}\| \leq 2M_0. \tag{13.61}$$

Thus we have shown that the following property holds:

(P7)   If an integer $p \in [0, \bar{q}]$ satisfies inequality (13.48), then [see relations (13.61), (13.46), (13.60)] we have

$$\|u_* - x_{(p+1)l}\| \leq 2M_0, \ \|u_* - x_i\| \leq 2M_0 + 1, \ i = pl, \dots, (p + 1)l$$

and

$$\|u_* - x_{pl}\|^2 - \|u_* - x_{(p+1)l}\|^2 > \bar{c}(1 - (\tau^* L)^2)\gamma_1^2/4. \tag{13.62}$$

In view of (13.37), property (P7), (13.62), and (13.16),

$$4M_0^2 \geq \|u_* - x_0\|^2$$
$$\geq \|u_* - x_0\|^2 - \|u_* - x_{(\tilde{q}+1)l}\|^2$$
$$= \sum_{p=0}^{\tilde{q}} [\|u_* - x_{pl}\|^2 - \|u_* - x_{(p+1)l}\|^2]$$
$$\geq (\tilde{q}+1)\bar{c}(1-(\tau^*L)^2)\gamma_1^2/4$$

and

$$\tilde{q}+1 \leq 16M_0^2 \bar{c}^{-1}(1-(\tau^*L)^2)^{-1}\gamma_1^{-2} < n_0. \tag{13.63}$$

We assumed that an integer $\tilde{q} \in [0, n_0 - 1]$ and that for every integer $p \in [0, \tilde{q}]$ property (P6) holds and proved that $\tilde{q} + 1 < n_0$.

This implies that there exists an integer $q \in [0, n_0 - 1]$ such that for every integer p satisfying $0 \leq p < q$, property (P6) holds and that the following property holds:

(P8) For every integer $i \in \{ql, \ldots, (q+1)l - 1\}$ properties (P3) and (P4) hold.

Property (P7) (with $\tilde{q} = q - 1$) implies that

$$\|u_* - x_{jl}\| \leq 2M_0, \ j = 0, \ldots, q. \tag{13.64}$$

In view of (13.64) and property (P5),

$$\|u_* - x_i\| \leq 2M_0 + 1, \ i = 0, \ldots, (q+1)l. \tag{13.65}$$

It follows from (13.64), (13.65), and (13.36) that

$$\|x_i\| \leq 3M_0 + 1, \ i = 0, \ldots, (q+1)l.$$

Assume that p is a nonnegative integer,

$$\|x_i\| \leq 3M_0 + 1, \ i = pl, \ldots, (p+1)l - 1 \tag{13.66}$$

and that for all integers $i = pl, \ldots, (p+1)l - 1$ properties (P3) and (P4) hold.
Let

$$i \in \{pl, \ldots, (p+1)l - 1\}. \tag{13.67}$$

There are two cases:

$$A(i) = T \in \mathcal{L}_2; \tag{13.68}$$

$$A(i) = P_{\tau, f, C} \text{ with } \tau \in [\tau_*, \tau^*], \ (f, C) \in \mathcal{L}_1. \tag{13.69}$$

Assume that relation (13.68) is valid. Then in view of (13.68) and property (P3),

$$\|x_{i+1} - x_i\| \le \gamma_1. \tag{13.70}$$

It follows from (13.68), (13.19), (13.70), and (13.17) that

$$\|T(x_i) - x_i\| \le \|T(x_i) - x_{i+1}\| + \|x_{i+1} - x_i\| \le \delta + \gamma_1 \le (5/4)\gamma_1,$$
$$x_i \in \text{Fix}_{(5/4)\gamma_1}(T). \tag{13.71}$$

Thus we have shown that the following property holds:

(P9)   If (13.68) is true, then relations (13.70) and (13.71) hold.

Assume that (13.69) holds. In view of (13.69), (13.67), property (P4), (13.20), and (13.21),

$$\|x_{i+1} - P_C(x_i - \tau f(v_i))\| \le \delta,$$
$$\|v_i - P_C(x_i - \tau f(x_i))\| \le \delta,$$
$$\|x_i - v_i\| \le \gamma_1. \tag{13.72}$$

Relations (13.17), (13.21), and (13.72) imply that

$$\|x_i - P_C(x_i - \tau f(x_i))\|$$
$$\le \|x_i - v_i\| + \|v_i - P_C(x_i - \tau f(x_i))\|$$
$$\le \delta + \gamma_1 \le (5/4)\gamma_1. \tag{13.73}$$

It follows from (13.11), (13.69), (13.12), (13.13), (13.14), (13.67), (13.66), (13.72), (13.73), and (13.33) that all the assumptions of Lemma 13.3 hold with

$$x = x_i, \ y = v_i, \ \tilde{x} = x_{i+1}$$

(and with the constants M_0, M_1, L as in Theorem 13.1), and this implies that

$$\|x_{i+1} - x_i\|$$
$$\le 2\delta + (1 + \tau L)\|x_i - P_C(x_i - \tau f(x_i))\|$$
$$\le 2\delta + (5/4)(1 + \tau L)\gamma_1$$
$$\le (5/4)(1 + \tau^* L)\gamma_1 + 2\delta < 5\gamma_1. \tag{13.74}$$

(Note that the second inequality in (13.74) follows from (13.73), the third one follows from (13.69) and the last inequality follows from (13.14) and (13.33).)

In view of Lemma 2.2, (13.73), (13.69), (13.66), and (13.12), for every point $\xi \in C$,

$$0 \geq \langle x_i - \tau f(x_i) - P_C(x_i - \tau f(x_i)), \xi - P_C(x_i - \tau f(x_i)) \rangle$$
$$= \langle x_i - P_C(x_i - \tau f(x_i)), \xi - P_C(x_i - \tau f(x_i)) \rangle$$
$$- \tau \langle f(x_i), \xi - P_C(x_i - \tau f(x_i)) \rangle$$
$$\geq -\|x_i - P_C(x_i - \tau f(x_i))\|(\|\xi - x_i\| + \|x_i - P_C(x_i - \tau f(x_i))\|)$$
$$- \tau \langle f(x_i), \xi - x_i \rangle - \tau \langle f(x_i), x_i - P_C(x_i - \tau f(x_i)) \rangle$$
$$\geq -2\gamma_1(\|\xi - x_i\| + 2\gamma_1) - \tau \langle f(x_i), \xi - x_i \rangle - 2\tau^* \|f(x_i)\| \gamma_1,$$
$$\tau \langle f(x_i), \xi - x_i \rangle \geq -2\gamma_1 \|\xi - x_i\| - 4\gamma_1^2 - 2\gamma_1 \tau^* M_1$$

and

$$\langle f(x_i), \xi - x_i \rangle$$
$$\geq -2\gamma_1 \tau_*^{-1} \|\xi - x_i\| - 4\tau_*^{-1} \gamma_1^2 - 2\gamma_1 \tau^* \tau_*^{-1} M_1$$
$$\text{for each } \xi \in C. \tag{13.75}$$

Set

$$\bar{y} = P_C(x_i - \tau f(x_i)). \tag{13.76}$$

Relations (13.76) and (13.73) imply that

$$\|x_i - \bar{y}\| \leq (5/4)\gamma_1. \tag{13.77}$$

It follows from (13.77), (13.66), (13.13), (13.12), and (13.10) that

$$\|f(x_i) - f(\bar{y})\| \leq L\|x_i - \bar{y}\| \leq L(5/4)\gamma_1,$$
$$\bar{y} \in B(x_i, 1) \subset B(0, 3M_0 + 2), \quad \|f(\bar{y})\| \leq M_1. \tag{13.78}$$

(Note that the inclusion in (13.78) follows from (13.76), the inequality $\gamma_1 < 1/2$ and (13.66), and the last inequality in (13.78) follows from (13.12).)

In view of (13.75), (13.77), (13.78), and (13.15), for every point $\xi \in C$,

$$\langle f(\bar{y}), \xi - \bar{y} \rangle$$
$$\geq \langle f(\bar{y}), \xi - x_i \rangle - \|f(\bar{y})\| \|x_i - \bar{y}\|$$
$$\geq \langle f(\bar{y}), \xi - x_i \rangle - 2M_1 \gamma_1$$
$$\geq \langle f(x_i), \xi - x_i \rangle - \|f(\bar{y}) - f(x_i)\| \|\xi - x_i\| - 2M_1 \gamma_1$$
$$\geq -2\gamma_1 \tau_*^{-1} \|\xi - x_i\| - 4\gamma_1 \tau_*^{-1}$$
$$- 2M_1 \gamma_1 \tau_*^{-1} \tau^* - 2L\gamma_1 \|\xi - x_i\| - 2M_1 \gamma_1$$

$$\geq -2\gamma_1 \tau_*^{-1} \|\xi - \bar{y}\| - 2\gamma_1 \tau_*^{-1} \|\bar{y} - x_i\|$$
$$-4\tau_*^{-1}\gamma_1 - 2\gamma_1 M_1 \tau^* \tau_*^{-1}$$
$$-2L\gamma_1 \|\xi - \bar{y}\| - 2L\gamma_1 \|x_i - \bar{y}\| - 2M_1\gamma_1$$
$$\geq -(2\gamma_1 \tau_*^{-1} + 2L\gamma_1)\|\xi - \bar{y}\| - 6\gamma_1 \tau_*^{-1}$$
$$-2\gamma_1 M_1 \tau^* \tau_*^{-1} - 4L\gamma_1$$
$$-2M_1\gamma_1 \geq -\epsilon\|\bar{y} - \xi\| - \epsilon.$$

Combined with relation (13.76) the relation above implies that

$$\bar{y} \in S_\epsilon(f, C).$$

It follows from the inclusion above and (13.77) that

$$d(x_i, S_\epsilon(f, C)) \leq (5/4)\gamma_1.$$

Thus we have shown that the following property holds:

(P10)   if (13.69) is valid, then the inequalities

$$\|x_i - x_{i+1}\| \leq 5\gamma_1$$

[see (13.74)] and

$$d(x_i, S_\epsilon(f, C)) \leq (5/4)\gamma_1$$

hold.

In view of properties (P9) and (P10), for every integer $i \in \{pl, \ldots, (p+1)l - 1\}$,

$$\|x_i - x_{i+1}\| \leq 5\gamma_1.$$

This implies that for every pair of integers $i, j \in \{pl, \ldots, (p+1)l\}$, we have

$$\|x_i - x_j\| \leq 5l\gamma_1 < \epsilon/4. \tag{13.79}$$

[see (13.15)].

Let $j \in \{pl, \ldots, (p+1)l\}$. Assume that $T \in \mathcal{L}_2$. In view of (13.18) and property (P1) there exists an integer $i \in \{pl, \ldots, (p+1)l - 1\}$ such that

$$A(i) = T.$$

It follows from (13.68) and property (P9) that

$$x_i \in \mathrm{Fix}_{(2\gamma_1)}(T)$$

and

$$\|x_i - T(x_i)\| \leq 2\gamma_1.$$

Combined with relations (13.79) and (13.15) this implies that

$$\|x_j - T(x_j)\| \leq \|x_j - x_i\| + \|x_i - T(x_i)\| + \|T(x_j) - T(x_i)\| \leq \epsilon/2 + 2\gamma_1 < \epsilon,$$

$$x_j \in \text{Fix}_\epsilon(T) \text{ for all } T \in \mathcal{L}_2. \tag{13.80}$$

Assume that $(f, C) \in \mathcal{L}_1$. In view of (13.18) and property (P2), there exists an integer $i \in \{pl, \ldots, (p+1)l - 1\}$ such that

$$A(i) = P_{\tau f, C} \text{ with } \tau \in [\tau_*, \tau^*].$$

It follows from (13.69) and property (P10) that

$$d(x_i, S_\epsilon(f, C)) \leq 2\gamma_1.$$

Combined with relations (13.79) and (13.15) (the choice of γ_1) the inequality above implies that

$$d(x_j, S_\epsilon(f, C)) \leq d(x_i, S_\epsilon(f, C)) + \|x_i - x_j\| \leq 2\gamma_1 + \epsilon/4 < \epsilon$$

for all $(f, C) \in \mathcal{L}_1$. Theorem 13.1 is proved.

13.4 Examples

In this section we present examples for which Theorem 13.1 can be used.

Example 13.5. Let $p \geq 1$ be an integer, C_i, $i = 1, \ldots, p$ be nonempty closed convex subsets of the Hilbert space X and let for every integer $i \in \{1, \ldots, p\}$, $g_i : X \to R^1$ be a convex Fréchet differentiable function and $g_i'(x) \in X$ be its Fréchet derivative at a point $x \in X$. We assume that for all integers $i = 1, \ldots, p$ the mapping $g_i' : X \to X$ is Lipschitzian on all bounded subsets of X.

Consider the following multi-objective minimization problem

Find $x \in \cap_{i=1}^p C_i$ such that

$$g_i(x) = \inf\{g_i(z) : z \in C_i\} \text{ for all } i = 1, \ldots, p.$$

It is clear that this problem is equivalent to the following problem which is a particular case of the problem discussed in this section with $f_i = g_i'$, $i = 1, \ldots, p$, and for which Theorem 13.1 was stated:

Find $x \in \cap_{i=1}^{p} C_i$ such that for all $i = 1, \ldots, p$,

$$\langle g_i'(x), y - x \rangle \geq 0 \text{ for all } y \in C_i.$$

Let S be the set of solutions of these two problems. We assume that $S \neq \emptyset$.

Now it is not difficult to see that all the assumptions needed for Theorem 13.1 hold with $f_i = g_i'$, $i = 1, \ldots, p$, $\mathcal{L}_1 = \{(g_i', C_i) : i = 1, \ldots, p\}$ and $\mathcal{L}_2 = \emptyset$.

The constants M_0, M_1, L can be found, in principle, using the analytic description of the functions g_i and the sets C_i, $i = 1, \ldots, p$ which is usually given. In many cases the set $\cap_{i=1}^{p} C_i$ is contained in a ball or one of the sets C_i, $i = 1, \ldots, p$ is contained in a ball and the radius of the ball can be found. Then we can found the constant M_1, L, choose a positive constant $\tau^* < L^{-1}$ and apply Theorem 13.1 with $A \in \mathcal{R}$ such that for each integer $i \geq 0$ and each integer $j \in [0, p-1]$,

$$A(ip + j) = P_{\tau^* f_{j+1}, C_{j+1}}.$$

Our next example is a particular case of Example 13.5.

Example 13.6. Let $X = R^4$, $p = 2$,

$$C_1 = \{x = (x_1, x_2, x_3, x_4) \in R^4 : |x_1| \leq 10, \ x_3 = 2\},$$

$$C_2 = \{x = (x_1, x_2, x_3, x_4) \in R^4 : |x_2| \leq 10, \ x_4 = 2\},$$

$$g_1(x) = (2x_1 + x_2 + x_3 - x_4 - 3)^2, \ x = (x_1, x_2, x_3, x_4) \in R^4,$$

$$g_2(x) = (x_1 + 2x_2 + x_3 - x_4 - 3)^2, \ x = (x_1, x_2, x_3, x_4) \in R^4.$$

Evidently, the functions g_1, g_2 are convex and $g_1, g_2 \in C^2$.

Consider the problem

Find $x \in C_1 \cap C_2$ such that

$$g_i(x) = \inf\{g_i(z) : z \in C_i\} \text{ for } i = 1, 2.$$

As it was shown in Example 13.5, we can apply Theorem 13.1 for this problem with $f_i = g_i'$, $i = 1, 2$.

Now we define the constants which appear in Theorem 13.1. Set $l = 2$. Clearly,

$$C_1 \cap C_2 \subset \{x = (x_1, x_2, x_3, x_4) \in R^4 :$$

$$|x_1| \leq 10, \ |x_2| \leq 10, \ x_3 = 2, \ x_4 = 2\} \subset B(0, 16).$$

Thus we can set $M_0 = 16$. It is easy to see that the set of solutions of our problem

$$S = \{x = (x_1, x_2, 2, 2) : |x_1| \leq 10, \ |x_2| \leq 10, \ g_1(x) = 0, \ g_2(x) = 0\}$$

$$= \{x = (x_1, x_2, 2, 2) : |x_1| \leq 10, \ |x_2| \leq 10, \ 2x_1 + x_2 - 3 = 0, \ x_1 + 2x_2 - 3 = 0\}$$

$$= \{(1, 1, 2, 2)\}.$$

For all points $x = (x_1, x_2, x_3, x_4) \in R^4$ we have

$$f_1(x) = g_1'(x) = 2(2x_1 + x_2 + x_3 - x_4 - 3)(2, 1, 1, -1),$$
$$f_2(x) = g_2'(x) = 2(x_1 + 2x_2 + x_3 - x_4 - 3)(1, 2, 1, -1).$$

These equalities imply that for $i = 1, 2$,

$$f_i(B(0, 50)) \subset B(0, 1530)$$

(thus $M_1 = 1530$) and that the functions f_1, f_2 are Lipschitzian on R^4 with the Lipschitz constant $L = 12$. Put $\tau^* = 16^{-1}, \bar{c} = 1/2$.

We apply Theorem 13.1 with these constants and with $\epsilon = 10^{-3}$. Then (13.15) implies that we can set $\gamma_1 = 4^{-1} 10^{-7}$. By (13.16), we have

$$n_0 > 2 \cdot 16^3 \cdot 10^{14}$$

and in view of (13.17) the following inequality holds:

$$\delta < (16 \cdot 34)^{-1} \cdot 16^{-3} \cdot 10^{-14}.$$

Note that this example can also be considered as an example of a convex feasibility problem

$$\text{Find } x \in C_1 \cap C_2 \cap \{z \in R^n : g_1(z) \le 0\} \cap \{z \in R^n : g_2(z) \le 0\},$$

or equivalently

$$\text{Find } x \in \{z \in C_1 : g_1(z) \le 0\} \cap \{z \in C_2 : g_2(z) \le 0\}.$$

Now we describe how the subgradient algorithm is applied for our example. First of all, note that for any $y = (y_1, y_2, y_3, y_4) \in R^4$,

$$P_{C_1}(y) = (\min\{\max\{y_1, -10\}, 10\}, y_2, 2, y_4),$$
$$P_{C_2}(y) = (y_1, \min\{\max\{y_2, -10\}, 10\}, y_3, 2).$$

We apply Theorem 13.1 with $x_0 = (0, 0, 0, 0)$ and $A \in \mathcal{R}$ such that for each integer $i \ge 0$,

$$A(2i) = P_{16^{-1} f_1, C_1}, \quad A(2i + 1) = P_{16^{-1} f_2, C_2}.$$

Then our algorithm generates two sequences $\{x_i\}_{i=0}^{\infty}, \{v_i\}_{i=0}^{\infty} \subset R^4$ such that for every nonnegative integer i,

$$\|v_{2i} - P_{C_1}(x_{2i} - 16^{-1} f_1(x_{2i}))\| \le \delta,$$
$$\|x_{2i+1} - P_{C_1}(x_{2i} - 16^{-1} f_1(v_{2i}))\| \le \delta$$

and

$$\|v_{2i+1} - P_{C_2}(x_{2i+1} - 16^{-1}f_2(x_{2i+1}))\| \le \delta,$$

$$\|x_{2i+2} - P_{C_2}(x_{2i+1} - 16^{-1}f_2(v_{2i+1}))\| \le \delta.$$

For every nonnegative integer p we calculate

$$\Delta_p = \max\{\|v_{2p} - x_{2p}\|, \; \|v_{2p+1} - x_{2p+1}\|\}$$

and find the smallest (first) integer $p \ge 0$ such that $\Delta_p \le \gamma_1$. In view of Theorem 13.1, this nonnegative integer p exists and satisfies $p < n_0$. Then it follows from Theorem 13.1 that $x_{2p}, x_{2p+1}, x_{2p+2} \in S_\epsilon$.

Example 13.7. Let $p, q \ge 1$ be integers, $C_i, i = 1, \ldots, p, D_i, i = 1, \ldots, q$ be nonempty closed convex subsets of the Hilbert space X and let for every integer $i \in \{1, \ldots, p\}$, $g_i : X \to R^1$ be a convex Frechet differentiable function and $g_i'(x) \in X$ be its Frechet derivative at a point $x \in X$. We assume that for every integer $i = 1, \ldots, p,$

$$g_i(z) \ge 0 \text{ for all } z \in C_i$$

and that the mapping $g_i' : X \to X$ is Lipschitzian on all bounded subsets of X.

Consider the following convex feasibility problem

Find a point belonging to $(\cap_{i=1}^p \{z \in C_i : g_i(z) \le 0\}) \cap (\cap_{j=1}^q D_j)$.

It is easy to show that this problem is equivalent to the following problem which is a particular case of the problem discussed in this section, and for which Theorem 13.1 was stated:

Find $x \in \cap_{j=1}^q D_j$ such that for all $i = 1, \ldots, p,$

$$x \in C_i \text{ and } \langle g_i'(x), y - x \rangle \ge 0 \text{ for all } y \in C_i.$$

Let S be the set of solutions of these two problems. We assume that $S \neq \emptyset$.

Now it is not difficult to see that all the assumptions needed for Theorem 13.1 hold with $f_i = g_i', i = 1, \ldots, p, \mathcal{L}_1 = \{(g_i', C_i) : i = 1, \ldots, p\}$ and $\mathcal{L}_2 = \{P_{D_i} : i = 1, \ldots, q\}, l = p + q$.

The constants M_0, M_1, L can be found as it was explained in Example 13.5, using the analytic description of the functions g_i and the sets $C_i, D_j, i = 1, \ldots, p, j = 1, \ldots, q$ which is usually given. Then we choose a positive constant $\tau^* < L^{-1}$ and apply Theorem 13.1 with $A \in \mathcal{R}$ such that

$$A(i + (p + q)) = A(i) \text{ for all integers } i \ge 0,$$

$$A(i) = P_{\tau^* f_{i+1}, C_{i+1}}, \; i = 0, \ldots, p - 1, \; A(p - 1 + j) = P_{D_j}, \; j = 1, \ldots, q.$$

Chapter 14
Continuous Subgradient Method

In this chapter we study the continuous subgradient algorithm for minimization of convex functions, under the presence of computational errors. We show that our algorithms generate a good approximate solution, if computational errors are bounded from above by a small positive constant. Moreover, for a known computational error, we find out what an approximate solution can be obtained and how much time one needs for this.

14.1 Bochner Integrable Functions

Let $(Y, \|\cdot\|)$ be a Banach space and $-\infty < a < b < \infty$. A function $x : [a,b] \to Y$ is strongly measurable on $[a,b]$ if there exists a sequence of functions $x_n : [a,b] \to Y$, $n = 1, 2, \ldots$ such that for any integer $n \geq 1$ the set $x_n([a,b])$ is countable and the set $\{t \in [a,b] : x_n(t) = y\}$ is Lebesgue measurable for any $y \in Y$, and $x_n(t) \to x(t)$ as $n \to \infty$ in $(Y, \|\cdot\|)$ for almost every $t \in [a,b]$.

The function $x : [a,b] \to Y$ is Bochner integrable if it is strongly measurable and there exists a finite $\int_a^b \|x(t)\| dt$.

If $x : [a,b] \to Y$ is a Bochner integrable function, then for almost every (a. e.) $t \in [a,b]$,

$$\lim_{\Delta t \to 0} (\Delta t)^{-1} \int_t^{t+\Delta t} \|x(\tau) - x(t)\| d\tau = 0$$

and the function

$$y(t) = \int_a^t x(s) ds, \ t \in [a,b]$$

is continuous and a. e. differentiable on $[a,b]$.

© Springer International Publishing Switzerland 2016
A.J. Zaslavski, *Numerical Optimization with Computational Errors*, Springer Optimization and Its Applications 108, DOI 10.1007/978-3-319-30921-7_14

Let $-\infty < \tau_1 < \tau_2 < \infty$. Denote by $W^{1,1}(\tau_1, \tau_2; Y)$ the set of all functions $x : [\tau_1, \tau_2] \to Y$ for which there exists a Bochner integrable function $u : [\tau_1, \tau_2] \to Y$ such that

$$x(t) = x(\tau_1) + \int_{\tau_1}^{t} u(s)ds, \ t \in (\tau_1, \tau_2]$$

(see, e.g., [11, 27]). It is known that if $x \in W^{1,1}(\tau_1, \tau_2; Y)$, then this equation defines a unique Bochner integrable function u which is called the derivative of x and is denoted by x'.

14.2   Convergence Analysis for Continuous Subgradient Method

The study of continuous subgradient algorithms is an important topic in optimization theory. See, for example, [6, 10, 23, 27, 28] and the references mentioned therein. In this chapter we analyze its convergence under the presence of computational errors.

We suppose that X is a Hilbert space equipped with an inner product denoted by $\langle \cdot, \cdot \rangle$ which induces a complete norm $\| \cdot \|$. For each $x \in X$ and each $r > 0$ set

$$B(x, r) = \{y \in X : \|x - y\| \le r\}.$$

For each $x \in X$ and each nonempty set $E \subset X$ set

$$d(x, E) = \inf\{\|x - y\| : y \in E\}.$$

Let D be a nonempty closed convex subset of X. Then for each $x \in X$ there is a unique point $P_D(x) \in D$ satisfying

$$\|x - P_D(x)\| = \inf\{\|x - y\| : y \in D\}$$

(see Lemma 2.2).

Suppose that $f : X \to R^1 \cup \{\infty\}$ is a convex, lower semicontinuous and bounded from below function such that

$$\mathrm{dom}(f) := \{x \in X : f(x) < \infty\} \ne \emptyset.$$

Set

$$\inf(f) = \inf(f(x) : x \in X)$$

and

$$\text{argmin}(f) = \{x \in X : f(x) = \inf(f)\}.$$

For each set $D \subset X$ put

$$\inf(f; D) = \inf\{f(z) : z \in D\},$$
$$\sup(f; D) = \sup\{f(z) : z \in D\}.$$

Recall that for each $x \in \text{dom}(f)$,

$$\partial f(x) = \{l \in X : \langle l, y - x \rangle \leq f(y) - f(x) \text{ for all } y \in X\}.$$

In Sect. 14.4 we will prove the following result.

Theorem 14.1. *Let* $\delta \in (0, 1]$, $0 < \mu_1 < \mu_2$, $M > 0$,

$$\epsilon_0 = 2\delta(2M + 1)\mu_1^{-1}, \quad T_0 > 4M^2\mu_1^{-1}\epsilon_0^{-1} \tag{14.1}$$

and let $\mu : [0, T_0] \to R^1$ *be a Lebesgue measurable function such that*

$$\mu_1 \leq \mu(t) \leq \mu_2 \text{ for all } t \in [0, T_0]. \tag{14.2}$$

Assume that $x \in W^{1,1}(0, T_0; X)$,

$$x(0) \in \text{dom}(f) \cap B(0, M) \tag{14.3}$$

and that for almost every $t \in [0, T_0]$,

$$x(t) \in \text{dom}(f) \tag{14.4}$$

and

$$B(x'(t), \delta) \cap (-\mu(t)\partial f(x(t))) \neq \emptyset. \tag{14.5}$$

Then

$$\min\{f(x(t)) : t \in [0, T_0]\} \leq \inf(f; B(0, M)) + \epsilon_0.$$

In Theorem 14.1 δ is the computational error. According to this result we can find a point $\xi \in X$ such that

$$f(\xi) \leq \inf(f; B(0, M)) + c_1\delta$$

during a period of time $c_2\delta^{-1}$, where $c_1, c_2 > 0$ are constants depending only on μ_1, M.

14.3   An Auxiliary Result

Let $V \subset X$ be an open convex set and $g : V \to R^1$ be a convex locally Lipschitzian function.

Let $T > 0$, $x_0 \in X$ and let $u : [0, T] \to X$ be a Bochner integrable function. Set

$$x(t) = x_0 + \int_0^t u(s)ds, \ t \in [0, T].$$

Then $x : [0, T] \to X$ is differentiable and $x'(t) = u(t)$ for almost every $t \in [0, T]$.

Assume that

$$x(t) \in V \text{ for all } t \in [0, T].$$

We claim that the restriction of g to the set $\{x(t) : \ t \in [0, T]\}$ is Lipschitzian. Indeed, since the set $\{x(t) : \ t \in [0, T]\}$ is compact, the closure of its convex hull C is both compact and convex, and so the restriction of g to C is Lipschitzian. Hence the function $(g \cdot x)(t) := g(x(t))$, $t \in [0, T]$, is absolutely continuous. It follows that for almost every $t \in [0, T]$, both the derivatives $x'(t)$ and $(g \cdot x)'(t)$ exist:

$$x'(t) = \lim_{h \to 0} h^{-1}[x(t + h) - x(t)], \qquad (14.6)$$

$$(g \cdot x)'(t) = \lim_{h \to 0} h^{-1}[g(x(t + h)) - g(x(t))]. \qquad (14.7)$$

We continue with the following fact (see Proposition 8.3 of [101]).

Proposition 14.2. *Assume that $t \in [0, T]$ and that both the derivatives $x'(t)$ and $(g \cdot x)'(t)$ exist. Then*

$$(g \cdot x)'(t) = \lim_{h \to 0} h^{-1}[g(x(t) + hx'(t)) - g(x(t))]. \qquad (14.8)$$

Proof. There exist a neighborhood $\mathcal{U}$ of $x(t)$ in X and a constant $L > 0$ such that

$$|g(z_1) - g(z_2)| \leq L||z_1 - z_2|| \text{ for all } z_1, z_2 \in \mathcal{U}. \qquad (14.9)$$

Let $\epsilon > 0$ be given. In view of (14.6), there exists $\delta > 0$ such that

$$x(t + h), \ x(t) + hx'(t) \in \mathcal{U} \text{ for each } h \in [-\delta, \delta] \cap [-t, T - t], \qquad (14.10)$$

and such that for each $h \in [(-\delta, \delta) \setminus \{0\}] \cap [-t, T - t]$,

$$\|x(t + h) - x(t) - hx'(t)\| < \epsilon|h|. \tag{14.11}$$

Let

$$h \in [(-\delta, \delta) \setminus \{0\}] \cap [-t, T - t]. \tag{14.12}$$

It follows from (14.10), (14.9), (14.11), and (14.12) that

$$|g(x(t + h)) - g(x(t) + hx'(t))| \le L\|x(t + h) - x(t) - hx'(t)\| < L\epsilon|h|. \tag{14.13}$$

Clearly,

$$[g(x(t + h)) - g(x(t))]h^{-1} = [g(x(t + h)) - g(x(t) + hx'(t))]h^{-1}$$
$$+[g(x(t) + hx'(t)) - g(x(t))]h^{-1}. \tag{14.14}$$

Relations (14.13) and (14.14) imply that

$$|[g(x(t + h)) - g(x(t))]h^{-1} - [g(x(t) + hx'(t)) - g(x(t))]h^{-1}|$$
$$\le |g(x(t + h)) - g(x(t) + hx'(t))||h^{-1}| \le L\epsilon. \tag{14.15}$$

Since ϵ is an arbitrary positive number, (14.7) and (14.15) imply (14.8).

Corollary 14.3. *Let $z \in X$ and $g(y) = \|z - y\|^2$ for all $y \in X$. Then for almost every $t \in [0, T]$, the derivative $(g \cdot x)'(t)$ exists and*

$$(g \cdot x)'(t) = 2\langle x'(t), x(t) - z \rangle.$$

14.4 Proof of Theorem 14.1

Assume that the theorem does not hold. Then there exists

$$z \in B(0, M) \tag{14.16}$$

such that

$$f(x(t)) > f(z) + \epsilon_0 \text{ for all } t \in [0, T_0]. \tag{14.17}$$

Set

$$\phi(t) = \|z - x(t)\|^2, \ t \in [0, T_0]. \tag{14.18}$$

In view of Corollary 14.3, for a. e. $t \in [0, T_0]$, there exist derivatives $x'(t)$, $\phi'(t)$ and

$$\phi'(t) = 2\langle x'(t), x(t) - z\rangle. \tag{14.19}$$

By (14.5), for a. e. $t \in [0, T_0]$, there exist

$$\xi(t) \in \partial f(x(t)) \tag{14.20}$$

such that

$$\|x'(t) + \mu(t)\xi(t)\| \leq \delta. \tag{14.21}$$

It follows from (14.19) that for almost every $t \in [0, T_0]$,

$$\phi'(t) = 2\langle x(t) - z, x'(t)\rangle$$
$$= 2\langle x(t) - z, -\mu(t)\xi(t)\rangle + 2\langle x(t) - z, x'(t) + \mu(t)\xi(t)\rangle. \tag{14.22}$$

In view of (14.21), for almost every $t \in [0, T_0]$,

$$|\langle z - x(t), x'(t) + \mu(t)\xi(t)\rangle| \leq \delta\|z - x(t)\|. \tag{14.23}$$

By (14.17) and (14.20), for almost every $t \in [0, T_0]$,

$$\langle z - x(t), \xi(t)\rangle \leq -f(x(t)) + f(z) \leq -\epsilon_0. \tag{14.24}$$

It follows from (14.2), (14.22), (14.23), and (14.24) that for almost every $t \in [0, T_0]$,

$$\phi'(t) \leq -2\epsilon_0\mu(t) + 2\delta\|z - x(t)\|$$
$$\leq -2\epsilon_0\mu_1 + 2\delta\|z - x(t)\|. \tag{14.25}$$

Relations (14.3), (14.16), and (14.18) imply that

$$\phi(0) \leq 4M^2. \tag{14.26}$$

We show that for all $t \in [0, T_0]$,

$$\|z - x(t)\| = \phi(t)^{1/2} \leq 2M + 1.$$

Assume the contrary. Then there exists

$$\tau \in (0, T]$$

such that

$$\|z - x(t)\| < 2M + 1, \; t \in [0, \tau),$$ (14.27)

$$\|z - x(\tau)\| = 2M + 1.$$ (14.28)

By (14.1) and (14.25)–(14.28),

$$(2M + 1)^2 - 4M^2 \le \|z - x(\tau)\|^2 - \|z - x(0)\|^2$$

$$= \phi(\tau) - \phi(0) = \int_0^\tau \phi'(t)dt$$

$$\le \int_0^\tau (-2\epsilon_0\mu_1 + 2\delta\|z - x(t)\|)dt$$

$$\le \tau(-2\epsilon_0\mu_1 + 2\delta(2M + 1)) \le -\tau\epsilon_0\mu_1,$$

a contradiction. The contradiction we have reached proves that

$$\|z - x(t)\| \le 2M + 1, \; t \in [0, T_0].$$ (14.29)

By (14.1), (14.25), and (14.29), for almost every $t \in [0, T_0]$,

$$\phi'(t) \le -2\epsilon_0\mu_1 + 2\delta(2M + 1) \le -\epsilon_0\mu_1.$$ (14.30)

It follows from (14.26) and (14.30) that

$$4M^2 \ge \phi(0) - \phi(T_0) = -\int_0^{T_0} \phi'(t)dt \ge T_0\epsilon_0\mu_1,$$

$$T_0 \le 4M^2(\mu_1\epsilon_0)^{-1}.$$

This contradicts the choice of T_0 (see (4.1)). The contradiction we have reached proves Theorem 14.1.

14.5 Continuous Subgradient Projection Method

We use the notation and definitions introduced in Sect. 14.2.

Let C be a nonempty, convex, and closed set in the Hilbert space X, U be an open and convex subset of X such that

$$C \subset U$$

and $f : U \to R^1$ be a convex locally Lipschitzian function.

Let $x \in U$ and $\xi \in X$. Set

$$f^0(x, \xi) = \lim_{t \to 0^+} t^{-1}[f(x + t\xi) - f(x)], \tag{14.31}$$

$$\partial f(x; \xi) = \{l \in \partial f(x) : \langle l, \xi \rangle = f^0(x, \xi)\}. \tag{14.32}$$

It is a well-known fact of the convex analysis that

$$\partial f(x; \xi) \neq \emptyset.$$

Let $M > 1, L > 0$ and assume that

$$C \subset B(0, M - 1), \ \{y \in X : d(y, C) \leq 1\} \subset U, \tag{14.33}$$

$$|f(v_1) - f(v_2)| \leq L\|v_1 - v_2\|$$

$$\text{for all } v_1, v_2 \in B(0, M + 1) \cap U. \tag{14.34}$$

We will prove the following result.

Theorem 14.4. *Let $\delta \in (0, 1]$,*

$$0 < \mu_1 < \mu_2, \ \mu_1 \leq 1, \tag{14.35}$$

$$\epsilon_0 = 2\delta(10M + \mu_2(L + 1))\mu_1^{-1}, \tag{14.36}$$

$$T_0 > \delta^{-1}(10M + \mu_2(L + 1))^{-1}[(2\mu_1)^{-1}4M^2 + L(2M + 2)]\mu_1 \tag{14.37}$$

and let

$$\mu_1 \leq \mu \leq \mu_2. \tag{14.38}$$

Assume that $x \in W^{1,1}(0, T_0; X)$,

$$d(x(0), C) \leq \delta \tag{14.39}$$

and that for almost every $t \in [0, T_0]$, there exists $\xi(t) \in X$ such that

$$B(\xi(t), \delta) \cap \partial f(x(t); x'(t)) \neq \emptyset, \tag{14.40}$$

$$P_C(x(t) - \mu\xi(t)) \in B(x(t) + x'(t), \delta). \tag{14.41}$$

Then

$$\min\{f(x(t)) : \ t \in [0, T_0]\} \leq \inf(f; C) + \epsilon_0. \tag{14.42}$$

Proof. Assume that (14.42) does not hold. Then there exists

$$z \in C \tag{14.43}$$

such that

$$f(x(t)) > f(z) + \epsilon_0, \ t \in [0, T_0]. \tag{14.44}$$

For almost every $t \in [0, T_0]$ set

$$\phi(t) = x(t) + x'(t). \tag{14.45}$$

It is clear that $\phi : [0, T]$ is a Bochner integrable function. In view of (14.41) and (14.45), for almost every $t \in [0, T_0]$,

$$B(\phi(t), \delta) \cap C \neq \emptyset. \tag{14.46}$$

Define

$$C_\delta = \{x \in X : \ d(x, C) \leq \delta\}. \tag{14.47}$$

Clearly, C_δ is a convex closed set, for each $x \in C_\delta$,

$$B(x, \delta) \cap C \neq \emptyset \tag{14.48}$$

and in view of (14.46),

$$\phi(t) \in C_\delta \text{ for almost every } t \in [0, T_0]. \tag{14.49}$$

Evidently, the function $e^s \phi(s)$, $s \in [0, T_0]$ is Bochner integrable.
   We claim that for all $t \in [0, T_0]$,

$$x(t) = e^{-t} x(0) + e^{-t} \int_0^t e^s \phi(s) ds. \tag{14.50}$$

Clearly, (14.50) holds for $t = 0$. For every $t \in (0, T_0]$ we have

$$\int_0^t e^s \phi(s) ds = \int_0^t e^s (x(s) + x'(s)) ds$$

$$= \int_0^t (e^s x(s))' ds = e^t x(t) - x(0).$$

This implies (14.50) for all $t \in [0, T_0]$.

By (14.50), for all $t \in [0, T_0]$,

$$x(t) = e^{-t}x(0) + (1 - e^{-t})(1 - e^{-t})^{-1}e^{-t}\int_0^t e^s\phi(s)ds$$

$$= e^{-t}x(0) + (1 - e^{-t})\int_0^t e^s(e^t - 1)^{-1}\phi(s)ds. \tag{14.51}$$

In view of (14.49), for all $t \in [0, T_0]$,

$$\int_0^t e^s(e^t - 1)^{-1}\phi(s)ds \in C_\delta. \tag{14.52}$$

Relations (14.39), (14.51), and (14.52) imply that

$$x(t) \in C_\delta \text{ for all } t \in [0, T_0]. \tag{14.53}$$

It follows from (14.48) and (14.53) that for every $t \in [0, T_0]$, there exists

$$\hat{x}(t) \in C \tag{14.54}$$

such that

$$\|x(t) - \hat{x}(t)\| \leq \delta. \tag{14.55}$$

By (14.55) and Lemma 2.2, for almost every $t \in [0, T_0]$,

$$\langle \hat{x}(t) - P_C(x(t) - \mu\xi(t)), x(t) - \mu\xi(t) - P_C(x(t) - \mu\xi(t)) \rangle \leq 0. \tag{14.56}$$

Inequality (14.56) implies that for almost every $t \in [0, T_0]$,

$$\langle x(t) - P_C(x(t) - \mu\xi(t)), x(t) - \mu\xi(t) - P_C(x(t) - \mu\xi(t)) \rangle$$
$$\leq \langle x(t) - \hat{x}(t), x(t) - \mu\xi(t) - P_C(x(t) - \mu\xi(t)) \rangle. \tag{14.57}$$

It follows from (14.43) and Lemma 2.2 that

$$\langle z - P_C(x(t) - \mu\xi(t)), x(t) - \mu\xi(t) - P_C(x(t) - \mu\xi(t)) \rangle \leq 0. \tag{14.58}$$

In view of (14.32) and (14.40), for almost every $t \in [0, T_0]$ there exists

$$\hat{\xi}(t) \in \partial f(x(t); x'(t)) \tag{14.59}$$

such that

$$f^0(x(t), x'(t)) = \langle \hat{\xi}(t), x'(t) \rangle, \tag{14.60}$$

$$\|\xi(t) - \hat{\xi}(t)\| \leq \delta. \tag{14.61}$$

In view of (14.32) and (14.59), for almost every $t \in [0, T_0]$,

$$f(z) \geq f(x(t)) + \langle \hat{\xi}(t), z - x(t) \rangle \geq 0$$

and

$$f(x(t)) - f(z) \leq \langle \hat{\xi}(t), x(t) - z + x'(t) \rangle - \langle \hat{\xi}(t), x'(t) \rangle. \qquad (14.62)$$

By (14.41), for almost every $t \in [0, T_0]$,

$$\|(x(t) + x'(t)) - P_C(x(t) - \mu \xi(t))\| \leq \delta. \qquad (14.63)$$

Relations (14.45) and (14.63) imply that for almost every $t \in [0, T_0]$,

$$\begin{aligned}
\langle z &- x(t) - x'(t), x(t) - \mu \xi(t) - x(t) - x'(t) \rangle \\
&= \langle z - \phi(t), x(t) - \mu \xi(t) - \phi(t) \rangle \\
&= \langle z - \phi(t), x(t) - \mu \xi(t) - P_C(x(t) - \mu \xi(t)) \rangle \\
&\quad + \langle z - \phi(t), P_C(x(t) - \mu \xi(t)) - \phi(t) \rangle \\
&\leq \langle z - \phi(t), x(t) - \mu \xi(t) - P_C(x(t) - \mu \xi(t)) \rangle \\
&\quad + \delta \|z - \phi(t)\|. \qquad (14.64)
\end{aligned}$$

In view of (14.33) and (14.53), for all $t \in [0, T_0]$,

$$\|x(t)\| \leq M. \qquad (14.65)$$

It follows from (14.33), (14.41), (14.45), and (14.65) that for almost every $t \in [0, T_0]$,

$$\|\phi(t)\| = \|x(t) + x'(t)\| \leq M, \qquad (14.66)$$

$$\|x'(t)\| \leq 2M. \qquad (14.67)$$

By (14.33), (14.43), (14.64), and (14.66),

$$\begin{aligned}
\langle z &- x(t) - x'(t), -\mu \xi(t) - x'(t) \rangle \\
&\leq \langle z - \phi(t), x(t) - \mu \xi(t) - P_C(x(t) - \mu \xi(t)) \rangle + 2M\delta. \qquad (14.68)
\end{aligned}$$

By (14.33), (14.34), (14.38), (14.45), (14.53), (14.58), (14.59), (14.61), (14.63), and (14.65),

$$\begin{aligned}
\langle z &- \phi(t), x(t) - \mu \xi(t) - P_C(x(t) - \mu \xi(t)) \rangle \\
&\leq \langle z - P_C(x(t) - \mu \xi(t)), x(t) - \mu \xi(t) - P_C(x(t) - \mu \xi(t)) \rangle \\
&\quad + \delta \|x(t) - \mu \xi(t) - P_C(x(t) - \mu \xi(t))\| \leq \delta(2M + \mu_2(L + 1)). \qquad (14.69)
\end{aligned}$$

In view of (14.68) and (14.69), for almost every $t \in [0, T_0]$,

$$\langle x(t) + x'(t) - z, \mu\xi(t) + x'(t)\rangle \le \delta(4M + \mu_2(L + 1)). \tag{14.70}$$

It follows from (14.45), (14.61), (14.62), and (14.66) that for almost every $t \in [0, T_0]$,

$$
\begin{aligned}
f&(x(t)) - f(z) \\
&\le \langle \xi(t), x(t) - z + x'(t)\rangle + \langle \hat{\xi}(t) - \xi(t), x(t) - z + x'(t)\rangle \\
&\quad - \langle \xi(t), x'(t)\rangle + \langle \xi(t) - \hat{\xi}(t), x'(t)\rangle \\
&\le \langle \xi(t), x(t) - z + x'(t)\rangle - \langle \xi(t), x'(t)\rangle + 4M\delta.
\end{aligned}\tag{14.71}
$$

Relations (14.70) and (14.71) imply that for almost every $t \in [0, T_0]$,

$$
\begin{aligned}
f&(x(t)) - f(z) \\
&\le \mu^{-1}\langle -x'(t), x(t) + x'(t) - z\rangle + \mu^{-1}(4M + \mu_2(L + 1))\delta \\
&\quad - \langle \xi(t), x'(t)\rangle + 4M\delta.
\end{aligned}\tag{14.72}
$$

By (14.44) and (14.72), for almost every $t \in [0, T_0]$,

$$
\begin{aligned}
\epsilon_0 &< f(x(t)) - f(z) \\
&\le -\mu^{-1}\|x'(t)\|^2 - \mu^{-1}\langle x'(t), x(t) - z\rangle \\
&\quad - \langle \xi(t), x'(t)\rangle + \delta(8M + \mu_2(L + 1))\mu_1^{-1}
\end{aligned}\tag{14.73}
$$

and

$$
\begin{aligned}
\mu^{-1}&\|x'(t)\|^2 + \mu^{-1}\langle x'(t), x(t) - z\rangle \\
&+ \langle \xi(t), x'(t)\rangle + f(x(t)) - f(z) \\
&\le \delta(8M + \mu_2(L + 1))\mu_1^{-1}.
\end{aligned}\tag{14.74}
$$

In view of (14.61), (14.67), and (14.74),

$$
\begin{aligned}
\mu^{-1}&\|x'(t)\|^2 + \mu^{-1}\langle x'(t), x(t) - z\rangle \\
&+ \langle \hat{\xi}(t), x'(t)\rangle + f(x(t)) - f(z) \\
&\le \delta(10M + \mu_2(L + 1))\mu_1^{-1}.
\end{aligned}\tag{14.75}
$$

It follows from (14.60), (14.75), and Corollary 14.3 that for almost every $t \in [0, T_0]$,

$$
\begin{aligned}
(2\mu)^{-1}&(d/dt)(\|x(t) - z\|^2) + f^0(x(t), x'(t)) \\
&+ f(x(t)) - f(z) + \mu^{-1}\|x'(t)\|^2 \le \delta(10M + \mu_2(L + 1))\mu_1^{-1}.
\end{aligned}
$$

Using Proposition 14.2, the equality

$$f^0(x(t), x'(t)) = (f \circ x)'(t)$$

and integrating the inequality above over the interval $[0, t]$, we obtain that for all $t \in [0, T_0]$,

$$(2\mu)^{-1}\|x(t) - z\|^2 - (2\mu)^{-1}\|x(0) - z\|^2$$

$$+ f(x(t)) - f(x(0)) + \int_0^t (f(x(s)) - f(z))ds \leq \delta t(10M + \mu_2(L+1))\mu_1^{-1}.$$

$$(14.76)$$

Relations (14.33), (14.34), (14.48), and (14.53) imply that for all $t \in [0, T_0]$,

$$f(x(t)) \geq \inf(f; C) - \delta L. \qquad (14.77)$$

By (14.36), (14.38), (14.73), (14.76), and (14.77), for all $t \in [0, T_0]$,

$$(2\mu_2)^{-1}\|x(t) - z\|^2 - (2\mu_1)^{-1}\|x(0) - z\|^2$$

$$+ \inf(f; C) - f(x(0)) - \delta L$$

$$\leq \delta t(10M + \mu_2(L+1))\mu_1^{-1} - \int_0^t (f(x(s)) - f(z))ds$$

$$\leq \delta t(10M + \mu_2(L+1))\mu_1^{-1} - \epsilon_0 t = -\delta t(10M + \mu_2(L+1))\mu_1^{-1}.$$

The relation above with $t = T_0$ implies that

$$-(2\mu_1)^{-1}\|x(0) - z\|^2 + \inf(f; C) - f(x(0)) - \delta L$$

$$\leq -\delta T_0(10M + \mu_2(L+1))\mu_1^{-1}. \qquad (14.78)$$

In view of (14.33), (14.34), (14.48), and (14.53),

$$f(x(0)) \leq \sup(f; C) + \delta L. \qquad (14.79)$$

Relations (14.33), (14.34), and (14.79) imply that

$$\inf(f; C) - f(x(0)) \geq \inf(f; C) - \sup(f; C) - \delta L \geq -L(2M + 1). \qquad (14.80)$$

It follows from (14.33), (14.43), (14.65), and (14.78) that

$$-(2\mu_1)^{-1}4M^2 - L(2M + 2) \leq -\delta T_0(10M + \mu_2(L+1))\mu_1^{-1},$$

$$\delta T_0 \leq ((2\mu_1)^{-1}4M^2 + L(2M + 2))(10M + \mu_2(L+1))^{-1}\mu_1.$$

This contradicts (14.37). The contradiction we have reached completes the proof of Theorem 14.4.

In Theorem 14.4 δ is the computational error. According to this result we obtain a point $\xi \in C_\delta$ (see (14.47), (14.53)) such that

$$f(\xi) \leq \inf(f; C) + c_1 \delta$$

[see (14.36), (14.42)], during a period of time $c_2 \delta^{-1}$ [see (14.37)], where $c_1, c_2 > 0$ are constants depending only on μ_1, μ_2, L, M.

Chapter 15
Penalty Methods

In this chapter we use the penalty approach in order to study constrained minimization problems in infinite dimensional spaces. A penalty function is said to have the exact penalty property if there is a penalty coefficient for which a solution of an unconstrained penalized problem is a solution of the corresponding constrained problem. Since we consider optimization problems in general Banach spaces, not necessarily finite-dimensional, the existence of solutions of original constrained problems and corresponding penalized unconstrained problems is not guaranteed. By this reason we deal with approximate solutions and with an approximate exact penalty property which contains the classical exact penalty property as a particular case. In our recent research we established the approximate exact penalty property for a large class of inequality-constrained minimization problems. In this chapter we improve this result and obtain an estimation of the exact penalty.

15.1 An Estimation of Exact Penalty in Constrained Optimization

Penalty methods are an important and useful tool in constrained optimization. See, for example, [25, 33, 43, 45, 49, 57, 80, 85, 117, 121] and the references mentioned there. In this chapter we use the penalty approach in order to study constrained minimization problems in infinite dimensional spaces. A penalty function is said to have the exact penalty property if there is a penalty coefficient for which a solution of an unconstrained penalized problem is a solution of the corresponding constrained problem.

The notion of exact penalization was introduced by Eremin [48] and Zangwill [114] for use in the development of algorithms for nonlinear constrained optimization. Since that time, exact penalty functions have continued to play an important role in the theory of mathematical programming.

© Springer International Publishing Switzerland 2016

A.J. Zaslavski, *Numerical Optimization with Computational Errors*, Springer Optimization and Its Applications 108, DOI 10.1007/978-3-319-30921-7_15

In our recent research, which was summarized in [121] we established the approximate exact penalty property for a large class of inequality-constrained minimization problems. This approximate exact penalty property can be used for approximate solutions and contains the classical exact penalty property as a particular case. In this chapter we obtain an estimation of the exact penalty.

We use the convention that $\lambda \cdot \infty = \infty$ for all $\lambda \in (0, \infty)$, $\lambda + \infty = \infty$ and $\max\{\lambda, \infty\} = \infty$ for every real number λ and that supremum over empty set is $-\infty$. For every real number λ put $\lambda_+ = \max\{\lambda, 0\}$.

We use the following notation and definitions.

Let $(X, \|\cdot\|)$ be a Banach space. For every point $x \in X$ and every positive number $r > 0$ put

$$B(x, r) = \{y \in X : \|x - y\| \le r\}.$$

For every function $f : X \to R^1 \cup \{\infty\}$ and every nonempty set $A \subset X$ define

$$\mathrm{dom}(f) = \{x \in X : f(x) < \infty\},$$
$$\inf(f) = \inf\{f(z) : z \in X\}$$

and

$$\inf(f; A) = \inf\{f(z) : z \in A\}.$$

For every point $x \in X$ and every nonempty set $B \subset X$ define

$$d(x, B) = \inf\{\|x - y\| : y \in B\}. \tag{15.1}$$

Let $n \ge 1$ be an integer. For every $\kappa \in (0, 1)$ denote by Ω_κ the set of all vectors $\gamma = (\gamma_1, \dots, \gamma_n) \in R^n$ such that

$$\kappa \le \min\{\gamma_i : i = 1, \dots, n\} \text{ and } \max\{\gamma_i : i = 1, \dots, n\} = 1. \tag{15.2}$$

Let $g_i : X \to R^1 \cup \{\infty\}$, $i = 1, \dots, n$ be convex lower semicontinuous functions and $c = (c_1, \dots, c_n) \in R^n$. Define

$$A = \{x \in X : g_i(x) \le c_i \text{ for all } i = 1, \dots, n\}. \tag{15.3}$$

Let $f : X \to R^1 \cup \{\infty\}$ be a bounded from below lower semicontinuous function which satisfies the following growth condition

$$\lim_{\|x\| \to \infty} f(x) = \infty. \tag{15.4}$$

We suppose that there exists a point $\tilde{x} \in X$ such that

$$g_j(\tilde{x}) < c_j \text{ for all } j = 1, \dots, n \text{ and } f(\tilde{x}) < \infty. \tag{15.5}$$

We consider the following constrained minimization problem

$$f(x) \to \min \text{ subject to } x \in A. \qquad (P)$$

By (15.5), $A \neq \emptyset$ and $\inf(f; A) < \infty$.

For every $\gamma = (\gamma_1, \ldots, \gamma_n) \in (0, \infty)^n$ define

$$\psi_\gamma(z) = f(z) + \sum_{i=1}^{n} \gamma_i \max\{g_i(z) - c_i, 0\}, \ z \in X. \qquad (15.6)$$

It is clear that for every vector $\gamma \in (0, \infty)^n$ the function $\psi_\gamma : X \to R^1 \cup \{\infty\}$ is bounded from below and lower semicontinuous and satisfies $\inf(\psi_\gamma) < \infty$. We associate with problem (P) the corresponding family of unconstrained minimization problems

$$\psi_\gamma(z) \to \min, \ z \in X \qquad (P_\gamma)$$

where $\gamma \in (0, \infty)^n$.

Assume that there exists a function $h : X \times \text{dom}(f) \to R^1 \cup \{\infty\}$ such that the following assumptions hold:

(A1) $h(z, y)$ is finite for every pair of points $y, z \in \text{dom}(f)$ and $h(y, y) = 0$ for every point $y \in \text{dom}(f)$.

(A2)   For every point $y \in \text{dom}(f)$ the function $h(\cdot, y) \to R^1 \cup \{\infty\}$ is convex.

(A3)   For every point $z \in \text{dom}(f)$ and every positive number r

$$\sup\{h(z, y) : \ y \in \text{dom}(f) \cap B(0, r)\} < \infty.$$

(A4)   For every positive number M there is $M_1 > 0$ such that for every point $y \in X$ satisfying $f(y) \leq M$ there exists a neighborhood V of y in X such that if $z \in V$, then

$$f(z) - f(y) \leq M_1 h(z, y).$$

Remark 15.1. Note that if the function f is convex, then assumptions (A1)–(A4) hold with $h(z, y) = f(z) - f(y)$, $z \in X$, $y \in \text{dom}(f)$. In this case $M_1 = 1$ for all $M > 0$. If the function f is finite-valued and Lipschitzian on all bounded subsets of X, then assumptions (A1)–(A4) hold with $h(z, y) = \|z - y\|$ for all $z, y \in X$.

Let $\kappa \in (0, 1)$. The main result of [118] (Theorem 15.2 stated below) imply that if λ is sufficiently large, then any solution of problem $(P_{\lambda\gamma})$ with $\gamma \in \Omega_\kappa$ is a solution of problem (P). Note that if the space X is infinite-dimensional, then the existence of solutions of problems $(P_{\lambda\gamma})$ and (P) is not guaranteed. In this case Theorem 15.2 implies that for each $\epsilon > 0$ there exists $\delta(\epsilon) > 0$ which depends only on ϵ such that the following property holds:

If $\lambda \geq \Lambda_0$, $\gamma \in \Omega_\kappa$ and if x is a δ-approximate solution of $(P_{\lambda\gamma})$, then there exists an ϵ-approximate solution y of (P) such that $\|y - x\| \leq \epsilon$.

Here Λ_0 is a positive constant which does not depend on ϵ.

It should be mentioned that we deal with penalty functions whose penalty parameters for constraints $g_1, \ldots, g_n$ are $\lambda\gamma_1, \ldots, \lambda\gamma_n$ respectively, where $\lambda > 0$ and $(\gamma_1, \ldots, \gamma_n) \in \Omega_\kappa$ for a given $\kappa \in (0, 1)$. Note that the vector $(1, 1, \ldots, 1) \in \Omega_\kappa$ for any $\kappa \in (0, 1)$. Therefore our results also includes the case $\gamma_1 = \cdots = \gamma_n = 1$ where one single parameter λ is used for all constraints. Note that sometimes it is an advantage from numerical consideration to use penalty coefficients $\lambda\gamma_1, \ldots, \lambda\gamma_n$ with different parameters γ_i, $i = 1, \ldots, n$. For example, in the case when some of the constrained functions are very "small" and some of the constraint functions are very "large."

The next theorem is the main result of [118].

Theorem 15.2. *Let $\kappa \in (0, 1)$. Then there exists a positive number Λ_0 such that for each $\epsilon > 0$ there exists $\delta \in (0, \epsilon)$ such that the following assertion holds:*
If $\gamma \in \Omega_\kappa$, $\lambda \geq \Lambda_0$ and if $x \in X$ satisfies

$$\psi_{\lambda\gamma}(x) \leq \inf(\psi_{\lambda\gamma}) + \delta,$$

then there exists $y \in A$ such that

$$\|y - x\| \leq \epsilon \text{ and } f(y) \leq \inf(f; A) + \epsilon.$$

Note that Theorem 15.2 is just an existence result and it does not provide any estimation of the constant Λ_0. In this chapter we prove the main result of [119] which improves Theorem 15.2 and provides an estimation of the exact penalty Λ_0.

In view of (15.4) and (15.5), there exists a positive number M such that

$$\text{if } y \in X \text{ satisfies } f(y) \leq |f(\tilde{x})| + 1, \text{ then } \|y\| < M. \tag{15.7}$$

By (15.7), we have

$$\|\tilde{x}\| < M. \tag{15.8}$$

In view of (A4), there exists a positive number M_1 such that the following property holds:

(P1)   for every point $y \in X$ satisfying $f(y) \leq |f(\tilde{x})| + 1$ there exists a neighborhood V of y in X such that $f(z) - f(y) \leq M_1 h(z, y)$ for all $z \in V$.

By (15.4), (15.5), and assumption (A3), there exists a positive number M_2 such that

$$\sup\{h(\tilde{x}, z) : \ z \in X \text{ and } f(z) \leq f(\tilde{x}) + 1\} \leq M_2.$$

Remark 15.3. If the function f is convex, then by Remark 15.1, we choose $h(z, y) = f(z) - f(y)$ for all $z \in X$ and all $y \in \text{dom}(f)$ with $M_1 = 1$ for all $M > 0$ and then

$$\sup\{h(\tilde{x}, z) : \ z \in X \text{ and } f(z) \leq f(\tilde{x}) + 1\}$$
$$\leq \sup\{f(\tilde{x}) - f(z) : \ z \in X \text{ and } f(z) \leq f(\tilde{x}) + 1\} = f(\tilde{x}) - \inf(f).$$

Thus in this case M_2 can be any positive number such that $M_2 \geq f(\tilde{x}) - \inf(f)$.

If the function f is finite-valued and Lipschitzian on bounded subsets of X, then by Remark 15.1, we choose $h(z, y) = \|z - y\|$ for all $z, y \in X$ and M_1 is a Lipschitz constant of the restriction of f to $B(0, M)$. In this case

$$\sup\{h(\tilde{x}, z) : \ z \in X \text{ and } f(z) \leq f(\tilde{x}) + 1\}$$
$$\leq \sup\{\|\tilde{x} - z\| : \ z \in B(0, M)\} \leq 2M$$

and $M_2 = M$.

Let $\kappa \in (0, 1)$. Fix a number $\Lambda_0 > 1$ such that

$$\kappa \sum_{i=1}^{n} (c_i - g_i(\tilde{x})) > \max\{2\Lambda_0^{-1} M_1 M_2, \ 8\Lambda_0^{-2} M^2\}. \tag{15.9}$$

We will prove the following result obtained in [119].

Theorem 15.4. *For each $\epsilon \in (0, 1)$, each $\gamma \in \Omega_\kappa$, each $\lambda \geq \Lambda_0$ and each $x \in X$ which satisfies*

$$\psi_{\lambda\gamma}(x) \leq \inf(\psi_{\lambda\gamma}) + (2\Lambda_0)^{-1}\epsilon$$

there exists $y \in A$ such that

$$\|y - x\| \leq \epsilon \text{ and } f(y) \leq \psi_{\lambda\gamma}(x) \leq \inf(f; A) + \epsilon.$$

15.2 Proof of Theorem 15.4

Assume that the theorem does not hold. Then there exist

$$\epsilon \in (0, 1), \ \gamma = (\gamma_1, \ldots, \gamma_n) \in \Omega_\kappa, \tag{15.10}$$
$$\lambda \geq \Lambda_0 \text{ and } \bar{x} \in X \tag{15.11}$$

such that

$$\psi_{\lambda\gamma}(\bar{x}) \leq \inf(\psi_{\lambda\gamma}) + 2^{-1}\epsilon\Lambda_0^{-1} \tag{15.12}$$

and

$$\{y \in B(\bar{x}, \epsilon) \cap A : \ \psi_{\lambda\gamma}(y) \leq \psi_{\lambda\gamma}(\bar{x})\} = \emptyset. \tag{15.13}$$

By (15.12) and Ekeland's variational principle [50] (see Theorem 15.19), there exists a point $\bar{y} \in X$ such that

$$\psi_{\lambda\gamma}(\bar{y}) \leq \psi_{\lambda\gamma}(\bar{x}), \tag{15.14}$$

$$\|\bar{y} - \bar{x}\| \leq 2^{-1}\epsilon \tag{15.15}$$

and

$$\psi_{\lambda\gamma}(\bar{y}) \leq \psi_{\lambda\gamma}(z) + \Lambda_0^{-1}\|z - \bar{y}\| \text{ for all } z \in X. \tag{15.16}$$

In view of (15.13)–(15.15),

$$\bar{y} \notin A. \tag{15.17}$$

Define

$$I_1 = \{i \in \{1, \ldots, n\} : \ g_i(\bar{y}) > c_i\}, \tag{15.18}$$

$$I_2 = \{i \in \{1, \ldots, n\} : \ g_i(\bar{y}) = c_i\},$$

$$I_3 = \{i \in \{1, \ldots, n\} : \ g_i(\bar{y}) < c_i\}.$$

By (15.17) and (15.18), we have

$$I_1 \neq \emptyset. \tag{15.19}$$

Relations from (15.3), (15.6), (15.10), (15.11), (15.12), (15.14), (15.18), and (15.19) imply that

$$\inf\{f(z) : \ z \in A\} = \inf\{\psi_{\lambda\gamma}(z) : \ z \in A\} \geq \inf(\psi_{\lambda\gamma})$$
$$\geq \psi_{\lambda\gamma}(\bar{x}) - 1 \geq \psi_{\lambda\gamma}(\bar{y}) - 1 = f(\bar{y})$$
$$+ \sum_{i \in I_1} \lambda\gamma_i(g_i(\bar{y}) - c_i) - 1. \tag{15.20}$$

By (15.20), (15.18), and (15.5),

$$f(\bar{y}) \leq \inf\{f(z) : \ z \in A\} + 1 \leq f(\bar{x}) + 1. \tag{15.21}$$

In view of (15.21) and (15.7),

$$\|\bar{y}\| < M. \tag{15.22}$$

Property (P1), (15.21), and (15.22) imply that there exists an open neighborhood V of the point $\bar{y}$ in X such that

$$V \subset B(0, M), \tag{15.23}$$

$$f(z) - f(\bar{y}) \le M_1 h(z, \bar{y}) \text{ for each } z \in V. \tag{15.24}$$

Since the functions g_i, $i = 1, \ldots, n$ are lower semicontinuous it follows from (15.18) that there exists a positive number $r < 1$ such that for every point $y \in B(\bar{y}, r)$,

$$g_i(y) > c_i \text{ for each } i \in I_1. \tag{15.25}$$

By (15.21), (15.5), (15.14), (15.12), (15.25), and (15.16), for every point $z \in B(\bar{y}, r) \cap \mathrm{dom}(f)$, we have

$$\sum_{i \in I_1} \lambda \gamma_i (g_i(z) - c_i) + \sum_{i \in I_2 \cup I_3} \lambda \gamma_i \max\{g_i(z) - c_i, 0\}$$

$$- \sum_{i \in I_1} \lambda \gamma_i (g_i(\bar{y}) - c_i) - \sum_{i \in I_2 \cup I_3} \lambda \gamma_i \max\{g_i(\bar{y}) - c_i, 0\}$$

$$= \psi_{\lambda \gamma}(z) - \psi_{\lambda \gamma}(\bar{y}) - f(z) + f(\bar{y}) \ge -\Lambda_0^{-1} \|\bar{y} - z\| - f(z) + f(\bar{y}).$$

Combined with (15.11) this relation implies that for every point $z \in B(\bar{y}, r)$,

$$\sum_{i \in I_1} \gamma_i g_i(z) + \sum_{i \in I_2 \cup I_3} \gamma_i \max\{g_i(z) - c_i, 0\}$$

$$- \sum_{i \in I_1} \gamma_i g_i(\bar{y}) - \sum_{i \in I_2 \cup I_3} \gamma_i \max\{g_i(\bar{y}) - c_i, 0\}$$

$$+ \lambda^{-1}(f(z) - f(\bar{y})) \ge -\Lambda_0^{-2} \|\bar{y} - z\|.$$

By this inequality, (15.23) and (15.24), for every point $z \in B(\bar{y}, r) \cap V$,

$$\sum_{i \in I_1} \gamma_i g_i(z) + \sum_{i \in I_2 \cup I_3} \gamma_i \max\{g_i(z) - c_i, 0\}$$

$$+ \lambda^{-1} M_1 h(z, \bar{y}) + \Lambda_0^{-2} \|z - \bar{y}\|$$

$$\ge \sum_{i \in I_1} \gamma_i g_i(\bar{y}) + \sum_{i \in I_2 \cup I_3} \gamma_i \max\{g_i(\bar{y}) - c_i, 0\}. \tag{15.26}$$

In view of (A2), the function

$$\sum_{i \in I_1} \gamma_i g_i(z) + \sum_{i \in I_2 \cup I_3} \gamma_i \max\{g_i(z) - c_i, 0\}$$

$$+\lambda^{-1} M_1 h(z, \bar{y}) + \Lambda_0^{-2} \|z - \bar{y}\|, \ z \in X$$

is convex. Combined with the equality $h(\bar{y}, \bar{y}) = 0$ [see (A1)] this implies that
(15.26) holds true for every point $z \in X$.

Since relation (15.26) is valid for $z = \tilde{x}$ relations (15.5), (15.10), (15.2), (15.11),
(15.18), and (15.19) imply that

$$\sum_{i \in I_1} \gamma_i g_i(\tilde{x}) + \lambda^{-1} M_1 h(\tilde{x}, \bar{y}) + \Lambda_0^{-2} \|\tilde{x} - \bar{y}\|$$

$$\geq \sum_{i \in I_1} \gamma_i g_i(\bar{y}) > \sum_{i \in I_1} \gamma_i c_i.$$

Combined with (15.21), (15.5), (15.8), (15.22), (15.10), (15.2), assumption (A1)
and the choice of M_2 (see Sect. 15.1) this implies that

$$4\Lambda_0^{-2} M^2 + \Lambda_0^{-1} M_1 \sup\{h(\tilde{x}, z) : \ z \in X \text{ and } f(z) \leq f(\tilde{x}) + 1\}$$

$$\geq \Lambda_0^{-2} 4M^2 + \Lambda_0^{-1} M_1 (h(\tilde{x}, \bar{y})_+) \geq \sum_{i \in I_1} \gamma_i (c_i - g_i(\tilde{x}))$$

$$\geq \kappa \sum_{i=1}^{n} (c_i - g_i(\tilde{x}))$$

and

$$\kappa \sum_{i=1}^{n} (c_i - g_i(\tilde{x})) \leq 4\Lambda_0^{-2} M^2 + \Lambda_0^{-1} M_1 M_2.$$

This contradicts (15.9). The contradiction we have reached proves Theorem 15.4.

15.3 Infinite-Dimensional Inequality-Constrained Minimization Problems

In this section we use the penalty approach in order to study inequality-constrained
minimization problems in infinite dimensional spaces. For these problems, a
constraint is a mapping with values in a normed ordered space. For this class of
problems we introduce penalty functions, prove the exact penalty property, and

obtain an estimation of the exact penalty. Using this exact penalty property we obtain necessary and sufficient optimality conditions for the constrained minimization problems.

Let X be a vector space, X' be the set of all linear functionals on X and let Y be a vector space ordered by a convex cone Y_+ such that

$$Y_+ \cap (-Y_+) = \{0\}, \ \alpha Y_+ \subset Y_+ \text{ for all } \alpha \geq 0, \ Y_+ + Y_+ \subset Y_+.$$

We say that $y_1, y_2 \in Y$ satisfy $y_1 \leq y_2$ if and only if $y_2 - y_1 \in Y_+$.

We add to the space Y the largest element ∞ and suppose that $y + \infty = \infty$ for all $y \in Y \cup \{\infty\}$ and that $\alpha \cdot \infty = \infty$ for all $\alpha > 0$.

For a mapping $F : X \to Y \cup \{\infty\}$ we set

$$\text{dom}(F) = \{x \in X : F(x) < \infty\}.$$

A mapping $F : X \to Y \cup \{\infty\}$ is called convex if for all $x_1, x_2 \in X$ and all $\alpha \in (0, 1)$,

$$F(\alpha x_1 + (1 - \alpha)x_2) \leq \alpha F(x_1) + (1 - \alpha)F(x_2).$$

A function $G : Y \to R^1$ is called increasing if $G(y_1) \leq G(y_2)$ for all $y_1, y_2 \in Y$ satisfying $y_1 \leq y_2$.

Assume that $p : X \to R^1 \cup \{\infty\}$ is a convex function.

Recall that for $\bar{x} \in \text{dom}(p)$,

$$\partial p(\bar{x}) = \{l \in X' : l(x - \bar{x}) \leq p(x) - p(\bar{x}) \text{ for all } x \in X\}. \tag{15.27}$$

The set $\partial p(\bar{x})$ is a subdifferential of p at the point $\bar{x}$.

Since we consider convex minimization problems we need to use the following two important facts of convex analysis (see Theorems 3.6.1 and 3.6.4 of [76] respectively).

Proposition 15.5. *Let $p_1 : X \to R^1 \cup \{\infty\}$ and $p_2 : X \to R^1$ be convex functions. Then for any $\bar{x} \in \text{dom}(p_1)$,*

$$\partial(p_1 + p_2)(\bar{x}) = \partial p_1(\bar{x}) + \partial p_2(\bar{x}).$$

Proposition 15.6. *Let $F : X \to Y \cup \{\infty\}$ be a convex mapping, $G : Y \to R^1$ be an increasing convex function, $G(\infty) = \infty$ and let $\bar{x} \in \text{dom}(F)$. Then*

$$\partial(G \circ F)(\bar{x}) = \cup\{\partial(l \circ F)(\bar{x}) : l \in \partial G(F(\bar{x}))\}.$$

In this paper we suppose that $(X, \| \cdot \|)$ is a Banach space, $(Y, \| \cdot \|)$ is a normed space and that $(X^*, \| \cdot \|_*)$ and $(Y^*, \| \cdot \|_*)$ are their dual spaces.

We also suppose that Y is ordered by a convex cone Y_+ which is a closed subset of $(Y, \|\cdot\|)$.

For each function $f : Z \to R^1 \cup \{\infty\}$ where the set Z is nonempty put $\inf(f) = \inf\{f(z) : z \in Z\}$ and for each nonempty subset $A \subset Z$ put

$$\inf(f; A) = \inf\{f(z) : z \in A\}.$$

For all $y \in Y$ put

$$\mu(y) = \inf\{\|z\| : z \in Y \text{ and } z \geq y\}. \tag{15.28}$$

It is not difficult to see that

$$\mu(y) \geq 0 \text{ for all } y \in Y,$$

$$\mu(y) = 0 \text{ if and only if } y \leq 0,$$

$$\mu(y) \leq \|y\| \text{ for all } y \in Y, \tag{15.29}$$

$$\mu(\alpha y) = \alpha\mu(y) \text{ for all } \alpha \in [0, \infty) \text{ and all } y \in Y, \tag{15.30}$$

for all $y_1, y_2 \in Y$

$$\mu(y_1 + y_2) \leq \mu(y_1) + \mu(y_2) \tag{15.31}$$

and

$$\text{if } y_1 \leq y_2, \text{ then } \mu(y_1) \leq \mu(y_2). \tag{15.32}$$

Set

$$\mu(\infty) = \infty.$$

The functional μ was used in [115, 116, 121] for the study of minimization problems with increasing objective functions. Here we use it in order to construct a penalty function.

The following auxiliary result is proved in Sect. 15.4.

Lemma 15.7. *Let $y \in Y \setminus (-Y_+)$ and $l \in \partial\mu(y)$. Then*

$$l(z) \geq 0 \text{ for all } z \in Y_+, \ l(y) = \mu(y) \text{ and } \|l\|_* = 1. \tag{15.33}$$

For each $x \in X$, each $y \in Y$, and each $r > 0$ set

$$B_X(x, r) = \{z \in X : \|x - z\| \leq r\},$$

$$B_Y(y, r) = \{z \in Y : \|y - z\| \leq r\}.$$

A set $E \subset Y$ is called $(*)$-bounded from above if the following property holds:

(P2)   there exists $M_E > 0$ such that for each $y \in E$ there is $z \in B_Y(0, M_E)$ for which $z \geq y$.

Let $f : X \to R^1 \cup \{\infty\}$ be a bounded from below lower semicontinuous function which satisfies the following growth condition

$$\lim_{\|x\| \to \infty} f(x) = \infty. \tag{15.34}$$

Assume that $G : X \to Y \cup \{\infty\}$ is a convex mapping and $c \in Y$. Set

$$A = \{x \in X : G(x) \leq c\}. \tag{15.35}$$

We suppose that there exist $\tilde{M} > 0$, $\tilde{r} > 0$ and a nonempty set $\Omega \subset X$ such that

$$G(x) \leq c \text{ and } f(x) \leq \tilde{M} \text{ for all } x \in \Omega \tag{15.36}$$

and that the following property holds:

(P3)   for each $h \in Y$ satisfying $\|h\| = \tilde{r}$ there is $x_h \in \Omega$ such that $G(x_h) + h \leq c$. By (15.36) and (15.34),

$$\sup\{\|x\| : x \in \Omega\} < \infty.$$

Remark 15.8. Property (P3) is an infinite-dimensional version of Slater condition [58]. In particular, it holds if there exists $\tilde{x} \in A$ such that $f(\tilde{x}) < \infty$ and $c - G(\tilde{x})$ is an interior point of Y_+ in the normed space Y. In this case $\tilde{M}$ is any positive constant satisfying $\tilde{M} \geq f(\tilde{x})$ and $\Omega = \{\tilde{x}\}$.

We assume that G possesses the following two properties.

(P4)   If $\{y_i\}_{i=1}^\infty \subset Y$, $y \in Y$, $\lim_{i \to \infty} y_i = y$ in $(Y, \|\cdot\|)$ and $G(y) = \infty$, then the sequence $\{G(y_i)\}_{i=1}^\infty$ is not $(*)$-bounded above.

(P5)   If $\{y_i\}_{i=1}^\infty \subset Y$, $y \in Y$, $\lim_{i \to \infty} y_i = y$ in $(Y, \|\cdot\|)$ and $G(y) < \infty$, then for each given $\epsilon > 0$ and all sufficiently large natural numbers i there exists $u_i \in Y$ such that

$$G(y_i) \geq u_i \text{ and } \|u_i - G(y)\| \leq \epsilon.$$

Remark 15.9. Clearly, G possesses (P1) and (P2) if $G(X) \subset Y$ and G is continuous. It is easy to see that G possesses (P4) and (P5) if $Y = R^n$, $Y_+ = \{y \in R^n : y_i \geq 0, \ i = 1, \ldots, n\}$, $G = (g_1, \ldots, g_n)$ and the functions $g_i : X \to R^1 \cup \{\infty\}$, $i = 1, \ldots, n$ are lower semicontinuous. In general properties (P4) and (P5) are an infinite-dimensional version of the lower semicontinuity property.

We consider the following constrained minimization problem

$$f(x) \to \min \text{ subject to } x \in A. \tag{P}$$

In view of (15.36)

$$A \neq \emptyset \text{ and } \inf(f; A) < \infty.$$

For each $\gamma > 0$ define

$$\psi_\gamma(z) = f(z) + \gamma \mu(G(z) - c), \ z \in X. \tag{15.37}$$

The set ψ_γ, $\gamma > 0$ is our family of penalty functions.

The following auxiliary result is proved in Sect. 15.4.

Lemma 15.10. *For each $\gamma > 0$ the function $\psi_\gamma : X \to R^1 \cup \{\infty\}$ is lower semicontinuous.*

By (15.34) there is $M_1 > 0$ such that

$$\|z\| < M_1 \text{ for each } z \in X \text{ satisfying } f(z) \leq \tilde{M} + 1. \tag{15.38}$$

We use the following assumption.

Assumption (A1) There is $M_0 > 0$ such that for each $x \in X$ satisfying $f(x) \leq \tilde{M} + 1$ there is a neighborhood V of x in $(X, \| \cdot \|)$ such that for each $z \in V, f(z)$ is finite and

$$|f(z) - f(x)| \leq M_0 \|x - z\|. \tag{15.39}$$

Remark 15.11. Note that assumption (A1) is a form of a local Lipschitz property for f on the sublevel set $f^{-1}((-\infty, \tilde{M} + 1])$.

The following theorem is the first main result of this section.

Theorem 15.12. *Assume that (A1) holds, $M_0 > 0$ is as guaranteed by (A1) and let $\Lambda_0 > 1$ satisfy*

$$\tilde{r} > 4(\Lambda_0^{-2} + M_0 \Lambda_0^{-1})(\sup\{\|z\| : z \in \Omega\} + M_1). \tag{15.40}$$

Then for each $\epsilon \in (0, 1)$, each $\lambda \geq \Lambda_0$ and each $x \in X$ which satisfies

$$\psi_\lambda(x) \leq \inf(\psi_\lambda) + (2\Lambda_0)^{-1}\epsilon$$

there exists $y \in A$ such that

$$\|y - x\| \leq \epsilon \text{ and } f(y) \leq \inf(\psi_\lambda) + \epsilon \leq \inf(f; A) + \epsilon.$$

Corollary 15.13. *Assume that (A1) holds, $M_0 > 0$ is as guaranteed by (A1) and let $\Lambda_0 > 1$ satisfy (15.40). Then for each $\lambda \geq \Lambda_0$ and each sequence $\{x_i\}_{i=1}^{\infty} \subset X$ satisfying*

$$\lim_{i \to \infty} \psi_\lambda(x_i) = \inf(\psi_\lambda) \tag{15.41}$$

there exists a sequence $\{y_i\}_{i=1}^{\infty} \subset A$ such that

$$\lim_{i \to \infty} \|y_i - x_i\| = 0 \text{ and } \lim_{i \to \infty} f(y_i) = \inf(f; A). \tag{15.42}$$

Moreover, for each $\lambda \geq \Lambda_0$

$$\inf(f; A) = \inf(\psi_\lambda).$$

Corollary 15.14. *Assume that (A1) holds, $M_0 > 0$ is as guaranteed by (A1) and let $\Lambda_0 > 1$ satisfy (15.40). Then if $\lambda \geq \Lambda_0$ and if $x \in X$ satisfies*

$$\psi_\lambda(x) = \inf(\psi_\lambda), \tag{15.43}$$

then $x \in A$ and

$$f(x) = \psi_\lambda(x) = \inf(\psi_\lambda) = \inf(f; A). \tag{15.44}$$

Theorem 15.12 is proved in Sect. 15.5. In our second main result of this section we do not assume (A1). Instead of it we assume that the function f is convex and that the mapping G is finite-valued.

Theorem 15.15. *Assume that $G(X) \subset Y$, the function f is convex and let $\Lambda_0 > 1$ satisfy*

$$\tilde{r} > 4\Lambda_0^{-2}(\sup\{\|x\| : x \in \Omega\} + M_1) + 4\Lambda_0^{-1}(\tilde{M} - \inf(f)). \tag{15.45}$$

Then for each $\epsilon > 0$, each $\lambda \geq \Lambda_0$ and each $x \in X$ which satisfies

$$\psi_\lambda(x) \leq \inf(\psi_\lambda) + (2\Lambda_0)^{-1}\epsilon$$

there exists $y \in A$ such that

$$\|y - x\| \leq \epsilon \text{ and } f(y) \leq \inf(\psi_\lambda) + \epsilon \leq \inf(f; A) + \epsilon.$$

Corollary 15.16. *Let the assumptions of Theorem 15.15 hold. Then for each $\lambda \geq \Lambda_0$ and each sequence $\{x_i\}_{i=1}^{\infty} \subset X$ satisfying (15.41) there exists a sequence $\{y_i\}_{i=1}^{\infty} \subset A$ such that (15.42) holds. Moreover, for each $\lambda \geq \Lambda_0$*

$$\inf(f; A) = \inf(\psi_\lambda).$$

Corollary 15.17. *Let the assumptions of Theorem 15.15 hold. Then if $\lambda \geq \Lambda_0$ and if $x \in X$ satisfies (15.43), then $x \in A$ and (15.44) holds.*

Theorem 15.15 is proved in Sect. 15.6.

Using our exact penalty results we obtain necessary and sufficient optimality conditions for constrained minimization problems (P) with the convex function f.

Theorem 15.18. *Assume that $G(X) \subset Y$, the function f is convex, $\Lambda_0 > 1$ satisfies (15.45), $\lambda \geq \Lambda_0$ and $\bar{x} \in X$. Then the following assertions are equivalent.*

1. $\bar{x} \in A$ and $f(\bar{x}) = \inf(f; A)$.
2. $\psi_\lambda(\bar{x}) = \inf(\psi_\lambda)$.
3. *There exist*

$$l_0 \in \partial\mu(G(\bar{x}) - c), \; l_1 \in \partial(l_0 \circ (G(\cdot) - c))(\bar{x})$$

and $l_2 \in \partial f(\bar{x})$ such that $\lambda l_1 + l_2 = 0$.

Proof. The equivalence of assertions 1 and 2 follows from Corollaries 15.16 and 15.17. Therefore it is sufficient to show that assertions 2 and 3 are equivalent. It is clear that if at least one of assertions 2 and 3 holds, then $\bar{x} \in \mathrm{dom}(f)$. Therefore we may assume that $\bar{x} \in \mathrm{dom}(f)$. Clearly, the function ψ_λ is convex. By Propositions 15.5 and 15.6,

$$\partial\psi_\lambda(\bar{x}) = \partial f(\bar{x}) + \lambda\partial(\mu \circ (G(\cdot) - c))(\bar{x})$$

$$= \partial f(\bar{x}) + \cup\lambda\{\partial(l \circ (G(\cdot) - c))(\bar{x}) : \; l \in \partial\mu(G(\bar{x}) - c)\}.$$

Now in order to complete the proof it is sufficient to note that assertion 2 holds if and only if

$$0 \in \partial\psi_\lambda(\bar{x}).$$

It should be mentioned that Theorem 15.18 is an infinite-dimensional version of the classical Karush–Kuhn–Tucker theorem [58].

Assume now that the assumptions of Theorem 15.18 hold and assertions 1, 2, and 3 hold. Let l_0, l_1, and l_2 be as guaranteed by assertion 3. Then

$$l_0(G(\bar{x}) - c) = \mu(G(\bar{x}) - x) = 0$$

because $G(\bar{x}) \leq c$. This is an infinite-dimensional version of the complementary slackness condition [58].

The results of this section were obtained in [123].

In the proof of Theorem 15.12 we use the following fundamental variational principle of Ekeland [50].

Theorem 15.19. *Assume that (Z, ρ) is a complete metric space and that $\phi : Z \to R^1 \cup \{\infty\}$ is a lower semicontinuous bounded from below function which is not identically ∞. Let $\epsilon > 0$ and $x_0 \in Z$ be given such that*

$$\phi(x_0) \leq \phi(x) + \epsilon \text{ for all } x \in Z.$$

Then for any $\lambda > 0$ there is $\bar{x} \in Z$ such that

$$\phi(\bar{x}) \leq \phi(x_0), \; \rho(\bar{x}, x_0) \leq \lambda,$$
$$\phi(x) + (\epsilon/\lambda)\rho(x, \bar{x}) > \phi(\bar{x}) \text{ for all } x \in Z \setminus \{\bar{x}\}.$$

15.4   Proofs of Auxiliary Results

Proof of Lemma 15.7.   It is not difficult to see that

$$\|l\|_* \leq 1, \; l(y) = \mu(y) > 0, \; l(z) \geq 0 \text{ for all } z \in Y_+. \tag{15.46}$$

We will show that $\|l\|_* = 1$. By (15.46) and (15.28),

$$l(y) = \mu(y) = \inf\{\|z\| : z \in Y \text{ and } z \geq y\}. \tag{15.47}$$

Let

$$\epsilon \in (0, \mu(y)/4)$$

[see (15.46)]. By (15.47) there exists $z \in Y$ such that

$$z \geq y \text{ and } \|z\| \leq \mu(y) + \epsilon. \tag{15.48}$$

By (15.46) and (15.28)

$$\|z\| > 0 \tag{15.49}$$

and

$$\mu(y) = l(y) \leq l(z). \tag{15.50}$$

It follows from (15.50) and (15.48) that

$$l(z) \geq \|z\| - \epsilon.$$

Together with (15.49), (15.48), and (15.47) this implies that

$$\|l\|_* \geq l(z)\|z\|^{-1} \geq (\|z\| - \epsilon)\|z\|^{-1} \geq 1 - \epsilon\|z\|^{-1} \geq 1 - \epsilon\mu(y)^{-1}.$$

Since ϵ is any positive number satisfying $\epsilon < \mu(y)/4$ we conclude that

$$\|l\|_* \geq 1.$$

Combined with (15.46) this implies that $\|l\|_* = 1$. Lemma 15.7 is proved.

Proof of Lemma 15.10. It is sufficient to show that the function $\mu(G(\cdot) - c) : X \to R^1 \cup \{\infty\}$ is lower semicontinuous.

Assume that $y \in Y$, $\{y_i\}_{i=1}^{\infty} \subset Y$ and

$$\lim_{i \to \infty} \|y_i - y\| = 0. \tag{15.51}$$

It is sufficient to show that

$$\liminf_{i \to \infty} \mu(G(y_i) - c) \geq \mu(G(y) - c).$$

Extracting a subsequence and re-indexing if necessary we may assume without loss of generality that there exists

$$\lim_{i \to \infty} \mu(G(y_i) - c) < \infty. \tag{15.52}$$

We may assume without loss of generality that $\mu(G(y_i) - c)$ is finite for all integers $i \geq 1$.

Let $\epsilon > 0$. By (15.28) for any integer $i \geq 1$ there exists $z_i \in Y$ such that

$$z_i \geq G(y_i) - c, \quad \|z_i\| \leq \mu(G(y_i) - c) + \epsilon/4. \tag{15.53}$$

In view of (15.52) and (15.53) the sequence $\{\|z_i\|\}_{i=1}^{\infty}$ is bounded. Together with (15.53) this implies that the sequence $\{G(y_i) - c\}_{i=1}^{\infty}$ is (*)-bounded from above [see (P2)]. It follows from (P4) and (15.53) that

$$G(y) < \infty. \tag{15.54}$$

By (15.51), (15.54), and (P5) there exists a natural number i_0 such that for each integer $i \geq i_0$ there is $u_i \in Y$ which satisfies

$$G(y_i) \geq u_i \text{ and } \|u_i - G(y)\| \leq \epsilon/4. \tag{15.55}$$

In view of (15.55) and (15.53) for all integers $i \geq i_0$

$$G(y) - c = (G(y) - u_i) + u_i - c$$
$$\leq (G(y) - u_i) + G(y_i) - c \leq (G(y) - u_i) + z_i. \tag{15.56}$$

It follows from (15.55) and (15.53) that for all integers $i \geq i_0$

$$\|G(y) - u_i + z_i\| \leq \|G(y) - u_i\| + \|z_i\|$$
$$\leq \epsilon/4 + \mu(G(y_i) - c) + \epsilon/4. \tag{15.57}$$

By (15.56) and (15.57), for all integers $i \geq i_0$,

$$\mu(G(y) - c) \leq \|G(y) - u_i + z_i\| \leq \mu(G(y_i) - c) + \epsilon/2$$

and

$$\mu(G(y) - c) \leq \lim_{i \to \infty} \mu(G(y_i) - c) + \epsilon/2.$$

Since ϵ is any positive number we conclude that

$$\mu(G(y) - c) \leq \lim_{i \to \infty} \mu(G(y_i) - c).$$

Lemma 15.10 is proved.

15.5   Proof of Theorem 15.12

We show that the following property holds:

(P6)   For each $\epsilon \in (0, 1)$, each $\lambda \geq \Lambda_0$ and each $x \in X$ which satisfies

$$\psi_\lambda(x) \leq \inf(\psi_\lambda) + (2\Lambda_0)^{-1}\epsilon$$

there exists $y \in A$ for which

$$\|y - x\| \leq \epsilon \text{ and } \psi_\lambda(y) \leq \psi_\lambda(x).$$

(It is easy to see that (P6) implies the validity of Theorem 15.2.)

Assume the contrary. Then there exist

$$\epsilon \in (0, 1), \ \lambda \geq \Lambda_0, \ \bar{x} \in X \tag{15.58}$$

such that

$$\psi_\lambda(\bar{x}) \leq \inf(\psi_\lambda) + 2^{-1}\epsilon\Lambda_0^{-1}, \tag{15.59}$$

$$\{y \in B_X(\bar{x}, \epsilon) \cap A : \psi_\lambda(y) \leq \psi_\lambda(\bar{x})\} = \emptyset. \tag{15.60}$$

It follows from (15.59), Lemma 15.10, and Theorem 15.19 that there is $\bar{y} \in X$ such that

$$\psi_\lambda(\bar{y}) \leq \psi_\lambda(\bar{x}), \tag{15.61}$$

$$\|\bar{y} - \bar{x}\| \leq 2^{-1}\epsilon, \tag{15.62}$$

$$\psi_\lambda(\bar{y}) \leq \psi_\lambda(z) + \Lambda_0^{-1}\|z - \bar{y}\| \text{ for all } z \in X. \tag{15.63}$$

By (15.60), (15.61), and (15.62),

$$\bar{y} \notin A. \tag{15.64}$$

It follows from (15.37), (15.35), (15.59), and (15.61) that

$$\inf\{f(z) : z \in A\} = \inf\{\psi_\lambda(z) : z \in A\} \geq \inf(\psi_\lambda) \geq \psi_\lambda(\bar{x}) - 1$$
$$\geq \psi_\lambda(\bar{y}) - 1 \geq f(\bar{y}) - 1$$

and in view of (15.36),

$$f(\bar{y}) \leq \inf\{f(z) : z \in A\} + 1 \leq \tilde{M} + 1. \tag{15.65}$$

By (15.65) and (15.38)

$$\|\bar{y}\| < M_1. \tag{15.66}$$

In view of (A1) and (15.65) there exists a neighborhood V of $\bar{y}$ in $(X, \|\cdot\|)$ such that for each $z \in V$,

$$f(z) \text{ is finite and } |f(z) - f(\bar{y})| \leq M_0 \|z - \bar{y}\|. \tag{15.67}$$

It follows from (15.61), (15.59), (15.67), (15.37), (15.63), and (15.58) that for each $z \in V$

$$\lambda \mu(G(z) - c) - \lambda \mu(G(\bar{y}) - c) = \psi_\lambda(z) - \psi_\lambda(\bar{y}) - f(z) + f(\bar{y})$$
$$\geq -\Lambda_0^{-1}\|z - \bar{y}\| - f(z) + f(\bar{y}) \geq -\Lambda_0^{-1}\|z - \bar{y}\| - M_0\|z - \bar{y}\|$$

and

$$\mu(G(z) - c) - \mu(G(\bar{y}) - c) \geq -\|z - \bar{y}\|(\Lambda_0^{-2} + M_0\Lambda_0^{-1}).$$

This implies that for each $z \in V$

$$\mu(G(z) - c) + (\Lambda_0^{-2} + M_0\Lambda_0^{-1})\|z - \bar{y}\| \geq \mu(G(\bar{y}) - c). \tag{15.68}$$

Clearly the function

$$\tilde{\mu}(z) = \mu(G(z) - c) + (\Lambda_0^{-2} + M_0\Lambda_0^{-1})\|z - \bar{y}\|, \; z \in X \tag{15.69}$$

is convex. By (15.68) and (15.69)

$$0 \in \partial\tilde{\mu}(\bar{y}). \tag{15.70}$$

By (15.70), (15.69), and Proposition 15.5 there is

$$l_0 \in \partial(\mu(G(\cdot) - c))(\bar{y}) \tag{15.71}$$

such that

$$\|l_0\|_* \leq (\Lambda_0^{-2} + M_0\Lambda_0^{-1}). \tag{15.72}$$

It follows from (15.71) and Proposition 15.6 that there exists

$$l_1 \in \partial \mu(G(\bar{y}) - c) \tag{15.73}$$

such that

$$l_0 \in \partial(l_1 \circ (G(\cdot) - c))(\bar{y}). \tag{15.74}$$

In view of (15.74) for each $z \in X$

$$l_0(z - \bar{y}) \le l_1(G(z) - c) - l_1(G(\bar{y}) - c). \tag{15.75}$$

By (15.73), (15.64), and Lemma 15.7

$$\|l_1\|_* = 1, \ l_1(z) \ge 0 \text{ for all } z \in Y_+, \tag{15.76}$$
$$l_1(G(\bar{y}) - c) = \mu(G(\bar{y}) - c).$$

Let

$$h \in Y \text{ and } \|h\| = \tilde{r}. \tag{15.77}$$

By (15.77) and property (P3) there is

$$x_h \in \Omega \tag{15.78}$$

such that

$$G(x_h) + h \le c. \tag{15.79}$$

It follows from (15.78), (15.66), (15.72), (15.75), (15.76), and (15.79) that

$$-(\Lambda_0^{-2} + M_0\Lambda_0^{-1})(M_1 + \sup\{\|x\| : x \in \Omega\}) \le -(\Lambda_0^{-2} + M_0\Lambda_0^{-1})(\|x_h\| + \|\bar{y}\|)$$
$$\le -\|l_0\|_*(\|x_h\| + \|\bar{y}\|) \le l_0(x_h - \bar{y})$$
$$\le l_1(G(x_h) - c) - l_1(G(\bar{y}) - c) \le l_1(G(x_h) - c) - \mu(G(\bar{y}) - c)$$
$$\le l_1(G(x_h) - c) \le l_1(-h). \tag{15.80}$$

Since (15.80) holds for all h satisfying (15.77) we conclude using (15.76) that

$$(\Lambda_0^{-2} + M_0\Lambda_0^{-1})(M_1 + \sup\{\|x\| : x \in \Omega\})$$
$$\ge \sup\{l_1(h) : h \in Y \text{ and } \|h\| = \tilde{r}\} = \tilde{r}\|l_1\|_* = \tilde{r}.$$

This contradicts (15.40). The contradiction we have reached proves (P6) and Theorem 15.12 itself.

15.6 Proof of Theorem 15.15

We show that property (P6) (see Sect. 15.5) holds. (Note that (P6) implies the validity of Theorem 15.15).

Assume the contrary. Then there exist

$$\epsilon \in (0, 1), \; \lambda \geq \Lambda_0, \; \bar{x} \in X \tag{15.81}$$

such that (15.59) and (15.60) hold. It follows from (15.59), Lemma 15.10 and Ekeland's variational principle [50] that there is $\bar{y} \in X$ such that (15.61)–(15.63) hold. By (15.60), (15.61) and (15.62),

$$\bar{y} \notin A. \tag{15.82}$$

Arguing as in the proof of Theorem 15.12 we show that (15.37), (15.35), (15.59), (15.61), (15.36), and (15.38) imply that

$$f(\bar{y}) \leq \tilde{M} + 1, \; \|\bar{y}\| < M_1. \tag{15.83}$$

It follows from (15.37) and (15.63) that for each $z \in X$

$$f(z) + \lambda\mu(G(z) - c) - f(\bar{y}) - \lambda\mu(G(\bar{y}) - c)$$
$$= \psi_\lambda(z) - \psi_\lambda(\bar{y}) \geq -\Lambda_0^{-1}\|z - \bar{y}\|$$

and

$$\lambda^{-1}(f(z) - f(\bar{y})) + \mu(G(z) - c) - \mu(G(\bar{y}) - c)$$
$$\geq -\lambda^{-1}\Lambda_0^{-1}\|z - \bar{y}\|.$$

This implies that for all $z \in X$

$$\lambda^{-1}f(z) + \mu(G(z) - c) + \lambda^{-1}\Lambda_0^{-1}\|z - \bar{y}\| \geq \lambda^{-1}f(\bar{y}) + \mu(G(\bar{y}) - c). \tag{15.84}$$

Put

$$\tilde{\mu}(z) = \mu(G(z) - c) + \lambda^{-1}f(z) + \lambda^{-1}\Lambda_0^{-1}\|z - \bar{y}\|, \; z \in X \tag{15.85}$$

In view of (15.85) the function $\tilde{\mu}$ is convex. By (15.84) and (15.85),

$$0 \in \partial\tilde{\mu}(\bar{y}). \tag{15.86}$$

By (15.85), (15.86), (15.81), and Proposition 15.5 there exist

$$l_1 \in \partial(\mu \circ (G(\cdot) - c))(\bar{y}), \; l_2 \in \partial f(\bar{y}) \tag{15.87}$$

such that

$$\|l_1 + \lambda^{-1} l_2\|_* \leq \Lambda_0^{-2}. \tag{15.88}$$

It follows from (15.87), (15.30), (15.31), and Proposition 15.6 that there exists

$$l_0 \in \partial \mu (G(\bar{y}) - c) \tag{15.89}$$

such that

$$l_1 \in \partial (l_0 \circ (G(\cdot) - c))(\bar{y}). \tag{15.90}$$

In view of (15.90) for each $z \in X$

$$l_1(z - \bar{y}) \leq l_0(G(z) - c) - l_0(G(\bar{y}) - c). \tag{15.91}$$

By (15.89), (15.82), and Lemma 15.7

$$\|l_0\|_* = 1, \; l_0(z) \geq 0 \text{ for all } z \in Y_+, \tag{15.92}$$
$$l_0(G(\bar{y}) - c) = \mu(G(\bar{y}) - c).$$

Let

$$h \in Y \text{ and } \|h\| = \tilde{r}. \tag{15.93}$$

By (15.93) and property (P3) there is

$$x_h \in \Omega \tag{15.94}$$

such that

$$G(x_h) + h \leq c. \tag{15.95}$$

It follows from (15.94), (15.83), (15.88), (15.91), (15.87), (15.92), (15.36), and (15.95) that

$$
\begin{aligned}
-\Lambda_0^{-2}(M_1 + \sup\{\|x\| : x \in \Omega\}) &\leq -\Lambda_0^{-2}(\|x_h\| + \|\bar{y}\|) \\
&\leq -\|l_1 + \lambda^{-1} l_2\|_*(\|x_h\| + \|\bar{y}\|) \leq (l_1 + \lambda^{-1} l_2)(x_h - \bar{y}) \\
&= l_1(x_h - \bar{y}) + \lambda^{-1} l_2(x_h - \bar{y}) \\
&\leq l_0(G(x_h) - c) - l_0(G(\bar{y}) - c) + \lambda^{-1}(f(x_h) - f(\bar{y})) \\
&\leq l_0(G(x_h) - c) - \mu(G(\bar{y}) - c) + \lambda^{-1}(\tilde{M} - \inf(f)) \\
&\leq -l_0(h) + \lambda^{-1}(\tilde{M} - \inf(f))
\end{aligned}
$$

and in view of (15.81)

$$l_0(h) \le \Lambda_0^{-1}(\tilde{M} - \inf(f)) + \Lambda_0^{-2}(M_1 + \sup\{\|x\| : x \in \Omega\}). \tag{15.96}$$

Since the inequality above holds for all h satisfying (15.93) it follows from (15.92) and (15.96) that

$$\tilde{r} = \tilde{r}\|l_0\|_* = \sup\{l_0(h) : h \in Y \text{ and } \|h\| = \tilde{r}\}$$
$$\le \Lambda_0^{-1}(\tilde{M} - \inf(f)) + \Lambda_0^{-2}(M_1 + \sup\{\|x\| : x \in \Omega\}).$$

This contradicts (15.45). The contradiction we have reached proves (P6) and Theorem 15.15 itself.

15.7 An Application

Let X be a Hilbert space equipped with an inner product $\langle \cdot, \cdot \rangle$ which induces the complete norm $\| \cdot \|$. We use the notation and definitions introduced in Sect. 15.1.

Let n be a natural number, $g_i : X \to R^1 \cup \{\infty\}$, $i = 1, \dots, n$ be convex lower semicontinuous functions and $c = (c_1, \dots, c_n) \in R^n$. Set

$$A = \{x \in X : g_i(x) \le c_i \text{ for all } i = 1, \dots, n\}.$$

Let $f : X \to R^1 \cup \{\infty\}$ be a bounded from below lower semicontinuous function which satisfies the following growth condition

$$\lim_{\|x\| \to \infty} f(x) = \infty.$$

We suppose that there is $\tilde{x} \in X$ such that

$$g_j(\tilde{x}) < c_j \text{ for all } j = 1, \dots, n \text{ and } f(\tilde{x}) < \infty. \tag{15.97}$$

We consider the following constrained minimization problem

$$f(x) \to \min \text{ subject to } x \in A. \tag{P}$$

For each vector $\gamma = (\gamma_1, \dots, \gamma_n) \in (0, \infty)^n$ define

$$\psi_\gamma(z) = f(z) + \sum_{i=1}^{n} \gamma_i \max\{g_i(z) - c_i, 0\}, \ z \in X. \tag{15.98}$$

Clearly for each $\gamma \in (0, \infty)^n$ the function $\psi_\gamma : X \to R^1 \cup \{\infty\}$ is bounded from below and lower semicontinuous and satisfies $\inf(\psi_\gamma) < \infty$.

We suppose that the function f is convex. By Remark 15.1, (A1)–(A4) hold with $h(z, y) = f(z) - f(y)$, $z \in X$, $y \in \text{dom}(f)$. In this case $M_1 = 1$ for all $M > 0$.

There is $M > 0$ such that [see (15.7)]

$$\text{if } y \in X \text{ satisfies } f(y) \leq |f(\tilde{x})| + 1, \text{ then } \|y\| < M. \tag{15.99}$$

Clearly, (P1) holds with $M_1 = 1$. In view of Remark 15.3, the constant M_2 can be any positive number such that

$$M_2 \geq f(\tilde{x}) - \inf(f).$$

We suppose that M_2 is given.

Let $\kappa \in (0, 1)$ Choose $\Lambda_0 > 1$ [see (15.9)] such that

$$\kappa \sum_{i=1}^{n} (c_i - g_i(\tilde{x})) > \max\{2\Lambda_0^{-1}M_2, \ 8\Lambda_0^{-2}M^2\}.$$

By Theorem 15.4, the following property holds:

(P7)   for each $\epsilon \in (0, 1)$, each $\gamma \in \Omega_\kappa$, each $\lambda \geq \Lambda_0$ and each $x \in X$ satisfies

$$\psi_{\lambda\gamma}(x) \leq \inf(\psi_{\lambda\gamma}) + 2^{-1}\epsilon\Lambda_0^{-1}$$

there exists $y \in A$ such that

$$\|y - x\| \leq \epsilon \text{ and } f(y) \leq \psi_{\lambda\gamma}(x) \leq \inf(f; A) + \epsilon.$$

Property (P7) implies that for each $\gamma \in \Omega_\kappa$ and each $\lambda \geq \Lambda_0$,

$$\inf(\psi_{\lambda\gamma}) = \inf(f; A). \tag{15.100}$$

In order to obtain an approximate solution of problem (P) we apply the subgradient projection method, studied in Chap. 2, for the minimization of the function $\psi_{\Lambda_0\gamma}$, where γ is a fixed element of the set Ω_κ.

We suppose that problem (P) has a solution

$$x_* \in A \tag{15.101}$$

such that

$$f(x_*) \leq f(x) \text{ for all } x \in A. \tag{15.102}$$

By (15.100), (15.101), and (15.102),

$$f(x_*) = \psi_{\Lambda_0\gamma}(x_*) = \inf(\psi_{\Lambda_0\gamma}). \tag{15.103}$$

It follows from (15.97), (15.99), and (15.102) that

$$\|x_*\| < M. \tag{15.104}$$

In view of (15.103) and (15.104), we consider the minimization of $\psi_{\Lambda_0\gamma}$ on $B(0, M)$. We suppose that there exist an open convex set $U \subset X$ and a number $L > 0$ such that

$$B(0, M + 1) \subset U, \tag{15.105}$$

the functions f and g_i, $i = 1, \dots, n$ are finite-valued on U and that for all $x, y \in U$ and all $i = 1, \dots, n$,

$$|f(x) - f(y)| \le L\|x - y\|, \ |g_i(x) - g_i(y)| \le L\|x - y\|. \tag{15.106}$$

In view of (15.98) and (15.106), the function $\psi_{\Lambda_0\gamma}$ is Lipschitzian on U and for all $x, y \in U$,

$$|\psi_{\Lambda_0\gamma}(x) - \psi_{\Lambda_0\gamma}(y)|$$
$$\le |f(x) - f(y)| + n\Lambda_0 L\|x - y\| \le L\|x - y\|(1 + \Lambda_0 n). \tag{15.107}$$

In this section we use the projection on the set $B(0, M)$ denoted by $P_{B(0,M)}$ and defined for each $z \in X$ by

$$P_{B(0,M)}(z) = z \text{ if } \|z\| \le M,$$
$$P_{B(0,M)}(z) = M\|z\|^{-1}z \text{ if } \|z\| > M.$$

We apply the subgradient projection method, studied in Chap. 2, for the minimization problem of the function $\psi_{\Lambda_0\gamma}$ on the set $B(0, M)$. For each $\delta > 0$ set

$$\alpha(\delta) = 2^{-1}(2M + 1)^2(1 + L(\Lambda_0 n + 1))(8M + 2)^{-1/2}\delta^{1/2}(1 + \Lambda_0 n)^{1/2}$$
$$+ \delta(1 + \Lambda_0 n)(2M + 1) + (8M + 2)^{1/2}(1 + L(\Lambda_0 n + 1))\delta^{1/2}(1 + \Lambda_0 n)^{1/2}$$
$$+ (4M + 1)(1 + L(\Lambda_0 n + 1))(8M + 2)^{-1/2}\delta^{1/2}(1 + \Lambda_0 n)^{1/2}. \tag{15.108}$$

Let $\delta \in (0, 1]$ be our computational error which satisfies

$$\delta(1 + n\Lambda_0) < 1, \ 2\alpha(\delta)\Lambda_0 < 1. \tag{15.109}$$

Set

$$\delta_0 = \delta(1 + n\Lambda_0) \tag{15.110}$$

and

$$a = (2\delta_0(4M + 1))^{1/2}(L(1 + \Lambda_0 n) + 1)^{-1}.$$

Let us describe our algorithm.

Subgradient Projection Algorithm

Initialization: select an arbitrary $x_0 \in B(0, M)$.

Iterative step: given a current iteration vector $x_t \in U$ calculate

$$\xi_t \in \partial \psi_{\Lambda_0 \gamma}(x_t) + B(0, \delta_0) \qquad (15.111)$$

and the next iteration vector $x_{t+1} \in X$ such that

$$\|x_{t+1} - P_{B(0,M)}(x_t - a\xi_t)\| \le \delta_0. \qquad (15.112)$$

Let $t \ge 0$ be an integer. Let us explain how one can calculate ξ_t satisfying (15.111). We find $\eta_0 \in X$ satisfying

$$\eta_0 \in \partial f(x_t) + B(0, \delta).$$

For every $i = 1, \dots, n$, if $g_i(x_t) \le c_i$, then set $\eta_i = 0$ and if $g_i(x_t) > c_i$, then we calculate

$$\eta_i \in \partial g_i(x_t) + B(0, \delta).$$

Set

$$\xi_t = \eta_0 + \Lambda_0 \gamma \sum_{i=1}^{n} \eta_i.$$

It follows from the equality above, the choice of η_i, $i = 0, \dots, n$, the subdifferential calculus in [84], (15.98) and (15.110) that

$$B(\xi_t, \delta_0) \cap \partial \psi_{\Lambda_0 \gamma}(x_t) = B(\xi_t, \delta(1 + n\Lambda_0)) \cap \partial \psi_{\Lambda_0 \gamma}(x_t) \neq \emptyset$$

and (15.111) is true.

By Theorem 2.6, applied to the function $\psi_{\Lambda_0 \gamma}$, for each natural number T,

$$\psi_{\Lambda_0 \gamma} \left((T+1)^{-1} \sum_{t=0}^{T} x_t \right) - \psi_{\Lambda_0 \gamma}(x_*),$$

$$\min\{\psi_{\Lambda_0 \gamma}(x_t) : \ t = 0, \dots, T\} - \psi_{\Lambda_0 \gamma}(x_*)$$

$$\le 2^{-1}(T+1)^{-1}(2M+1)^2(L(1+\Lambda_0 n)+1)(2\delta_0(4M+1))^{-1/2}$$

$$+\delta_0(2M+1)$$

$$+2^{-1}(2\delta_0(4M+1))^{1/2}(L(1+\Lambda_0 n)+1)$$

$$+\delta_0(4M+1)(L(1+\Lambda_0 n)+1)(2\delta_0(4M+1))^{-1/2}.$$

Now we can think about the best choice of T. It was explained in Chap. 2 that it should be at the same order as

$$\lfloor \delta_0^{-1} \rfloor = \lfloor \delta^{-1}(1 + n\Lambda_0)^{-1} \rfloor.$$

Put $T = \lfloor \delta_0^{-1} \rfloor$ and obtain from (15.108) and (15.110) that

$$\psi_{\Lambda_0 \gamma}\left((T+1)^{-1} \sum_{t=0}^{T} x_t \right) - \psi_{\Lambda_0 \gamma}(x_*),$$

$$\min\{\psi_{\Lambda_0 \gamma}(x_t) : t = 0, \dots, T\} - \psi_{\Lambda_0 \gamma}(x_*) \le \alpha(\delta) \qquad (15.113)$$

By (15.100), (15.101), (15.103), and (15.113),

$$\psi_{\Lambda_0 \gamma}\left((T+1)^{-1} \sum_{t=0}^{T} x_t \right) \le \inf(\psi_{\Lambda_0 \gamma}) + \alpha(\delta). \qquad (15.114)$$

Let $\tau \in \{0, \dots, T\}$ satisfy

$$\psi_{\Lambda_0 \gamma}(x_\tau) = \min\{\psi_{\Lambda_0 \gamma}(x_t) : t = 0, \dots, T\}.$$

In view of (15.101), (15.103), (15.106), and (15.113),

$$\psi_{\Lambda_0 \gamma}(x_\tau) \le \inf(\psi_{\Lambda_0 \gamma}) + \alpha(\delta). \qquad (15.115)$$

It follows from (15.109), (15.114), (15.115), and property (P7) that there exist $y_0, y_1 \in A$ such that

$$\left\| y_0 - (T+1)^{-1} \sum_{t=0}^{T} x_t \right\| \le 2\alpha(\delta)\Lambda_0, \ \|y_1 - x_\tau\| \le 2\alpha(\delta)\Lambda_0,$$

$$f(y_0) \le \psi_{\Lambda_0 \gamma}\left((T+1)^{-1} \sum_{t=0}^{T} x_t \right) \le \inf(f; A) + \alpha(\delta),$$

$$f(y_1) \le \psi_{\Lambda_0 \gamma}(x_\tau) \le \inf(f; A) + \alpha(\delta).$$

The analogous analysis can be also done for the mirror descent method.

Chapter 16
Newton's Method

In this chapter we study the convergence of Newton's method for nonlinear equations and nonlinear inclusions in a Banach space. Nonlinear mappings, which appear in the right-hand side of the equations, are not necessarily differentiable. Our goal is to obtain an approximate solution in the presence of computational errors. In order to meet this goal, in the case of inclusions, we study the behavior of iterates of nonexpansive set-valued mappings in the presence of computational errors.

16.1 Pre-differentiable Mappings

Newton's method is an important and useful tool in optimization and numerical analysis. See, for example, [8, 21, 22, 24, 32, 41, 46, 47, 63–65, 88, 94, 97, 99, 102] and the references mentioned therein. We study equations with nonlinear mappings which are not necessarily differentiable. In this section we consider this class of mappings.

Let $(X, \| \cdot \|)$ and $(Y, \| \cdot \|)$ be normed spaces. For each $x \in X$, each $y \in Y$, and each $r > 0$ set

$$B_X(x, r) = \{u \in X : \|u - x\| \le r\},$$

$$B_Y(y, r) = \{v \in Y : \|v - y\| \le r\}.$$

Let $I_X(x) = x$ for all $x \in X$ and let $I_Y(y) = y$ for all $y \in Y$. Denote by $\mathcal{L}(X, Y)$ the set of all linear continuous operators $A : X \to Y$. For each $A \in \mathcal{L}(X, Y)$ set

$$\|A\| = \sup\{\|A(x)\| : x \in B_X(0, 1)\}.$$

Let $U \subset X$ be a nonempty open set, $F : U \to Y$, $x \in U$ and $\gamma > 0$. We say that the mapping F is (γ)-pre-differentiable at x if there exists $A \in \mathcal{L}(X, Y)$ such that

© Springer International Publishing Switzerland 2016
A.J. Zaslavski, *Numerical Optimization with Computational Errors*, Springer
Optimization and Its Applications 108, DOI 10.1007/978-3-319-30921-7_16

$$\gamma \geq \limsup_{h \to 0} \|F(x + h) - F(x) - A(h)\| \|h\|^{-1}$$

$$= \lim_{\epsilon \to 0^+} \sup\{\|F(x + h) - F(x) - A(h)\| \|h\|^{-1} : h \in B_X(0, \epsilon) \setminus \{0\}\}. \quad (16.1)$$

If $A \in \mathcal{L}(X, Y)$ satisfies (16.1), then A is called a (γ)-pre-derivative of F at x.

We denote by $\partial_\gamma F(x)$ the set of all (γ)-pre-derivatives of F at x. Note that the set $\partial_\gamma F(x)$ can be empty. We say that the mapping F is (γ)-pre-differentiable if it is (γ)-pre-differentiable at every $x \in U$.

If $G : U \to Y$, $x \in U$ and G is Frechet differentiable at x, then we denote by $G'(x)$ the Frechet derivative of G at x.

Proposition 16.1. *Let* $G : U \to Y$, $g : U \to Y$, $x \in U$, $\gamma > 0$, G *be Frechet differentiable at* x *and let*

$$\|g(z_2) - g(z_1)\| \leq \gamma \|z_2 - z_1\| \textit{ for all } z_1, z_2 \in U.$$

Then $G'(x)$ *is the* (γ)*-pre-derivative of* $G + g$ *at* $x \in X$.

Proof. For every $h \in X \setminus \{0\}$ such that $\|h\|$ is sufficiently small,

$$\|h\|^{-1} \|(G + g)(x + h) - (G + g)(x) - (G'(x))(h)\|$$

$$\leq \|h\|^{-1} \|G(x + h) - G(x) - (G'(x))(h)\| + \|h\|^{-1} \|g(x + h) - g(x)\|$$

$$\leq \|h\|^{-1} \|G(x + h) - G(x) - (G'(x))(h)\| + \gamma.$$

This implies that

$$\limsup_{h \to 0} \|h\|^{-1} \|(G + g)(x + h) - (G + g)(x) - (G'(x))(h)\| \leq \gamma.$$

Proposition 16.1 is proved.

In our analysis of Newton's method we need the following mean-valued theorem.

Theorem 16.2. *Assume that* $U \subset X$ *is a nonempty open set,* $\gamma > 0$, *a mapping* $F : U \to Y$ *is* (γ)*-pre-differentiable at every point of* U *and that* $x, y \in U$ *satisfy* $x \neq y$ *and*

$$\{tx + (1 - t)y : t \in [0, 1]\} \subset U.$$

Then there exists $t_0 \in (0, 1)$ *such that for every*

$$A \in \partial_\gamma F(x + t_0(y - x))$$

the following inequality holds:

$$\|F(x) - F(y)\| \leq \|A(y - x)\| + \gamma \|y - x\|.$$

Proof. For each $t \in [0, 1]$ set

$$\phi(t) = \|F(x) - F(x + t(y - x))\|, \ t \in [0, 1]. \tag{16.2}$$

Since the mapping F is (γ)-pre-differentiable, the function ϕ is continuous on $[0, 1]$. For every $t \in [0, 1]$ set

$$\psi(t) = \phi(t) - t(\phi(1) - \phi(0)), \ t \in [0, 1]. \tag{16.3}$$

Clearly, the function ψ is continuous on $[0, 1]$ and

$$\psi(0) = \phi(0) = 0, \ \psi(1) = 0. \tag{16.4}$$

By (16.4), there exists $t_0 \in (0, 1)$ such that either

(a) t_0 is a point of minimum of ψ on $[0, 1]$

or

(b) t_0 is a point of maximum of ψ on $[0, 1]$.

It is easy to see that in the case (a)

$$\liminf_{t \to t_0^+}(\psi(t) - \psi(t_0))(t - t_0)^{-1} \geq 0, \ \limsup_{t \to t_0^-}(\psi(t) - \psi(t_0))(t - t_0)^{-1} \leq 0 \tag{16.5}$$

and in the case (b)

$$\limsup_{t \to t_0^+}(\psi(t) - \psi(t_0))(t - t_0)^{-1} \leq 0, \ \liminf_{t \to t_0^-}(\psi(t) - \psi(t_0))(t - t_0)^{-1} \geq 0. \tag{16.6}$$

Assume that

$$A \in \partial_\gamma F(x + t_0(y - x)). \tag{16.7}$$

Let $t \in (0, 1)$. By (16.2),

$$|\phi(t) - \phi(t_0)|$$
$$||\|F(x) - F(x + t(y - x))\| - \|F(x) - F(x + t_0(y - x))\|||$$
$$\leq \|F(x + t(y - x)) - F(x + t_0(y - x))\|$$
$$\leq \|F(x + t(y - x)) - F(x + t_0(y - x)) - (t - t_0)A(y - x)\|$$
$$+ |t - t_0| \|A(y - x)\|. \tag{16.8}$$

Let $\epsilon > 0$. Since the mapping F is (γ)-pre-differentiable on U, it follows from (16.1) that there exists $\delta(\epsilon) \in (0, \epsilon)$ such that

$$B_X(x + t_0(y - x), \delta(\epsilon)) \subset U \tag{16.9}$$

and that for each $h \in B_X(0, \delta(\epsilon))$,

$$\|F(x + t_0(y - x) + h) - F(x + t_0(y - x)) - A(h)\| \leq (\gamma + \epsilon)\|h\|. \tag{16.10}$$

Assume that $t \in (0, 1)$ satisfies

$$|t - t_0| \leq \delta(\epsilon)(\|x\| + \|y\| + 1)^{-1}. \tag{16.11}$$

Set

$$h = (t - t_0)(y - x). \tag{16.12}$$

In view of (16.11) and (16.12), inequality (16.10) holds. By (6.8), (6.10), and (6.12),

$$|\phi(t) - \phi(t_0)| \leq |t - t_0|\|A(y - x)\| + (\gamma + \epsilon)|t - t_0|\|y - x\| \tag{16.13}$$

for every $t \in (0, 1)$ satisfying (16.11).

Assume that the case (a) holds. Then (16.3), (16.5), and (16.13) imply that

$$0 \leq \liminf_{t \to t_0^+}(\psi(t) - \psi(t_0))(t - t_0)^{-1}$$

$$= \liminf_{t \to t_0^+}(\phi(t) - t(\phi(1) - \phi(0)) - (\phi(t_0) - t_0(\phi(1) - \phi(0))))(t - t_0)^{-1}$$

$$= \liminf_{t \to t_0^+}(\phi(t) - \phi(t_0))(t - t_0)^{-1} - \phi(1)$$

$$\leq \|A(y - x)\| + (\gamma + \epsilon)\|y - x\| - \|F(x) - F(y)\|.$$

Since ϵ is any positive number we conclude that

$$\|F(x) - F(y)\| \leq \|A(y - x)\| + \gamma\|y - x\|.$$

Assume that the case (b) holds. Then (16.2), (16.3), (16.6), and (16.13) imply that

$$0 \leq \liminf_{t \to t_0^-}(\psi(t) - \psi(t_0))(t - t_0)^{-1}$$

$$= \liminf_{t \to t_0^-}(\phi(t) - t(\phi(1) - \phi(0)) - (\phi(t_0) - t_0(\phi(1) - \phi(0))))(t - t_0)^{-1}$$

$$= \liminf_{t \to t_0^-} (\phi(t) - \phi(t_0))(t - t_0)^{-1} - \phi(1)$$

$$= \liminf_{t \to t_0^-} |\phi(t) - \phi(t_0)| |t - t_0|^{-1} - \|F(x) - F(y)\|$$

$$\leq \|A(y - x)\| + (\gamma + \epsilon)\|y - x\| - \|F(x) - F(y)\|.$$

Since ϵ is any positive number we conclude that

$$\|F(x) - F(y)\| \leq \|A(y - x)\| + \gamma \|y - x\|.$$

This completes the proof of Theorem 16.2.

Let $x \in X$ and $r > 0$. Define $P_{x,r}(z) \in X$ for every $z \in X$ by

$$P_{x,r}(z) = z \text{ if } z \in B_X(x, r)$$

and

$$P_{x,r}(z) = x + r\|z - x\|^{-1}(z - x) \text{ if } z \in X \setminus B_X(x, r).$$

Clearly, for each $z \in X$,

$$\|z - P_{x,r}(z)\| = \inf\{\|z - y\| : y \in B_X(x, r)\}.$$

16.2   Convergence of Newton's Method

We use the notation and definitions of Sect. 16.1. Suppose that the normed space X is Banach.

Let $\gamma > 0, r > 0, \bar{x} \in X, U$ be a nonempty open subset of X such that

$$B_X(\bar{x}, r) \subset U \tag{16.14}$$

and let $F : U \to Y$ be a (γ)-pre-differentiable mapping. Let

$$L > 0, \quad \gamma_1 \in (0, 1). \tag{16.15}$$

Suppose that for each $z \in U$ there exists

$$A(z) \in \partial_\gamma F(z) \tag{16.16}$$

such that for each $z_1, z_2 \in B_X(\bar{x}, r)$,

$$\|A(z_1) - A(z_2)\| \leq L\|z_1 - z_2\|. \tag{16.17}$$

Let $A \in \mathcal{L}(Y, X)$ satisfy

$$\|I_X - A \circ A(\bar{x})\| \leq \gamma_1, \tag{16.18}$$

$$M = \|A\| \tag{16.19}$$

and let a positive number K satisfy

$$\|A(F(\bar{x}))\| \le K \le 4^{-1}r. \tag{16.20}$$

In view of (16.15), (16.18), and (16.19),

$$M > 0.$$

Set

$$h = MLK. \tag{16.21}$$

Let us consider the equation

$$ht^2 - (1 - \gamma_1 - M\gamma)t + 1 = 0 \tag{16.22}$$

with respect to the variable $t \in R^1$. We suppose that

$$\gamma_1 + M\gamma \le 2^{-1}, \ MLK = h < 4^{-1}(1 - \gamma_1 - M\gamma)^2. \tag{16.23}$$

Equation (16.22) has solutions

$$(2h)^{-1}(1 - \gamma_1 - M\gamma + ((1 - \gamma_1 - M\gamma)^2 - 4h)^{1/2})$$

and

$$(2h)^{-1}(1 - \gamma_1 - M\gamma - ((1 - \gamma_1 - M\gamma)^2 - 4h)^{1/2}).$$

Set

$$t_0 = (2h)^{-1}(1 - \gamma_1 - M\gamma - ((1 - \gamma_1 - M\gamma)^2 - 4h)^{1/2}). \tag{16.24}$$

For every $x \in U$ define

$$T(x) = x - A(F(x)). \tag{16.25}$$

The following result is proved in Sect. 16.4.

Theorem 16.3. *For each $x, y \in B_X(\bar{x}, Kt_0)$,*

$$\|T(x) - T(y)\| \le (3/4)\|x - y\|,$$
$$T(B_X(\bar{x}, Kt_0)) \subset B_X(\bar{x}, Kt_0),$$

there exists a unique point $x_* \in B_X(\bar{x}, Kt_0)$ such that $A(F(x_*)) = 0$ and for each $x \in B_X(\bar{x}, Kt_0)$,

$$\|T^i(x) - x_*\| \le 2(3/4)^i Kt_0, \ i = 0, 1, \ldots.$$

If the operator A is injective, then $F(x_*) = 0$. Moreover, the following assertions hold.

1. *Let $\delta > 0$, a natural number n_0 satisfy*

$$(3 \cdot 4^{-1})^{n_0} 2Kt_0 \leq \delta \tag{16.26}$$

and let a sequence $\{x_i\}_{i=0}^{\infty} \subset B_X(\bar{x}, Kt_0)$ satisfy

$$\|T(x_i) - x_{i+1}\| \leq \delta \text{ for all integers } i \geq 0.$$

Then for all integers $n \geq n_0$,

$$\|x_n - x_*\| \leq 5\delta.$$

2. *Let $\delta > 0$, a natural number $n_0 > 4$ satisfy (16.26) and let sequences*

$$\{x_i\}_{i=0}^{\infty} \subset B_X(\bar{x}, Kt_0), \ \{y_i\}_{i=0}^{\infty} \subset X$$

satisfy for all integers $i \geq 0$,

$$\|y_{i+1} - T(x_i)\| \leq \delta, \ x_{i+1} = P_{\bar{x}, Kt_0}(y_{i+1}). \tag{16.27}$$

Then for all integers $n \geq n_0$,

$$\|x_n - x_*\| \leq 10\delta.$$

3. *Let $\epsilon > 0$, an integer $n_0 > 2$ satisfy*

$$(3 \cdot 4^{-1})^{n_0} 16Kt_0 \leq \epsilon \tag{16.28}$$

and let a positive number δ satisfy

$$\delta < 128^{-1}\epsilon, \ 24(n_0 + 1)\delta < K, \ K \geq 6\|A(F(\bar{x}))\|. \tag{16.29}$$

Assume that $\{x_i\}_{i=0}^{n_0} \subset X$, $x_0 = \bar{x}$ and that if an integer $i \in [0, n_0 - 1]$ satisfies

$$x_i \in B_X(\bar{x}, Kt_0),$$

then

$$\|T(x_i) - x_{i+1}\| \leq \delta. \tag{16.30}$$

Then

$$\{x_i\}_{i=0}^{n_0} \subset B_X(\bar{x}, Kt_0),$$
$$\|x_{n_0-1} - x_*\| \leq \epsilon \text{ and } \|x_{n_0-1} - x_{n_0}\| < \epsilon/4.$$

16.3   Auxiliary Results

We use the notation, assumptions, and definitions introduced in Sects. 16.1 and 16.2.

Lemma 16.4. *The mapping* $T : U \to X$ *is* (γM)-*pre-differentiable at every point of* U *and for every* $x \in U$,

$$I_X - A \circ A(x) \in \partial_{M\gamma} T(x).$$

Proof. Let $x \in U$ and $\epsilon > 0$. In view of (16.1) and (16.16), there exists $\delta > 0$ such that

$$B_X(x, \delta) \subset U \tag{16.31}$$

and for each $h \in B_X(0, \delta)$,

$$\|F(x + h) - F(x) - (A(x))(h)\|$$
$$\leq (\gamma + \epsilon(\|A\| + 1)^{-1})\|h\|. \tag{16.32}$$

By (16.19), (16.25), (16.31), and (16.32), for each $h \in B_X(0, \delta)$,

$$\|T(x + h) - T(x) - (I_X - A \circ A(x))(h)\|$$
$$= \|x + h - A(F(x + h)) - x + A(F(x)) - h + A((A(x))(h))\|$$
$$= \|(-A)(F(x + h) - F(x) - (A(x))(h))\|$$
$$\leq \|A\|\|F(x + h) - F(x) - (A(x))(h)\|$$
$$\leq \|A\|(\gamma + \epsilon(\|A\| + 1)^{-1})\|h\|$$
$$\leq (\|A\|\gamma + \epsilon)\|h\| = (M\gamma + \epsilon)\|h\|.$$

Since ϵ is any positive number, this completes the proof of Lemma 16.4.

Lemma 16.5. *Let* $r_0 \in (0, r]$ *and* $x \in B_X(\bar{x}, r_0)$. *Then*

$$\|I_X - A \circ A(x)\| \leq \gamma_1 + MLr_0.$$

Proof. By (16.17), (16.18), and (16.19),

$$\|I_X - A \circ A(x)\|$$
$$= \|I_X - A \circ A(\bar{x}) + A \circ A(\bar{x}) - A \circ A(x)\|$$
$$\leq \|I_X - A \circ A(\bar{x})\| + \|A\|\|A(\bar{x}) - A(x)\|$$
$$\leq \gamma_1 + ML\|x - \bar{x}\| \leq \gamma_1 + MLr_0.$$

Lemma 16.5 is proved.

Lemma 16.6. *Let $r_0 \in (0, r]$ and*

$$x, y \in B_X(\bar{x}, r_0).$$

Then

$$\|T(x) - T(y)\| \leq (\gamma_1 + MLr_0 + M\gamma)\|y - x\|.$$

Proof. By (16.14), Theorem 16.2, and Lemmas 16.4 and 16.5, there exists $\tau_0 \in (0, 1)$ such that

$$\|T(x) - T(y)\|$$
$$\leq \|(I_X - A \circ A(x + \tau_0(y - x)))(y - x)\| + M\gamma\|y - x\|$$
$$\leq (\gamma_1 + MLr_0 + M\gamma)\|y - x\|.$$

Lemma 16.6 is proved.

Lemma 16.7. *Let $r_0 \in (0, r]$. Then for each $x \in B_X(\bar{x}, r_0)$,*

$$\|T(x) - \bar{x}\| \leq K + \gamma_1 r_0 + M(\gamma r_0 + Lr_0^2).$$

Proof. Let

$$x \in B_X(\bar{x}, r_0). \tag{16.33}$$

By (16.19), (16.20), and (16.25),

$$\|T(x) - \bar{x}\| = \|x - A(F(x)) - \bar{x}\|$$
$$= \|A[(A(\bar{x}))(x - \bar{x}) - F(x) + F(\bar{x})]$$
$$-A(F(\bar{x})) + (I_X - A \circ A(\bar{x}))(x - \bar{x})\|$$
$$\leq \|A\|\|F(x) - F(\bar{x}) - (A(\bar{x}))(x - \bar{x})\|$$
$$+\|A(F(\bar{x}))\| + \|I_X - A \circ A(\bar{x})\|\|x - \bar{x}\|$$
$$\leq M\|F(x) - F(\bar{x}) - (A(\bar{x}))(x - \bar{x})\| + K + \gamma_1\|x - \bar{x}\|. \tag{16.34}$$

For every $x \in U$ define

$$\phi(x) = F(x) - F(\bar{x}) - (A(\bar{x}))(x - \bar{x}). \tag{16.35}$$

We show that the mapping ϕ is (γ)-pre-differentiable on U. By (16.35), for each $x \in U$ and each $h \in X$ satisfying $x + h \in U$,

$$\phi(x + h) - \phi(x) - (A(x) - A(\bar{x}))(h)$$
$$= F(x + h) - F(\bar{x}) - (A(\bar{x}))(x + h - \bar{x})$$
$$-F(x) + F(\bar{x}) + (A(\bar{x}))(x - \bar{x}) - (A(x) - A(\bar{x}))(h)$$

$$= F(x + h) - F(x) - (A(\bar{x}))(h) - (A(x) - A(\bar{x}))(h)$$
$$= F(x + h) - F(x) - (A(x))(h).$$

Together with (16.1) and (16.16) this implies that the mapping ϕ is (γ)-pre-differentiable at every point of U and that for all $z \in U$,

$$A(z) - A(\bar{x}) \in \partial_\gamma \phi(z). \tag{16.36}$$

In view of (16.17), for all $z \in B_X(\bar{x}, r_0)$,

$$\|A(z) - A(\bar{x})\| \le L\|z - \bar{x}\|. \tag{16.37}$$

It follows from (16.35), (16.36), and Theorem 16.2 that for every $z \in B_X(\bar{x}, r_0)$,

$$\|\phi(z)\| = \|\phi(z) - \phi(\bar{x})\| \le \gamma\|z - \bar{x}\| + L\|z - \bar{x}\|^2. \tag{16.38}$$

By (16.33)–(16.35) and (16.38), for every $x \in B_X(\bar{x}, r_0)$,

$$\|T(x) - \bar{x}\| \le K + \gamma_1 r_0 + M(\gamma r_0 + L r_0^2).$$

Lemma 16.7 is proved.

16.4   Proof of Theorem 16.3

Set

$$r_0 = K t_0. \tag{16.39}$$

We show that

$$r_0 \le r. \tag{16.40}$$

Indeed, in view of (16.20), (16.23), (16.24), and (16.39),

$$r_0 = K t_0$$
$$= 2K[(1 - \gamma_1 - M\gamma) + ((1 - \gamma_1 - M\gamma)^2 - 4h)^{1/2}]^{-1}$$
$$\le 2(1 - \gamma_1 - M\gamma)^{-1} K \le 4K \le r.$$

By (16.21), (16.39), (16.40), and Lemma 16.6, for each $x, y \in B_X(\bar{x}, K t_0)$,

$$\|T(x) - T(y)\| \le (\gamma_1 + MLK t_0 + M\gamma)\|x - y\|$$
$$\le (\gamma_1 + h t_0 + M\gamma)\|x - y\|. \tag{16.41}$$

It follows from (16.24) that

$$t_0 \le 2(1 - \gamma_1 - M\gamma)^{-1}. \tag{16.42}$$

Relations (16.23) and (16.42) imply that

$$
\begin{aligned}
\gamma_1 &+ ht_0 + M\gamma \\
&\le \gamma_1 + M\gamma + 2(1 - \gamma_1 - M\gamma)^{-1}(1 - \gamma_1 - M\gamma)^2/4 \\
&= \gamma_1 + M\gamma + 2^{-1}(1 - \gamma_1 - M\gamma) \\
&= 2^{-1} + 2^{-1}(\gamma_1 + M\gamma) \le 3/4.
\end{aligned} \tag{16.43}
$$

By (16.41) and (16.43), for each $x, y \in B_X(\bar{x}, Kt_0)$,

$$\|T(x) - T(y)\| \le (3 \cdot 4^{-1})\|x - y\|. \tag{16.44}$$

Let

$$x \in B_X(\bar{x}, r_0). \tag{16.45}$$

It follows from (16.21), (16.22), (16.24), (16.39), (16.40), (16.45), and Lemma 16.7 that

$$
\begin{aligned}
\|T(x) - \bar{x}\| &\le K + \gamma_1 Kt_0 + M(\gamma Kt_0 + LK^2 t_0^2) \\
&\le K(1 + \gamma_1 t_0 + M\gamma t_0 + MLKt_0^2) \\
&= K(1 + \gamma_1 t_0 + M\gamma t_0 + ht_0^2) = Kt_0,
\end{aligned}
$$

$$T(x) \in B_X(\bar{x}, r_0)$$

and

$$T(B_X(\bar{x}, r_0)) \subset B_X(\bar{x}, r_0). \tag{16.46}$$

Relations (16.45) and (16.46) imply that there exists a unique point

$$x_* \in B_X(\bar{x}, r_0)$$

such that

$$T(x_*) = x_*.$$

In order to complete the proof of Theorem 16.3 it is sufficient to show that assertions 1, 2, and 3 hold.

Let us prove assertion 1. For each integer $i \ge 0$,

$$\|x_{i+1} - x_*\| \le \|x_{i+1} - T(x_i)\| + \|T(x_i) - x_*\| \le \delta + (3 \cdot 4^{-1})\|x_i - x_*\|. \tag{16.47}$$

By induction we show that for all integers $p \geq 1$,

$$\|x_p - x_*\| \leq (3 \cdot 4^{-1})^p \|x_0 - x_*\| + \delta \sum_{i=0}^{p-1} (3 \cdot 4^{-1})^i. \tag{16.48}$$

In view of (16.47), inequality (16.48) holds for $p = 1$.

Assume that an integer $p \geq 1$ and that (16.48) holds. It follows from (16.47) and (16.48) that

$$\|x_{p+1} - x_*\| \leq \delta + (3/4)\|x_p - x_*\|$$

$$\leq (3 \cdot 4^{-1})^{p+1} \|x_0 - x_*\| + \delta \sum_{i=0}^{p} (3 \cdot 4^{-1})^i.$$

Thus we showed by induction that (16.48) holds for all integers $p \geq 1$. By (16.48) and the choice of n_0 [see (16.26)], for all integer $n \geq n_0$,

$$\|x_n - x_*\| \leq (3 \cdot 4^{-1})^{n_0} 2Kt_0 + 4\delta \leq 5\delta.$$

Assertion 1 is proved.

Let us prove assertion 2. In view of (16.27), for each integer $i \geq 0$,

$$\delta \geq \|y_{i+1} - T(x_i)\| \geq \|y_{i+1} - x_{i+1}]\|$$

and

$$\|x_{i+1} - T(x_i)\| \leq 2\delta.$$

By the relation above, (16.26) and assertion 1, for all integers $n \geq n_0$,

$$\|x_n - x_*\| \leq 10\delta.$$

Assertion 2 is proved.

Let us prove assertion 3. In view of (16.24),

$$t_0 = 2((1 - \gamma_1 - M\gamma) + ((1 - \gamma_1 - M\gamma)^2 - 4h)^{1/2})^{-1}$$

$$\geq (1 - \gamma_1 - M\gamma)^{-1}. \tag{16.49}$$

By (16.23),(16.25), (16.29), (16.30), and (16.49),

$$\|x_1 - x_0\| = \|x_1 - \bar{x}\| \leq \|x_1 - T(x_0)\| + \|T(x_0) - x_0\|$$

$$\leq \delta + \|A(F(\bar{x}))\| \leq \delta + 6^{-1}K$$

$$< K < K(1 - \gamma_1 - M\gamma)^{-1} \leq Kt_0 \tag{16.50}$$

and

$$x_1 \in B_X(\bar{x}, Kt_0). \tag{16.51}$$

Relations (16.30), (16.44), (16.50), and (16.51) imply that

$$\begin{aligned}
\|x_2 - x_1\| &\leq \|x_2 - T(x_1)\| + \|T(x_1) - x_1\| \\
&\leq \delta + \|T(x_1) - T(x_0)\| + \|T(x_0) - x_1\| \\
&\leq 2\delta + (3 \cdot 4^{-1})\|x_1 - x_0\| \\
&\leq 2\delta + (3 \cdot 4^{-1})\delta + 6^{-1}(3/4)K. \tag{16.52}
\end{aligned}$$

It follows from (16.23), (16.29), (16.49), (16.50), and (16.52),

$$\begin{aligned}
\|x_2 - \bar{x}\| &\leq \|x_2 - x_1\| + \|x_1 - x_0\| \\
&\leq 2\delta + (3 \cdot 4^{-1})(\delta + 6^{-1}K) + (\delta + 6^{-1}K) \\
&\leq 2\delta + 2(\delta + 6^{-1}K) \leq 4\delta + K/3 < K \leq Kt_0. \tag{16.53}
\end{aligned}$$

Assume that an integer p satisfies

$$2 \leq p < n_0,$$

$$x_i \in B_X(\bar{x}, Kt_0) \tag{16.54}$$

for all $i = 1, \ldots, p$ and for all $q = 2, \ldots, p$,

$$\|x_q - x_{q-1}\| \leq (3 \cdot 4^{-1})^{q-1}\|x_1 - x_0\| + 2\delta \sum_{i=0}^{q-2}(3 \cdot 4^{-1})^i. \tag{16.55}$$

(In view of (16.51), (16.52), and (16.53), our assumption holds for $p = 2$.) By (16.30), (16.44), (16.54), and (16.55),

$$\begin{aligned}
\|x_{p+1} - x_p\| &\leq \|x_{p+1} - T(x_p)\| + \|T(x_p) - T(x_{p-1})\| + \|T(x_{p-1}) - x_p\| \\
&\leq 2\delta + 3 \cdot 4^{-1}\|x_p - x_{p-1}\| \\
&\leq (3 \cdot 4^{-1})^p\|x_1 - x_0\| + (3 \cdot 2^{-1})\delta \sum_{i=0}^{p-2}(3 \cdot 4^{-1})^i + 2\delta \\
&= (3 \cdot 4^{-1})^p\|x_1 - x_0\| + 2\delta \sum_{i=0}^{p-1}(3 \cdot 4^{-1})^i
\end{aligned}$$

and (16.55) holds for $q = p + 1$. By (16.55) which holds for all $q = 2, \ldots, p + 1$, (16.29), and (16.50),

$$\|x_{p+1} - x_0\| \leq \sum_{q=1}^{p+1} \|x_q - x_{q-1}\|$$

$$\leq \sum_{q=1}^{p+1} (3 \cdot 4^{-1})^{q-1} \|x_1 - x_0\| + 8\delta p$$

$$\leq 4\|x_1 - x_0\| + 8n_0\delta \leq 4\delta + 2 \cdot 3^{-1}K + 8n_0\delta$$

$$\leq 8(n_0 + 1)\delta + 2 \cdot 3^{-1}K \leq K \leq Kt_0$$

and (16.54) holds for $i = p + 1$.

Thus we showed by induction that our assumption holds for $p = n_0$. Together with (16.54) and (16.55) holding for $p = n_0$, (16.28) and (16.29) this implies that

$$x_{n_0} \in B_X(\bar{x}, Kt_0),$$

$$\|x_{n_0} - x_{n_0-1}\| \leq 8\delta + (3 \cdot 4^{-1})^{n_0-1} 2Kt_0 \leq \epsilon/8 + \epsilon/16. \tag{16.56}$$

Set

$$\tilde{x}_0 = x_{n_0-1} \tag{16.57}$$

and for all integer $i \geq 0$ set

$$\tilde{x}_{i+1} = T(\tilde{x}_i). \tag{16.58}$$

By (16.28), (16.30), (16.44), and (16.56)–(16.58),

$$\|\tilde{x}_0 - \tilde{x}_1\| = \|x_{n_0-1} - T(x_{n_0-1})\| \leq \|x_{n_0-1} - x_{n_0}\| + \delta \leq \epsilon/4 \tag{16.59}$$

and for all integers $i \geq 0$,

$$\|\tilde{x}_{i+2} - \tilde{x}_{i+1}\| = \|T(\tilde{x}_{i+1}) - T(\tilde{x}_i)\| \leq (3 \cdot 4^{-1})\|\tilde{x}_{i+1} - \tilde{x}_i\|.$$

Together with (16.59) this implies that for all integers $i \geq 0$,

$$\|\tilde{x}_{i+1} - \tilde{x}_i\| \leq (3 \cdot 4^{-1})^i \epsilon/4. \tag{16.60}$$

Clearly,

$$x_* = \lim_{i \to \infty} \tilde{x}_i.$$

By (16.57) and (16.60),

$$\|x_* - x_{n_0-1}\| = \|x_* - \tilde{x}_0\| \le \lim_{q \to \infty} \|\tilde{x}_q - \tilde{x}_0\| \le \sum_{q=0}^{\infty} \|\tilde{x}_q - \tilde{x}_{q+1}\| \le \epsilon.$$

Assertion 3 is proved. This completes the proof of Theorem 16.3.

16.5 Set-Valued Mappings

Let (X, ρ) be a complete metric space. For each $z \in X$ and each $r > 0$ set

$$B(z, r) = \{y \in X : \rho(z, y) \le r\}.$$

For each $x \in X$ and each nonempty set $C \subset X$ define

$$\rho(x, C) = \inf\{\rho(x, y) : y \in C\}.$$

In Sect. 16.7 we prove the following result which is important in our study of Newton's method for nonlinear inclusions.

Theorem 16.8. *Suppose that $\phi : X \to 2^X$, $a > 0$, $\theta \in (0, 1)$, $\bar{x} \in X$,*

$$\phi(x) \ne \emptyset \text{ for all } x \in B(\bar{x}, a), \tag{16.61}$$

$$\rho(\bar{x}, \phi(\bar{x})) < a(1 - \theta), \tag{16.62}$$

for all $u, v \in B(\bar{x}, a)$,

$$\sup\{\rho(z, \phi(v)) : z \in \phi(u) \cap B(\bar{x}, a)\} \le \theta \rho(u, v) \tag{16.63}$$

and that the set

$$graph(\phi; B(\bar{x}, a)) := \{(x, y) \in B(\bar{x}, a) \times B(\bar{x}, a) : y \in \phi(x)\}$$

is closed. Then the following assertions hold.

1. Assume that a sequence $\{\epsilon_i\}_{i=0}^{\infty} \subset (0, \infty)$ satisfies $\sum_{i=0}^{\infty} \epsilon_i < \infty$,

$$2(1 - \theta)^{-1} \sum_{i=0}^{\infty} \epsilon_i + (1 - \theta)^{-1}(\max\{\epsilon_i : i = 0, 1, \dots\} + \rho(\bar{x}, \phi(\bar{x}))) \le a \tag{16.64}$$

and that a sequence $\{x_i\}_{i=0}^{\infty} \subset X$ satisfies

$$x_0 = \bar{x} \tag{16.65}$$

and for each integer $i \geq 0$ satisfying $x_i \in B(\bar{x}, a)$, the inequalities

$$\rho(x_{i+1}, \phi(x_i)) \leq \epsilon_i, \tag{16.66}$$

$$\rho(x_i, x_{i+1}) \leq \rho(x_i, \phi(x_i)) + \epsilon_i \tag{16.67}$$

hold. Then

$$\rho(x_i, \bar{x}) < a - \epsilon_{i-1} \text{ for all integers } i \geq 1, \tag{16.68}$$

for each integer $k \geq 1$,

$$\rho(x_k, x_{k+1}) \leq \theta^k(\rho(\bar{x}, \phi(\bar{x})) + \epsilon_0) + \sum_{j=0}^{k-1} \epsilon_j \theta^{k-1-j} + \sum_{j=1}^{k} \epsilon_j \theta^{k-j}, \tag{16.69}$$

for each integer $k \geq 0$,

$$\sum_{i=k}^{\infty} \rho(x_i, x_{i+1}) \leq (1-\theta)^{-1} \rho(x_k, x_{k+1}) + 2(1-\theta)^{-1} \sum_{i=k}^{\infty} \epsilon_i \tag{16.70}$$

and there exists

$$x_* = \lim_{n \to \infty} x_n \in B(\bar{x}, a)$$

satisfying $x_* \in \phi(x_*)$.

2. *Let $\epsilon \in (0, 1)$, a natural number $n_0 > 2$ satisfy*

$$4\theta^{n_0}(a(1-\theta) + 1) < \epsilon(1-\theta) \tag{16.71}$$

and let a number $\delta \in (0, \epsilon)$ satisfy

$$\delta < (8n_0)^{-1}\epsilon(1-\theta) \tag{16.72}$$

and

$$\delta < [a - (1-\theta)^{-1}\rho(\bar{x}, \phi(\bar{x}))](1-\theta)(2n_0 + 1)^{-1}. \tag{16.73}$$

Assume that a sequence $\{x_i\}_{i=0}^{n_0} \subset X$ satisfies

$$x_0 = \bar{x}$$

and for each integer $i \in [0, n_0 - 1]$ satisfying $x_i \in B(\bar{x}, a)$, the inequalities

$$\rho(x_{i+1}, \phi(x_i)) \leq \delta, \tag{16.74}$$

$$\rho(x_i, x_{i+1}) \leq \rho(x_i, \phi(x_i)) + \delta \tag{16.75}$$

hold. Then

$$\rho(x_i, \bar{x}) < a - \delta \text{ for all } i = 1, \dots, n_0$$

and there exists $x_* \in B(\bar{x}, a)$ *such that*

$$x_* \in \phi(x_*) \text{ and } \rho(x_{n_0}, x_*) < \epsilon.$$

A prototype of Theorem 16.8, for which $\phi : X \to 2^X \setminus \{\emptyset\}$ is a strict contraction, was proved in Chap. 9 of [121].

16.6 An Auxiliary Result

Lemma 16.9. *Suppose that* $\phi : X \to 2^X$, $a > 0$, $\theta \in (0, 1)$, $\bar{x} \in X$, (16.61)–(16.63) *hold and let a sequence* $\{\epsilon_i\}_{i=0}^{\infty} \subset (0, \infty)$ *satisfy* $\sum_{i=0}^{\infty} \epsilon_i < \infty$ *and* (16.64). *Then the following assertions hold.*

1. *Let* $x_0 = \bar{x}$ *and* $x_1 \in X$ *satisfy*

$$\rho(x_1, \phi(x_0)) \le \epsilon_0, \ \rho(x_0, x_1) \le \rho(x_0, \phi(x_0)) + \epsilon_0. \tag{16.76}$$

Then $\rho(x_1, \bar{x}) < a - \epsilon_0$.
2. *Assume that* $n \ge 1$ *is an integer,* $\{x_i\}_{i=0}^{n+1} \subset X$,

$$x_0 \in B(\bar{x}, a), \tag{16.77}$$

$$\rho(x_i, \bar{x}) < a - \epsilon_{i-1}, \ i = 1, \dots, n \tag{16.78}$$

and that for each integer $i \in \{0, \dots, n\}$ *the inequalities*

$$\rho(x_{i+1}, \phi(x_i)) \le \epsilon_i, \tag{16.79}$$

$$\rho(x_i, x_{i+1}) \le \rho(x_i, \phi(x_i)) + \epsilon_i \tag{16.80}$$

hold. Then for each integer $k \in [0, n-1]$ *there exists*

$$y_{k+1} \in \phi(x_k) \cap B(\bar{x}, a) \tag{16.81}$$

such that

$$\rho(x_{k+1}, y_{k+1}) < 2\epsilon_k, \tag{16.82}$$

for all integers $k = 0, \dots, n-1$,

$$\rho(x_{k+2}, x_{k+1}) \le \theta\rho(x_k, x_{k+1}) + \epsilon_k + \epsilon_{k+1} \tag{16.83}$$

for each integer s satisfying $0 \leq s < n$ and each integer k satisfying $s < k \leq n$,

$$\rho(x_k, x_{k+1}) \leq \theta^{k-s} \rho(x_s, x_{s+1}) + \sum_{i=s}^{k-1} \theta^{i-s}(\epsilon_{k-i-1+s} + \epsilon_{k-i+s}) \qquad (16.84)$$

and

$$\sum_{p=0}^{n} \rho(x_p, x_{p+1}) \leq \sum_{p=0}^{n} \theta^p \rho(x_0, x_1) + \sum_{i=0}^{n-1} \epsilon_i \left(\sum_{j=0}^{n-1-i} \theta^j \right) + \sum_{i=1}^{n} \epsilon_i \left(\sum_{j=0}^{n-i} \theta^j \right).$$

$$(16.85)$$

Moreover, if $x_0 = \bar{x}$, then $\rho(x_{n+1}, \bar{x}) < a - \epsilon_n$.

Proof. Let us prove assertion 1. By (16.64) and (16.76),

$$\rho(x_1, \bar{x}) \leq \rho(\bar{x}, \phi(\bar{x})) + \epsilon_0 < a - \epsilon_0.$$

Assertion 1 is proved.

Let us prove assertion 2. Assume that an integer

$$k \in [0, n-1]. \qquad (16.86)$$

By (16.77)–(16.80) and (16.86),

$$x_k, x_{k+1} \in B(\bar{x}, a), \ x_{k+2} \in X,$$

$$\rho(x_{k+2}, \phi(x_{k+1})) \leq \epsilon_{k+1}, \ \rho(x_{k+1}, \phi(x_k)) \leq \epsilon_k, \qquad (16.87)$$

$$\rho(x_{k+2}, x_{k+1}) \leq \rho(x_{k+1}, \phi(x_{k+1})) + \epsilon_{k+1}. \qquad (16.88)$$

We show that

$$\rho(x_{k+2}, x_{k+1}) \leq \theta \rho(x_k, x_{k+1}) + \epsilon_k + \epsilon_{k+1}.$$

Let a positive number Δ satisfy

$$\Delta < \min\{\epsilon_k, \ a - \epsilon_k - \rho(x_{k+1}, \bar{x})\} \qquad (16.89)$$

[see (16.78)]. In view of (16.87), there exists

$$y_{k+1} \in \phi(x_k) \qquad (16.90)$$

such that

$$\rho(x_{k+1}, y_{k+1}) < \epsilon_k + \Delta. \qquad (16.91)$$

By (16.89) and (16.91),

$$\rho(y_{k+1}, \bar{x}) < \epsilon_k + \Delta + \rho(x_{k+1}, \bar{x}) < a. \tag{16.92}$$

Relations (16.89)–(16.92) imply that

$$y_{k+1} \in \phi(x_k) \cap B(\bar{x}, a), \quad \rho(x_{k+1}, y_{k+1}) < 2\epsilon_k. \tag{16.93}$$

Thus (16.81) and (16.82) hold. It follows from (16.63), (16.88), (16.91), and (16.93) that

$$
\begin{aligned}
\rho(x_{k+2}, x_{k+1}) &\le \epsilon_{k+1} + \rho(x_{k+1}, \phi(x_{k+1})) \\
&\le \epsilon_{k+1} + \rho(x_{k+1}, y_{k+1}) + \rho(y_{k+1}, \phi(x_{k+1})) \\
&\le \epsilon_{k+1} + \epsilon_k + \Delta + \sup\{\rho(z, \phi(x_{k+1})) : z \in \phi(x_k) \cap B(\bar{x}, a)\} \\
&\le \epsilon_{k+1} + \epsilon_k + \Delta + \theta\rho(x_k, x_{k+1}).
\end{aligned}
$$

Since Δ is an arbitrary positive number satisfying (16.89) we conclude that

$$\rho(x_{k+2}, x_{k+1}) \le \theta\rho(x_k, x_{k+1}) + \epsilon_k + \epsilon_{k+1} \tag{16.94}$$

for all integers $k = 0, \dots, n-1$.

Let an integer s satisfy

$$0 \le s < n.$$

We show by induction that for each integer k satisfying $s < k \le n$, (16.84) holds. In view of (16.94),

$$\rho(x_{s+2}, x_{s+1}) \le \theta\rho(x_s, x_{s+1}) + \epsilon_s + \epsilon_{s+1}$$

and (16.84) holds for $k = s + 1$.

Assume that an integer k satisfies

$$s < k < n$$

and that (16.84) holds. By (16.84) and (16.94),

$$
\begin{aligned}
\rho(x_{k+2}, x_{k+1}) &\le \theta\rho(x_k, x_{k+1}) + \epsilon_k + \epsilon_{k+1} \\
&\le \theta^{k+1-s}\rho(x_s, x_{s+1}) + \sum_{i=s}^{k-1} \theta^{i-s+1}(\epsilon_{k-i-1+s} + \epsilon_{k-i+s}) + \epsilon_k + \epsilon_{k+1} \\
&= \theta^{k+1-s}\rho(x_s, x_{s+1}) + \sum_{i=s}^{k} \theta^{i-s}(\epsilon_{k-i+s} + \epsilon_{k+1-i+s}).
\end{aligned}
$$

Thus by induction we showed that (16.84) holds for all integers k satisfying $s < k \le n$.

In particular, for all integers $p = 1, \ldots, n$,

$$\rho(x_p, x_{p+1}) \leq \theta^p \rho(x_0, x_1) + \sum_{i=0}^{p-1} \theta^i (\epsilon_{p-i-1} + \epsilon_{p-i}). \tag{16.95}$$

It follows from (16.95) that

$$\sum_{p=0}^{n} \rho(x_p, x_{p+1}) \leq \left(\sum_{p=0}^{n} \theta^p \right) \rho(x_0, x_1) + \sum_{p=1}^{n} \sum_{i=0}^{p-1} \theta^i (\epsilon_{p-i-1} + \epsilon_{p-i})$$

$$= \left(\sum_{p=0}^{n} \theta^p \right) \rho(x_0, x_1) + \sum_{p=1}^{n} \sum_{i=0}^{p-1} \epsilon_i \theta^{p-i-1} + \sum_{p=1}^{n} \sum_{i=1}^{p} \epsilon_i \theta^{p-i}$$

$$= \left(\sum_{p=0}^{n} \theta^p \right) \rho(x_0, x_1) + \sum_{i=0}^{n-1} \epsilon_i \left(\sum_{j=0}^{n-1-i} \theta^j \right) + \sum_{i=1}^{n} \epsilon_i \left(\sum_{j=0}^{n-i} \theta^j \right).$$

Thus (16.85) holds.

' Assume that

$$x_0 = \bar{x}.$$

By (16.64), (16.80), and (16.85),

$$\rho(\bar{x}, x_{n+1}) = \rho(x_0, x_{n+1}) \leq \sum_{p=0}^{n} \rho(x_p, x_{p+1})$$

$$\leq (1 - \theta)^{-1} \rho(x_0, x_1) + (1 - \theta)^{-1} \sum_{i=0}^{n-1} \epsilon_i + (1 - \theta)^{-1} \sum_{i=1}^{n} \epsilon_i$$

$$\leq (1 - \theta)^{-1} (\rho(\bar{x}, \phi(\bar{x})) + \epsilon_0) + 2(1 - \theta)^{-1} \sum_{i=0}^{n} \epsilon_i - (1 - \theta)^{-1} \epsilon_n$$

$$< a - (1 - \theta)^{-1} \epsilon_n$$

and

$$\rho(\bar{x}, x_{n+1}) < a - \epsilon.$$

This completes the proof of Lemma 16.9.

16.7 Proof of Theorem 16.8

Let us prove assertion 1. By (16.66), (16.67), and assertion 1 of Lemma 16.9,

$$\rho(x_1, \bar{x}) < a - \epsilon_0. \tag{16.96}$$

We show that for all integers $p \geq 1$,

$$\rho(x_p, \bar{x}) < a - \epsilon_{p-1}. \tag{16.97}$$

In view of (16.96), inequality (16.97) holds for $p = 1$.

Assume that $n \geq 1$ is an integer and (16.97) holds for all integers $p = 1, \ldots, n$. By (16.66), (16.67), (16.97), and assertion 2 of Lemma 16.9,

$$\rho(x_{n+1}, \bar{x}) < a - \epsilon_n.$$

Thus (16.97) holds for all integers $p \geq 1$.

It follows from (16.66), (16.67), (16.85), (16.97), and Lemma 16.9 that

$$\sum_{p=0}^{\infty} \rho(x_p, x_{p+1}) \leq (1 - \theta)^{-1} \rho(\bar{x}, x_1) + 2(1 - \theta)^{-1} \sum_{i=0}^{\infty} \epsilon_i < \infty.$$

This implies that there exists

$$x_* = \lim_{p \to \infty} x_p \in B(\bar{x}, a) \tag{16.98}$$

and

$$\lim_{p \to \infty} \rho(x_p, x_{p+1}) = 0. \tag{16.99}$$

Lemma 16.9, (16.81), (16.82), and (16.97) imply that for each integer $p \geq 0$ there exists

$$y_{p+1} \in \phi(x_p) \cap B(\bar{x}, a) \tag{16.100}$$

such that

$$\rho(x_{p+1}, y_{p+1}) < 2\epsilon_p. \tag{16.101}$$

By (16.97), (16.98), (16.100), and (16.101),

$$\lim_{p \to \infty} \rho(x_p, y_p) = 0, \quad \lim_{p \to \infty} \rho(y_p, x_*) = 0$$

and since graph$(\phi; B(\bar{x}, a))$ is closed we conclude that

$$x_* \in \phi(x_*).$$

Lemma 16.9, (16.67), (16.84), and (16.97) imply that for each integer $k > 0$,

$$\rho(x_k, x_{k+1}) \le \theta^k \rho(x_0, x_1) + \sum_{j=0}^{k-1} \epsilon_j \theta^{k-1-j} + \sum_{j=1}^{k} \epsilon_j \theta^{k-j}$$

$$\le \theta^k (\rho(\bar{x}, \phi(\bar{x})) + \epsilon_0) + \sum_{j=0}^{k-1} \epsilon_j \theta^{k-1-j} \sum_{j=1}^{k} \epsilon_j \theta^{k-j}. \quad (16.102)$$

Let $k \ge 0$ be an integer. In view of (16.64), (16.66), (16.67), and (16.97), we apply Lemma 16.9 to the sequences $\{x_{k+i}\}_{i=0}^{\infty}$, $\{\epsilon_{k+i}\}_{i=0}^{\infty}$ and obtain from (16.85) that

$$\sum_{p=0}^{\infty} \rho(x_{k+p}, x_{k+p+1}) \le (1 - \theta)^{-1} \rho(x_k, x_{k+1}) + 2(1 - \theta)^{-1} \sum_{p=k}^{\infty} \epsilon_p.$$

Assertion 1 is proved.

Let us prove assertion 2. For $i = 0, \ldots, n_0 - 1$ set

$$\epsilon_i = \delta. \quad (16.103)$$

By (16.73) and (16.103), for every integer $i \ge n_0$ there exists $\epsilon_i > 0$ such that (16.64) holds,

$$2(1 - \theta)^{-1} \sum_{i=n_0}^{\infty} \epsilon_i < \epsilon/4, \ \epsilon_i \le \delta \text{ for all integers } i \ge 0. \quad (16.104)$$

Clearly, for every integer $i \ge n_0 + 1$, there exists $x_i \in X$ such that the following property holds:

if an integer $i \ge 0$ satisfies $x_i \in B(\bar{x}, a)$, then (16.66) and (16.67) hold.

It follows from (16.62), (16.64), (16.66)–(16.69), (16.71), (16.72), (16.103), (16.104), and assertion 1 that

$$\rho(x_i, \bar{x}) < a - \epsilon_{i-1} \text{ for all integers } i \ge 1,$$

$$\rho(x_i, \bar{x}) < a - \delta \text{ for all integers } i = 1, \ldots, n_0$$

and

$$\rho(x_{n_0}, x_{n_0+1}) \le \theta^{n_0} (\rho(\bar{x}, \phi(\bar{x})) + \delta) + 2n_0\delta$$
$$< 4^{-1}\epsilon(1 - \theta) + 4^{-1}\epsilon(1 - \theta). \quad (16.105)$$

By assertion 1, (16.64), (16.66), (16.67), (16.70), (16.104), and (16.105),

$$\sum_{i=n_0}^{\infty} \rho(x_i, x_{i+1}) \leq (1 - \theta)^{-1} \rho(x_{n_0}, x_{n_0+1}) + 2(1 - \theta)^{-1} \sum_{i=n_0}^{\infty} \epsilon_i < \epsilon/2 + \epsilon/4$$

(16.106)

and there exists

$$x_* = \lim_{n \to \infty} x_n \in B(\bar{x}, a)$$

(16.107)

satisfying $x_* \in \phi(x_*)$. It follows from (16.106) and (16.107) that

$$\rho(x_{n_0}, x_*) = \lim_{n \to \infty} \rho(x_{n_0}, x_n) \leq \sum_{i=n_0}^{\infty} \rho(x_i, x_{i+1}) < \epsilon.$$

Assertion 2 is proved. This completes the proof of Theorem 16.8.

16.8 Pre-differentiable Set-Valued Mappings

Let $(Z, \| \cdot \|)$ be a normed space. For each $x \in Z$ and each nonempty set $C \subset Z$ define

$$d(x, C) = \inf\{\|x - z\| : z \in C\}.$$

For each $z \in Z$ and each $r > 0$ set

$$B_Z(z, r) = \{y \in Z : \|z - y\| \leq r\}.$$

Let $I_Z(z) = z$ for all $z \in Z$.

For each pair of nonempty sets $C_1, C_2 \subset Z$ define

$$H(C_1, C_2) = \max\{\sup\{d(x, C_2) : x \in C_1\}, \sup\{d(y, C_1) : y \in C_2\}\}.$$

Let $(X, \| \cdot \|)$ and $(Y, \| \cdot \|)$ be normed spaces. Denote by $\mathcal{L}(X, Y)$ the set of all linear continuous operators $A : X \to Y$. For each $A \in \mathcal{L}(X, Y)$ set

$$\|A\| = \sup\{\|A(x)\| : x \in B_X(0, 1)\}.$$

Let $U \subset X$ be a nonempty open set, $F : U \to 2^Y \setminus \{\emptyset\}$, $x \in U$ and $\gamma > 0$. We say that the mapping F is (γ)-pre-differentiable at x if there exists $A \in \mathcal{L}(X, Y)$ such that the following property holds:

(P1)   for each $\epsilon > 0$ there exists $\delta(\epsilon) > 0$ such that

$$B_X(x, \delta(\epsilon)) \subset U$$

(16.108)

and that for each $h \in B_X(0, \delta(\epsilon))$,

$$F(x) + A(h) \subset F(x + h) + (\gamma + \epsilon)\|h\|B_Y(0, 1), \tag{16.109}$$

$$F(x + h) \subset F(x) + A(h) + (\gamma + \epsilon)\|h\|B_Y(0, 1), \tag{16.110}$$

If $A \in \mathcal{L}(X, Y)$ and (P1) holds, then A is called a (γ)-pre-derivative of F at x.

We denote by $\partial_\gamma F(x)$ the set of all (γ)-pre-derivatives of F at x. Note that the set $\partial_\gamma F(x)$ can be empty.

Clearly, $A \in \mathcal{L}(X, Y)$ satisfies $A \in \partial_\gamma F(x)$ if and only if for each $\epsilon > 0$ there exists $\delta(\epsilon) > 0$ such that

$$B_X(x, \delta(\epsilon)) \subset U$$

and that for each $h \in B_X(0, \delta(\epsilon))$,

$$H(F(x) + A(h), F(x + h)) \le (\gamma + \epsilon)\|h\|. \tag{16.111}$$

We say that the mapping F is (γ)-pre-differentiable if it is (γ)-pre-differentiable at every $x \in U$.

Recall that if $G : U \to Y$, $x \in U$ and G is Frechet differentiable at x, then we denote by $G'(x)$ the Frechet derivative of G at x.

Proposition 16.10. *Let $G : U \to Y$, $g : U \to 2^Y \setminus \{\emptyset\}$, $x \in U$, $\gamma > 0$, G be Frechet differentiable at x and let*

$$H(g(z_2), g(z_1)) \le \gamma\|z_2 - z_1\| \text{ for all } z_1, z_2 \in U. \tag{16.112}$$

Then $G'(x)$ is the (γ)-pre-derivative of $G + g$ at $x \in X$.

Proof. Let $\epsilon > 0$. There exists $\delta(\epsilon) > 0$ such that

$$B_X(x, \delta(\epsilon)) \subset U$$

and that for each $h \in B_X(0, \delta(\epsilon)) \setminus \{0\}$,

$$\|h\|^{-1}\|G(x + h) - G(x) - (G'(x))(h)\| < \epsilon/2. \tag{16.113}$$

By (16.112) and (16.113), for each $h \in B_X(0, \delta(\epsilon)) \setminus \{0\}$,

$$G(x + h) + g(x + h)$$
$$\subset G(x) + (G'(x))(h) + 2^{-1}\epsilon\|h\|B_Y(0, 1) + g(x + h)$$
$$\subset G(x) + (G'(x))(h) + 2^{-1}\epsilon\|h\|B_Y(0, 1) + g(x) + (\gamma + 4^{-1}\epsilon)\|h\|B_Y(0, 1)$$
$$\subset G(x) + g(x) + (G'(x))(h) + (\gamma + \epsilon)\|h\|B_Y(0, 1)$$

and

$$G(x) + g(x) + (G'(x))(h)$$
$$\subset G(x + h) + 2^{-1}\epsilon \|h\| B_Y(0, 1) + g(x)$$
$$\subset G(x + h) + 2^{-1}\epsilon \|h\| B_Y(0, 1) + g(x + h) + (\gamma + 4^{-1}\epsilon)\|h\| B_Y(0, 1)$$
$$\subset G(x + h) + g(x + h) + (\gamma + \epsilon)\|h\| B_Y(0, 1).$$

Proposition 16.10 is proved.

The next result is a mean-value theorem for pre-differentiable set-valued mappings.

Theorem 16.11. *Assume that $U \subset X$ is a nonempty open set, $\gamma > 0$, a mapping $F : U \to 2^Y \setminus \{\emptyset\}$ is (γ)-pre-differentiable at every point of U, $x, y \in U$ satisfy $x \neq y$ and*

$$\{tx + (1 - t)y : t \in [0, 1]\} \subset U$$

and that

$$\tilde{x} \in F(x). \tag{16.114}$$

Then there exists $t_0 \in (0, 1)$ such that for every

$$A \in \partial_\gamma F(x + t_0(y - x))$$

the following inequality holds:

$$d(\tilde{x}, F(y)) \leq \|A(y - x)\| + \gamma \|y - x\|.$$

Proof. For each $t \in [0, 1]$ set

$$\phi(t) = d(\tilde{x}, F(x + t(y - x))), \ t \in [0, 1]. \tag{16.115}$$

Since the mapping F is (γ)-pre-differentiable, the function ϕ is continuous on $[0, 1]$. For every $t \in [0, 1]$ set

$$\psi(t) = \phi(t) - t(\phi(1) - \phi(0)), \ t \in [0, 1]. \tag{16.116}$$

Clearly, the function ψ is continuous on $[0, 1]$ and in view of (16.114)

$$\psi(0) = \phi(0) = 0, \ \psi(1) = 0.$$

Therefore there exists $t_0 \in (0, 1)$ such that either

(a) t_0 is a point of minimum of ψ on $[0, 1]$

or

$$\text{(b) } t_0 \text{ is a point of maximum of } \psi \text{ on } [0, 1].$$

It is easy to see that in the case (a)

$$\liminf_{t \to t_0^+}(\psi(t)-\psi(t_0))(t-t_0)^{-1} \geq 0, \ \limsup_{t \to t_0^-}(\psi(t)-\psi(t_0))(t-t_0)^{-1} \leq 0 \qquad (16.117)$$

and in the case (b)

$$\limsup_{t \to t_0^+}(\psi(t) - \psi(t_0))(t - t_0)^{-1} \leq 0, \ \liminf_{t \to t_0^-}(\psi(t) - \psi(t_0))(t - t_0)^{-1} \geq 0.$$
$$(16.118)$$

Assume that

$$A \in \partial_\gamma F(x + t_0(y - x)). \qquad (16.119)$$

By (16.115), for every $t \in [0, 1]$,

$$\phi(t) - \phi(t_0)$$
$$= d(\tilde{x}, F(x + t(y - x))) - d(\tilde{x}, F(x + t_0(y - x)))$$
$$\leq H(F(x + t(y - x)), F(x + t_0(y - x))),$$
$$\phi(t_0) - \phi(t)$$
$$= d(\tilde{x}, F(x + t_0(y - x))) - d(\tilde{x}, F(x + t(y - x)))$$
$$\leq H(F(x + t(y - x)), F(x + t_0(y - x)))$$

and

$$|\phi(t) - \phi(t_0)| \leq H(F(x + t(y - x)), F(x + t_0(y - x))). \qquad (16.120)$$

Let $\epsilon > 0$. Since the mapping F is (γ)-pre-differentiable on U, it follows from (16.119) that there exists $\delta(\epsilon) \in (0, \epsilon)$ such that

$$B_X(x + t_0(y - x), \delta(\epsilon)) \subset U \qquad (16.121)$$

and that for each

$$h \in B_X(0, \delta(\epsilon)) \qquad (16.122)$$

we have

$$H(F(x + t_0(y - x)) + A(h), F(x + t_0(y - x) + h)) \leq (\gamma + \epsilon/4)\|h\|. \qquad (16.123)$$

Assume that $t \in [0, 1]$ satisfies

$$|t - t_0| \leq \delta(\epsilon)(\|x\| + \|y\| + 1)^{-1}. \tag{16.124}$$

Set

$$h = (t - t_0)(y - x). \tag{16.125}$$

In view of (16.124) and (16.125), relations (16.122) and (16.123) hold. By (16.123) and (16.125),

$$H(F(x + t_0(y - x)) + (t - t_0)A(y - x), F(x + t(y - x)))$$
$$\leq (\gamma + 4^{-1}\epsilon)|t - t_0|\|y - x\|.$$

It follows from the relation above and (16.120) that

$$|\phi(t) - \phi(t_0)|$$
$$\leq H(F(x + t_0(y - x)), F(x + t_0(y - x)) + (t - t_0)A(y - x))$$
$$+ H(F(x + t_0(y - x)) + (t - t_0)A(y - x), F(x + t(y - x)))$$
$$\leq |t - t_0|\|A(y - x)\| + (\gamma + 4^{-1}\epsilon)|t - t_0|\|y - x\|. \tag{16.126}$$

Assume that the case (a) holds. Then (16.115)–(16.117), (16.124), and (16.126) imply that

$$0 \leq \liminf_{t \to t_0^+}(\psi(t) - \psi(t_0))(t - t_0)^{-1}$$
$$= \liminf_{t \to t_0^+}(\phi(t) - t(\phi(1) - \phi(0)) - (\phi(t_0) - t_0(\phi(1) - \phi(0))))(t - t_0)^{-1}$$
$$= \liminf_{t \to t_0^+}(\phi(t) - \phi(t_0))(t - t_0)^{-1} - (\phi(1) - \phi(0))$$
$$\leq \|A(y - x)\| + (\gamma + 4^{-1}\epsilon)\|y - x\| - d(\tilde{x}, F(y)).$$

Since ϵ is any positive number we conclude that

$$d(\tilde{x}, F(y)) \leq \|A(y - x)\| + \gamma\|y - x\|. \tag{16.127}$$

Assume that the case (b) holds. Then (16.114)–(16.116), (16.118), and (16.126) imply that

$$0 \leq \liminf_{t \to t_0^-}(\psi(t) - \psi(t_0))(t - t_0)^{-1}$$
$$= \liminf_{t \to t_0^-}(\phi(t) - t(\phi(1) - \phi(0)) - (\phi(t_0) - t_0(\phi(1) - \phi(0))))(t - t_0)^{-1}$$

$$= \lim_{t \to t_0^-} \inf(\phi(t) - \phi(t_0))(t - t_0)^{-1} - (\phi(1) - \phi(0))$$

$$= \lim_{t \to t_0^-} \inf |\phi(t) - \phi(t_0)||t - t_0|^{-1} - d(\tilde{x}, F(y))$$

$$\leq \|A(y - x)\| + (\gamma + 4^{-1}\epsilon)\|y - x\| - d(\tilde{x}, F(y)).$$

Since ϵ is any positive number we conclude that

$$d(\tilde{x}, F(y)) \leq \|A(y - x)\| + \gamma\|y - x\|.$$

Thus (16.127) holds in both cases. This completes the proof of Theorem 16.11.

16.9   Newton's Method for Solving Inclusions

We use the notation and definitions of Sect. 16.8. Suppose that the normed space X is Banach.

Let $\gamma > 0, r > 0, \tilde{x} \in X$, U be a nonempty open subset of X such that

$$B_X(\tilde{x}, r) \subset U$$

and let $F : U \to 2^Y \setminus \{\emptyset\}$ be a (γ)-pre-differentiable mapping at all points of U such that $F(x)$ is a closed set for all $x \in U$.

Let

$$L > 0, \ \gamma_1 \in (0, 1).$$

Suppose that for each $z \in U$ there exists

$$A(z) \in \partial_\gamma F(z) \tag{16.128}$$

such that for each $z_1, z_2 \in B_X(\tilde{x}, r)$,

$$\|A(z_1) - A(z_2)\| \leq L\|z_1 - z_2\|. \tag{16.129}$$

Let $A \in \mathcal{L}(Y, X)$ satisfy

$$\|I_X - A \circ A(\tilde{x})\| \leq \gamma_1, \tag{16.130}$$

there exists a continuous operator $A^{-1} : X \to Y$, a positive number K satisfy

$$\inf\{\|A(z)\| : z \in F(\tilde{x})\} \leq K \leq 4^{-1}r \text{ and } M := \|A\|. \tag{16.131}$$

In view of (16.130),

$$M > 0.$$

For every $x \in U$ define

$$T(x) = x - A(F(x)) \tag{16.132}$$

and for every $x \in X \setminus U$ set $T(x) = \emptyset$. The following result is proved in Sect. 16.11.

Theorem 16.12. *Let*

$$\gamma_1 + M\gamma \le 4^{-1}, \quad K \le (16ML)^{-1}, \quad r_0 := \min\{(4ML)^{-1}, r\}. \tag{16.133}$$

Then for each $x, y \in B_X(\bar{x}, r_0)$,

$$H(T(x), T(y)) \le 2^{-1}\|x - y\|,$$

$$d(\bar{x}, T(\bar{x})) \le 4^{-1}r_0$$

and there exists $x_* \in B_X(\bar{x}, r_0)$ such that $x_* \in T(x_*)$ and $0 \in F(x_*)$. Moreover, the following assertions hold.

1. Assume that a sequence $\{\epsilon_i\}_{i=0}^{\infty} \subset (0, \infty)$ satisfies $\sum_{i=0}^{\infty} \epsilon_i < \infty$,

$$4\sum_{i=0}^{\infty} \epsilon_i + 2\max\{\epsilon_i : i = 0, 1, \ldots\} \le 2^{-1}r_0$$

and that a sequence $\{x_i\}_{i=0}^{\infty} \subset X$ satisfies

$$x_0 = \bar{x}$$

and for each integer $i \ge 0$ satisfying $x_i \in B_X(\bar{x}, r_0)$, the inequalities

$$d(x_{i+1}, T(x_i)) \le \epsilon_i,$$

$$\|x_i - x_{i+1}\| \le d(x_i, T(x_i)) + \epsilon_i$$

hold. Then

$$\|x_i - \bar{x}\| < r_0 - \epsilon_{i-1} \text{ for all integers } i \ge 1,$$

for each integer $k \ge 1$,

$$\|x_k - x_{k+1}\| \le 2^{-k}(4^{-1}r_0 + \epsilon_0) + \sum_{j=0}^{k-1} 2^{-k+1+j}\epsilon_j + \sum_{j=1}^{k} 2^{-k+j}\epsilon_j,$$

for each integer $k \ge 0$,

$$\sum_{i=k}^{\infty} \|x_i - x_{i+1}\| \le 2\|x_k - x_{k+1}\| + 4\sum_{i=k}^{\infty} \epsilon_i$$

and there exists

$$\lim_{n\to\infty} x_n \in B_X(\bar{x}, r_0)$$

satisfying $\lim_{n\to\infty} x_n \in T(\lim_{n\to\infty} x_n)$ *and* $0 \in F(\lim_{n\to\infty} x_n)$.

2. *Let* $\epsilon \in (0, 1)$, *a natural number* $n_0 > 2$ *satisfy*

$$2^{-n_0}(2^{-1}r_0 + 1) < 8^{-1}\epsilon$$

and let a number $\delta \in (0, \epsilon)$ *satisfy*

$$\delta < (16n_0)^{-1}\epsilon,$$

and

$$\delta < 4^{-1}(2n_0 + 1)^{-1}r_0.$$

Assume that a sequence $\{x_i\}_{i=0}^{n_0} \subset X$ *satisfies*

$$x_0 = \bar{x}$$

and for each integer $i \in [0, n_0 - 1]$ *satisfying* $x_i \in B_X(\bar{x}, r_0)$, *the inequalities*

$$d(x_{i+1}, T(x_i)) \le \delta,$$

$$\|x_i - x_{i+1}\| \le d(x_i, T(x_i)) + \delta$$

hold. Then

$$\|x_i - \bar{x}\| < r_0 - \delta \text{ for all } i = 1, \dots, n_0$$

and there exists $x_* \in B_X(\bar{x}, r_0)$ *such that*

$$F(x_*) = 0 \text{ and } \|x_{n_0} - x_*\| < \epsilon.$$

16.10 Auxiliary Results for Theorem 16.12

Lemma 16.13. *The mapping* $T : U \to 2^X \setminus \{\emptyset\}$ *is* (γM)-*pre-differentiable at every point of* U *and for every* $x \in U$,

$$I_X - A \circ A(x) \in \partial_{M\gamma} T(x).$$

Proof. Let $x \in U$ and $\epsilon > 0$. In view of (16.128), there exists $\delta > 0$ such that

$$B_X(x, \delta) \subset U$$

and for each $h \in B_X(0, \delta)$,

$$H(F(x) + (A(x))(h), F(x + h)) \le (\gamma + \epsilon(\|A\| + 1)^{-1})\|h\|.$$

By the inclusion above, (16.131) (16.132), for each $h \in B_X(0, \delta)$,

$$H(T(x) + (I_X - A \circ A(x))(h), T(x + h))$$
$$= H(x - A(F(x)) + h - A((A(x))(h)), x + h - A(F(x + h)))$$
$$= H(-A(F(x) + (A(x))(h)), -A(F(x + h)))$$
$$\leq \|A\|(\gamma + \epsilon(\|A\| + 1)^{-1})\|h\| \leq (M\gamma + \epsilon)\|h\|.$$

Since ϵ is any positive number, this completes the proof of Lemma 16.13.

Lemma 16.14. *Let $r_0 \in (0, r]$ and $x \in B_X(\bar{x}, r_0)$. Then*

$$\|I_X - A \circ A(x)\| \leq \gamma_1 + MLr_0.$$

Proof. By (16.129) and (16.130),

$$\|I_X - A \circ A(x)\|$$
$$= \|I_X - A \circ A(\bar{x}) + A \circ A(\bar{x}) - A \circ A(x)\|$$
$$\|I_X - A \circ A(\bar{x})\| + \|A\|\|A(\bar{x}) - A(x)\|$$
$$\leq \gamma_1 + ML\|x - \bar{x}\| \leq \gamma_1 + MLr_0.$$

Lemma 16.14 is proved.

Lemma 16.15. *Let $r_0 \in (0, r]$ and*

$$x, y \in B_X(\bar{x}, r_0).$$

Then

$$H(T(x), T(y)) \leq (\gamma_1 + MLr_0 + M\gamma)\|y - x\|.$$

Proof. By Lemma 16.13, the mapping T is (γM)-pre-differentiable and for every $x \in U$,

$$I_X - A \circ A(x) \in \partial_{\gamma M} T(x).$$

We may assume that $x \neq y$.
    Let

$$\tilde{x} \in T(x).$$

By Theorem 16.11 and Lemmas 16.13 and 16.14, there exists $t_0 \in (0, 1)$ such that

$$d(\tilde{x}, T(y)) \leq \|(I_X - A \circ A(x + t_0(y - x)))(y - x)\| + M\gamma\|y - x\|$$
$$\leq (\gamma_1 + MLr_0 + M\gamma)\|y - x\|.$$

Sine $\tilde{x}$ is an arbitrary element of $T(x)$ this implies that

$$\sup\{d(\tilde{x}, T(y)) : \tilde{x} \in T(x)\| \le \|y - x\|(\gamma_1 + MLr_0 + M\gamma).$$

Analogously,

$$\sup\{d(\tilde{y}, T(x)) : \tilde{y} \in T(y)\| \le \|y - x\|(\gamma_1 + MLr_0 + M\gamma).$$

Therefore

$$H(T(x), T(y)) \le \|y - x\|(\gamma_1 + MLr_0 + M\gamma).$$

Lemma 16.15 is proved.

16.11   Proof of Theorem 16.12

By (16.131)–(16.133),

$$d(\tilde{x}, T(\tilde{x})) = \inf\{\|A(z)\| : z \in F(\tilde{x})\| \le K \le 4^{-1}r_0.$$

Let

$$x, y \in B_X(\tilde{x}, r_0).$$

Lemma 16.15 imply that

$$H(T(x), T(y)) \le \|y - x\|(\gamma_1 + MLr_0 + M\gamma) \le 2^{-1}\|x - y\|.$$

Clearly, the set

$$\{(x, y) \in B_X(\tilde{x}, r_0) \times B_X(\tilde{x}, r_0) : y \in T(x)\}$$

is closed. It is not difficult to see that Theorem 16.12 follows from Theorem 16.8 applied to the mapping T.

References

1. Alber YI (1971) On minimization of smooth functional by gradient methods. USSR Comput Math Math Phys 11:752–758
2. Alber YI, Iusem AN, Solodov MV (1997) Minimization of nonsmooth convex functionals in Banach spaces. J Convex Anal 4:235–255
3. Alber YI, Iusem AN, Solodov MV (1998) On the projected subgradient method for nonsmooth convex optimization in a Hilbert space. Math Program 81:23–35
4. Alvarez F, Lopez J, Ramirez CH (2010) Interior proximal algorithm with variable metric for second-order cone programming: applications to structural optimization and support vector machines. Optim Methods Softw 25:859–881
5. Ansari QH, Yao JC (1999) A fixed point theorem and its applications to a system of variational inequalities. Bull Aust Math Soc 59:433–442
6. Antipin AS (1994) Minimization of convex functions on convex sets by means of differential equations. Differ Equ 30:1365–1375
7. Aragon Artacho FJ, Geoffroy MH (2007) Uniformity and inexact version of a proximal method for metrically regular mappings. J Math Anal Appl 335:168–183
8. Aragon Artacho FJ, Dontchev AL, Gaydu M, Geoffroy MH, Veliov VM (2011) Metric regularity of Newtons iteration. SIAM J Control Optim 49:339–362
9. Attouch H, Bolte J (2009) On the convergence of the proximal algorithm for nonsmooth functions involving analytic features. Math Program Ser B 116:5–16
10. Baillon JB (1978) Un Exemple Concernant le Comportement Asymptotique de la Solution du Probleme $0 \in du/dt + \partial\phi(u)$. J Funct Anal 28:369–376
11. Barbu V, Precupanu T (2012) Convexity and optimization in Banach spaces. Springer, Heidelberg, London, New York
12. Barty K, Roy J-S, Strugarek C (2007) Hilbert-valued perturbed subgradient algorithms. Math Oper Res 32:551–562
13. Bauschke HH, Borwein JM (1996) On projection algorithms for solving convex feasibility problems. SIAM Rev 38:367–426
14. Bauschke HH, Combettes PL (2011) Convex analysis and monotone operator theory in Hilbert spaces. Springer, New York
15. Bauschke HH, Borwein JM, Combettes PL (2003) Bregman monotone optimization algorithms. SIAM J Control Optim 42:596–636
16. Bauschke HH, Goebel R, Lucet Y, Wang X (2008) The proximal average: basic theory. SIAM J Optim 19:766–785
17. Bauschke H, Moffat S, Wang X (2012) Firmly nonexpansive mappings and maximally monotone operators: correspondence and duality. Set-Valued Var Anal 20:131–153

© Springer International Publishing Switzerland 2016

A.J. Zaslavski, *Numerical Optimization with Computational Errors*, Springer Optimization and Its Applications 108, DOI 10.1007/978-3-319-30921-7

18. Beck A, Sabach S (2015) Weiszfeld's method: old and new results. J Optim Theory Appl 164:1–40

19. Beck A, Teboulle M (2003) Mirror descent and nonlinear projected subgradient methods for convex optimization. Oper Res Lett 31:167–175

20. Beck A, Teboulle M (2009) A fast iterative shrinkage-thresholding algorithm for linear inverse problems. SIAM J Imag Sci 2:183–202

21. Ben-Israel A (1966) A Newton-Raphson method for the solution of equations. J Math Anal Appl 15:243–253

22. Ben-Israel A, Greville TNE (1974) Generalized inverses: theory and applications. Wiley, New York

23. Bolte J (2003) Continuous gradient projection method in Hilbert spaces. J Optim Theory Appl 119:235–259

24. Bonnans JF (1994) Local analysis of Newton-type methods for variational inequalities and nonlinear programming. Appl Math Optim 29:161–186

25. Boukari D, Fiacco AV (1995) Survey of penalty, exact-penalty and multiplier methods from 1968 to 1993. Optimization 32:301–334

26. Bregman LM (1967) A relaxation method of finding a common point of convex sets and its application to the solution of problems in convex programming. Z Vycisl Mat Mat Fiz 7:620–631

27. Brezis H (1973) Opérateurs maximaux monotones. North Holland, Amsterdam

28. Bruck RE (1974) Asymptotic convergence of nonlinear contraction semigroups in a Hilbert space. J Funct Anal 18:15–26

29. Burachik RS, Iusem AN (1998) A generalized proximal point algorithm for the variational inequality problem in a Hilbert space. SIAM J Optim 8:197–216

30. Burachik RS, Grana Drummond LM, Iusem AN, Svaiter BF (1995) Full convergence of the steepest descent method with inexact line searches. Optimization 32:137–146

31. Burachik RS, Lopes JO, Da Silva GJP (2009) An inexact interior point proximal method for the variational inequality problem. Comput Appl Math 28:15–36

32. Burachik RS, Kaya CY, Sabach S (2012) A generalized univariate Newton method motivated by proximal regularization. J Optim Theory Appl 155:923–940

33. Burke JV (1991) An exact penalization viewpoint of constrained optimization. SIAM J Control Optim 29:968–998

34. Butnariu D, Kassay G (2008) A proximal-projection method for finding zeros of set-valued operators. SIAM J Control Optim 47:2096–2136

35. Ceng LC, Mordukhovich BS, Yao JC (2010) Hybrid approximate proximal method with auxiliary variational inequality for vector optimization. J Optim Theory Appl 146:267–303

36. Censor Y, Zenios SA (1992) The proximal minimization algorithm with D-functions. J. Optim. Theory Appl. 73:451–464

37. Censor Y, Gibali A, Reich S (2011) The subgradient extragradient method for solving variational inequalities in Hilbert space. J Optim Theory Appl 148:318–335

38. Censor Y, Gibali A, Reich S (2012) A von Neumann alternating method for finding common solutions to variational inequalities. Nonlinear Anal 75:4596–4603

39. Censor Y, Gibali A, Reich S, Sabach S (2012) Common solutions to variational inequalities. Set-Valued Var Anal 20:229–247

40. Chen Z, Zhao K (2009) A proximal-type method for convex vector optimization problem in Banach spaces. Numer Funct Anal Optim 30:70–81

41. Chen X, Nashed Z, Qi L (1997) Convergence of Newtons method for singular smooth and nonsmooth equations using adaptive outer inverses. SIAM J Optim 7:445–462

42. Chuong TD, Mordukhovich BS, Yao JC (2011) Hybrid approximate proximal algorithms for efficient solutions in for vector optimization. J Nonlinear Convex Anal 12:861–864

43. Clarke FH (1983) Optimization and nonsmooth analysis. Willey Interscience, New York

44. Demyanov VF, Vasilyev LV (1985) Nondifferentiable optimization. Optimization Software, New York

45. Di Pillo G, Grippo L (1989) Exact penalty functions in constrained optimization. SIAM J Control Optim 27:1333–1360
46. Dontchev AL, Rockafellar RT (2010) Newton's method for generalized equations: a sequential implicit function theorem. Math Program 123:139–159
47. Dontchev AL, Rockafellar RT (2013) Convergence of inexact Newton methods for generalized equations. Math Program Ser B 139:115–137
48. Eremin II (1966) The penalty method in convex programming. Sov Math Dokl 8:459–462
49. Eremin II (1971) The penalty method in convex programming. Cybernetics 3:53–56
50. Ekeland I (1974) On the variational principle. J Math Anal Appl 47, 324–353
51. Ermoliev YM (1966) Methods for solving nonlinear extremal problems. Cybernetics 2:1–17
52. Facchinei F, Pang J-S (2003) Finite-dimensional variational inequalities and complementarity problems, volume I and volume II. Springer, New York
53. Guler O (1991) On the convergence of the proximal point algorithm for convex minimization. SIAM J Control Optim 29:403–419
54. Gwinner J, Raciti F (2009) On monotone variational inequalities with random data. J Math Inequal 3:443–453
55. Hager WW, Zhang H (2007) Asymptotic convergence analysis of a new class of proximal point methods. SIAM J Control Optim 46:1683–1704
56. Hager WW, Zhang H (2008) Self-adaptive inexact proximal point methods. Comput Optim Appl 39:161–181
57. Han S-P, Mangasarian OL (1979) Exact penalty function in nonlinear programming. Math Program 17:251–269
58. Hiriart-Urruty J-B, Lemarechal C (1993) Convex analysis and minimization algorithms. Springer, Berlin
59. Iiduka H, Takahashi W, Toyoda M (2004) Approximation of solutions of variational inequalities for monotone mappings. Pan Am Math J 14:49–61
60. Ioffe AD, Zaslavski AJ (2000) Variational principles and well-posedness in optimization and calculus of variations. SIAM J Control Optim 38:566–581
61. Iusem A, Nasri M (2007) Inexact proximal point methods for equilibrium problems in Banach spaces. Numer Funct Anal Optim 28:1279–1308
62. Iusem A, Resmerita E (2010) A proximal point method in nonreflexive Banach spaces. Set-Valued Var Anal 18:109–120
63. Izmailov AF, Solodov MV (2014) Newton-type methods for optimization and variational problems. Springer International Publishing, Cham
64. Kantorovich LV (1948) Functional analysis and applied mathematics. Usp Mat Nauk 3:89–185
65. Kantorovich LV, Akilov GP (1982) Functional analysis. Pergamon Press, Oxford, New York
66. Kaplan A, Tichatschke R (1994) Stable methods for ill-posed variational problems. Akademie Verlag, Berlin
67. Kaplan A, Tichatschke R (1998) Proximal point methods and nonconvex optimization. J Global Optim 13:389–406
68. Kaplan A, Tichatschke R (2007) Bregman-like functions and proximal methods for variational problems with nonlinear constraints. Optimization 56:253–265
69. Kassay G (1985) The proximal points algorithm for reflexive Banach spaces. Stud Univ Babes-Bolyai Math 30:9–17
70. Kiwiel KC (1996) Restricted step and Levenberg–Marquardt techniques in proximal bundle methods for nonconvex nondifferentiable optimization. SIAM J Optim 6:227–249
71. Konnov IV (1997) On systems of variational inequalities. Russ Math (Iz VUZ) 41:79–88
72. Konnov IV (2001) Combined relaxation methods for variational inequalities. Springer, Berlin, Heidelberg
73. Konnov IV (2008) Nonlinear extended variational inequalities without differentiability: applications and solution methods. Nonlinear Anal. 69:1–13
74. Konnov IV (2009) A descent method with inexact linear search for mixed variational inequalities. Russ Math (Iz VUZ) 53:29–35

75. Korpelevich GM (1976) The extragradient method for finding saddle points and other problems. Ekon Matem Metody 12:747–756
76. Kutateladze SS (1979) Convex operators. Usp Math Nauk 34:167–196
77. Lemaire B (1989) The proximal algorithm. In: Penot JP (ed) International series of numerical mathematics, vol 87. Birkhauser-Verlag, Basel, pp 73–87
78. Lotito PA, Parente LA, Solodov MV (2009) A class of variable metric decomposition methods for monotone variational inclusions. J Convex Anal 16:857–880
79. Mainge P-E (2008) Strong convergence of projected subgradient methods for nonsmooth and nonstrictly convex minimization. Set-Valued Anal 16:899–912
80. Mangasarian OL, Pang J-S (1997) Exact penalty functions for mathematical programs with linear complementary constraints. Optimization 42:1–8
81. Martinet B (1978) Pertubation des methodes d'optimisation: application. RAIRO Anal Numer 12:153–171
82. Minty GJ (1962) Monotone (nonlinear) operators in Hilbert space. Duke Math J 29:341–346
83. Minty GJ (1964) On the monotonicity of the gradient of a convex function. Pac J Math 14:243–247
84. Mordukhovich BS (2006) Variational analysis and generalized differentiation, I: I: basic theory. Springer, Berlin
85. Mordukhovich BS (2006) Variational analysis and generalized differentiation, II: applications. Springer, Berlin
86. Mordukhovich BS, Nam NM (2014) An easy path to convex analysis and applications. Morgan&Clayton Publishes, San Rafael, CA
87. Moreau JJ (1965) Proximite et dualite dans un espace Hilbertien. Bull Soc Math Fr 93:273–299
88. Nashed MZ, Chen X (1993) Convergence of Newton-like methods for singular operator equations using outer inverses. Numer Math 66:235–257
89. Nedic A, Ozdaglar A (2009) Subgradient methods for saddle-point problems. J Optim Theory Appl 142:205–228
90. Nemirovski A, Yudin D (1983) Problem complexity and method efficiency in optimization. Wiley, New York
91. Nesterov Yu (1983) A method for solving the convex programming problem with convergence rate $O(1/k^2)$. Dokl Akad Nauk 269:543–547
92. Nesterov Yu (2004) Introductory lectures on convex optimization. Kluwer, Boston
93. Pang J-S (1985) Asymmetric variational inequality problems over product sets: applications and iterative methods. Math Program 31:206–219
94. Pang J-S (1990) Newton's method for B-differentiable equations. Math Oper Res 15:311–341
95. Polyak BT (1967) A general method of solving extremum problems. Dokl Akad Nauk 8:593–597
96. Polyak BT (1987) Introduction to optimization. Optimization Software, New York
97. Polyak BT (2007) Newtons method and its use in optimization. Eur J Oper Res 181:1086–1096
98. Polyak RA (2015) Projected gradient method for non-negative least squares. Contemp Math 636:167–179
99. Qi L, Sun J (1993) A nonsmooth version of Newton's method. Math Program 58:353–367
100. Reich S, Sabach S (2010) Two strong convergence theorems for Bregman strongly nonexpansive operators in reflexive Banach spaces. Nonlinear Anal 73:122–135
101. Reich S, Zaslavski AJ (2014) Genericity in nonlinear analysis. Springer, New York
102. Robinson SM (1994) Newtons method for a class of nonsmooth functions. Set-Valued Anal 2:291–305
103. Rockafellar RT (1976) Augmented Lagrangians and applications of the proximal point algorithm in convex programming. Math Oper Res 1:97–116
104. Rockafellar RT (1976) Monotone operators and the proximal point algorithm. SIAM J Control Optim 14:877–898
105. Shor NZ (1985) Minimization methods for non-differentiable functions. Springer, Berlin

106. Solodov MV, Svaiter BF (2000) Error bounds for proximal point subproblems and associated inexact proximal point algorithms. Math Program 88:371–389
107. Solodov MV, Svaiter BF (2001) A unified framework for some inexact proximal point algorithms. Numer Funct Anal Optim 22:1013–1035
108. Solodov MV, Zavriev SK (1998) Error stability properties of generalized gradient-type algorithms. J Optim Theory Appl 98:663–680
109. Su M, Xu H-K (2010) Remarks on the gradient-projection algorithm. J Nonlinear Anal Optim 1:35–43
110. Weiszfeld EV (1937) Sur le point pour lequel la somme des distances de n points donnes est minimum. Tohoku Math J 43:355–386
111. Xu H-K (2006) A regularization method for the proximal point algorithm. J Global Optim 36:115–125
112. Xu H-K (2011) Averaged mappings and the gradient-projection algorithm. J Optim Theory Appl 150:360–378
113. Yamashita N, Kanzow C, Morimoto T, Fukushima M (2001) An infeasible interior proximal method for convex programming problems with linear constraints. J Nonlinear Convex Anal 2:139–156
114. Zangwill WI (1967) Nonlinear programming via penalty functions. Manage Sci 13:344–358
115. Zaslavski AJ (2003) Existence of solutions of minimization problems with an increasing cost function and porosity. Abstr Appl Anal 2003:651–670
116. Zaslavski AJ (2003) Generic existence of solutions of minimization problems with an increasing cost function. J. Nonlinear Funct Anal Appl 8:181–213
117. Zaslavski AJ (2005) A sufficient condition for exact penalty in constrained optimization. SIAM J Optim 16:250–262
118. Zaslavski AJ (2007) Existence of approximate exact penalty in constrained optimization. Math Oper Res 32:484–495
119. Zaslavski AJ (2010) An estimation of exact penalty in constrained optimization. J Nonlinear Convex Anal 11:381–389
120. Zaslavski AJ (2010) Convergence of a proximal method in the presence of computational errors in Hilbert spaces. SIAM J Optim 20:2413–2421
121. Zaslavski AJ (2010) Optimization on metric and normed spaces. Springer, New York
122. Zaslavski AJ (2010) The projected subgradient method for nonsmooth convex optimization in the presence of computational errors. Numer Funct Anal Optim 31:616–633
123. Zaslavski AJ (2011) An estimation of exact penalty for infinite-dimensional inequality-constrained minimization problems. Set-Valued Var Anal 19:385–398
124. Zaslavski AJ (2011) Inexact proximal point methods in metric spaces. Set-Valued Var Anal 19:589–608
125. Zaslavski AJ (2011) Maximal monotone operators and the proximal point algorithm in the presence of computational errors. J. Optim Theory Appl 150:20–32
126. Zaslavski AJ (2012) The extragradient method for convex optimization in the presence of computational errors. Numer Funct Anal Optim 33:1399–1412
127. Zaslavski AJ (2012) The extragradient method for solving variational inequalities in the presence of computational errors. J Optim Theory Appl 153:602–618
128. Zaslavski AJ (2013) The extragradient method for finding a common solution of a finite family of variational inequalities and a finite family of fixed point problems in the presence of computational errors. J Math Anal Appl 400:651–663
129. Zeng LC, Yao JC (2006) Strong convergence theorem by an extragradient method for fixed point problems and variational inequality problems. Taiwan J Math 10:1293–1303

Index

A
Absolutely continuous function, 154
Algorithm, 11, 13, 14
Approximate solution, 1, 44

B
Banach space, 225
Bochner integrable function, 225

C
Cardinality of a set, 137
Collinear vectors, 86
Compact set, 228
Concave function, 26, 36
Continuous subgradient algorithm, 225
Convex–concave function, 11
Convex cone, 247
Convex function, 1, 4, 6, 11, 20, 26, 35
Convex hull, 228
Convex minimization problem, 105
Convex set, 11, 12

E
Ekelands variational principle, 244, 252
Euclidean norm, 167
Euclidean space, 86, 169
Exact penalty, 239
Extragradient method, 183, 205

F
Fermat–Weber location problem, 85, 86

F
Fréchet derivative, 45, 59
Fréchet diferentiable function, 45, 59, 63

G
Gâteaux derivative, 106
Gâteaux differential function, 106
Gradient-type method, 8, 73

H
Hilbert space, 1, 4, 6, 11, 20

I
Increasing function, 247
Inner product, 1, 4, 6, 20

K
Karush–Kuhn–Tucker theorem, 252

L
Lebesgue measurable function, 225, 227
Linear functional, 246
Linear inverse problem, 74
Lower semicontinuous function, 6, 137

M
Maximal monotone operator, 169
Metric space, 149
Minimization problem, 2
Minimizer, 15, 16, 22, 42

© Springer International Publishing Switzerland 2016
A.J. Zaslavski, *Numerical Optimization with Computational Errors*, Springer
Optimization and Its Applications 108, DOI 10.1007/978-3-319-30921-7

Printed in the United States
By Bookmasters